Rotating Machinery Handbook

Volume I

Rotating Machinery Handbook
Volume I

Edited by **George Rivera**

NY RESEARCH PRESS
New York

Published by NY Research Press,
23 West, 55th Street, Suite 816,
New York, NY 10019, USA
www.nyresearchpress.com

Rotating Machinery Handbook: Volume I
Edited by George Rivera

International Standard Book Number: 978-1-63238-403-4 (Hardback)

Printed in China.

Contents

Preface

Rotating Machinery can be defined in a manner by its name itself. Any mechanism which has in its design a rotating member and usually generates, transforms, modifies or converts power or adds kinetic energy to a process. There must be very few machines and apparatus as well as mechanical systems that do not incorporate rotating components. Wheels, bearings, shafts, gears and seals can be called rotating machinery. Turbines can be called rotating machines as well, as they usually require special mechanical designs, monitoring and operations and includes gas, wind and tidal, steam, generators, internal combustion engines as well as others. Indeed, most of the mechanical devices used by the individual have some form of rotating machinery in its design. An extensive array of industrial applications is available for rotating machinery in fields that include oil, power and processing industries. The need for more and more qualified personnel is increasing constantly as there is a continuous expansion in this field and graduates with a higher skill set are in demand. The demand for better and more efficient rotating machinery is pushing the envelope of research on advancements to a higher level. The economic ramifications of the technological advancements in the field of rotating machinery are also not to be ignored.

This book is an attempt to understand as well as compile and collate the available current and proposed research topics under the branch of rotating machinery. I would like to thank all those who put their hard work and effort into these research subjects. I would also like to thank my friends and family for supporting me at every step in my life.

Editor

1

Aerothermal Analysis of a Turbine Casing Impingement Cooling System

Riccardo Da Soghe, Bruno Facchini, Mirko Micio, and Antonio Andreini

Energy Engineering Department "S. Stecco", University of Florence, Street S. Marta 3, 50139 Florence, Italy

Correspondence should be addressed to Riccardo Da Soghe, dasoghe@brun.de.unifi.it

Academic Editor: N. Sitaram

Heat transfer and pressure drop for a representative part of a turbine active cooling system were numerically investigated by means of an in-house code. This code has been developed in the framework of an internal research program and has been validated by experiments and CFD. The analysed system represents the classical open bird cage arrangement that consists of an air supply pipe with a control valve and the present system with a collector box and pipes, which distribute cooling air in circumferential direction of the casing. The cooling air leaves the ACC system through small holes at the bottom of the tubes. These tubes extend at about 180° around the casing and may involve a huge number of impinging holes; as a consequence, the impinging jets mass flow rate may vary considerably along the feeding manifold with a direct impact on the achievable heat transfer levels. This study focuses on the performance, in terms of heat transfer coefficient and pressure drop, of several impinging tube geometries. As a result of this analysis, several design solutions have been compared and discussed.

1. Introduction

Impingement with high velocity jets has become an established method for surface cooling or heating in a wide variety of processes and thermal control applications. The use of impingement jets for the cooling of modern aero-engine components is widespread, especially within the hot stationary parts. Since the cooling performance of impinging jets is very high, this method provides an efficient way to manage a component heat load when a sufficient pressure head and geometrical characteristics are available for its implementation. The cooling jets are usually arranged as arrays [1].

Aero-engine casing temperature control is a very effective way to reduce aerodynamic losses and specific fuel consumption due to blade tip clearance. Because of the significant variations of centrifugal and thermal loads which occur at different engine operating conditions, the tip clearance can be extremely variable; such dimensional variation may worsen engine performance and reduce the components' life span [2]. To overcome these issues, the Active thermal Clearance Control (ACC), generally based on impingement cooling, has been successfully introduced in several applications as described in Halila et al. [3], Beck and Fasching [4], and more recently by Justak and Doux [2]. In such systems, impinging jets are directed towards the external turbine casing by means of a series of circumferential feeding pipes with the final aim of keeping the clearance between blade tip and casing as constant as possible under different engine operating conditions (Figures 1 and 2). Depending on the engine architecture, the cooling air could be bled from the compressor or extracted from the fan flow. The impinging system pressure ratio is then a consequence of the ACC configuration: lower than 1.1 in case of cooling air extracted from the fan flow, while higher β when the coolant comes from the compressor.

The correct evaluation of both the impinging jet mass flow rate (i.e., the characterization of the holes discharge coefficient) and the related heat transfer coefficient is a fundamental activity.

The discharge coefficient (C_d) is defined as the ratio of the actual mass flow rate through a hole and the isentropic flow rate. It summarizes all the losses that limit the actual

FIGURE 1: ACC system, Ahmed et al. [5].

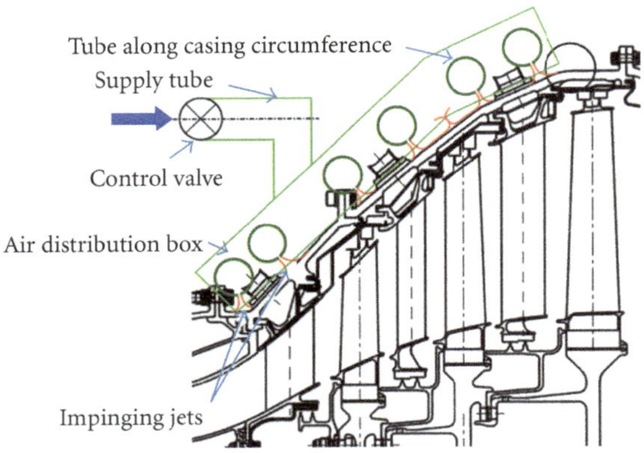

FIGURE 2: Scheme of a LPT ACC system, Ahmed et al. [6].

mass flow rate through a hole: entry pressure losses, internal losses due to friction, and exit losses.

Many parameters may influence the discharge coefficient [7]: geometrical, such as hole shape, hole angle, space between holes, and length to diameter ratio, and fluid-dynamical, such as pressure ratio across the hole and Reynolds and Mach number of the two cross-flows and inside the hole. For this reason, several studies have been carried out on different geometries of holes subjected to a wide range of fluid-dynamics conditions; extensive reviews can be found in Hay and Lampard [8] and in Andreini and DaSoghe [9]. Gritsch et al. [10] have investigated the behaviour of a single hole of large diameter (10 mm); they proposed a method for correlating the discharge coefficients, assuming that pressure losses inside the hole and those related to the hole entry and exit are independent. Internal losses are found to be dependent on the pressure ratio across the hole, while the entry and exit losses depend on the jet to cross-flow momentum ratio. With these hypotheses in mind, Gritsch et al. [11] have studied the influence of the internal cross-flow on shaped holes, while Rowbury et al. [12] proposed a method to quantify the influence of the external cross-flow on the C_d. More recently, Schulz et al. [13] have performed several experiments, analysing the behaviour of the C_d under varying internal and external cross-flow conditions and geometrical angles and using the jet-to-cross-flow momentum ratio to reduce data.

Referring to the heat transfer, measurements of multiple jet impingement arrays can be found in a relevant number of existing publications. Comprehensive reviews on this field have been provided by Martin [15] and Han et al. [16]. Some recent and very interesting contributions have been made by Ahmed and coworkers [5, 6] who have performed some numerical simulations of the flow in a short tube section of an ACC system for a low pressure turbine. The length-to-diameter ratio of the sharp-edged cylindrical nozzles, ranging from 0.25 to 2, was also accounted for. The authors reveal that increasing the Mach number by simultaneously reducing the orifice diameters led to slightly decreasing Nusselt numbers, with average deviations of the order of 14%. The predicted discharge coefficients increased significantly by augmenting the Mach number. The main limiting aspect of these two interesting contributions consists in the fact that the authors do not evaluate the effects of the undercowl flow on the impingement jets.

As mentioned above, many published works deal with heat transfer coefficient due to impinging jets; most of them analyse, by means of experiments or numerical simulations, the behaviour of several hole geometries and arrangements, but it also appears a complete deficiency of studies conduced on a real-engine ACC geometry. Similar conclusions could be pointed out referring to impinging holes discharge coefficient prediction. Indeed, there seems to be a lack of general studies, and therefore of useful design correlations concerning the effects on the discharge coefficient due to the interaction among adjacent holes drilled in cylindrical feeding pipes [9].

Hence, the University of Florence has promoted an internal research project aimed at the characterization of the holes discharge coefficient and heat transfer phenomena of a real-engine ACC geometry. The main goals of this project are to provide a general understanding of the fluid flow and heat transfer characteristics, to develop useful correlations for both the C_d and the HTC coefficient and, finally, to validate CFD with high quality experimental data. In the framework of the project two test rigs have been developed. The first one, aims to evaluate the discharge coefficient of several jet holes drilled into a circular duct. The tested geometries replicate the final part of a feeding manifold impingement tube used in active clearance control systems. By using the provided experimental data (see [17]) and validated CFD calculations, Andreini and DaSoghe [9] have developed an empirical correlation for the prediction of the impinging holes discharge coefficient: it expresses the C_d of each hole as a function of the ratio between the hole and the manifold mass velocity and the local value of the pressure ratio. Recently, the authors have revised this correlation to make it sensible to the nozzle length-to-diameter ratio [18]. The second test rig, aims to evaluate the heat transfer coefficient and the adiabatic effectiveness of a multijet impingement array (for which the undercowl flow is considered) which reproduces an active clearance control system of a commercial aircraft. The test rig has

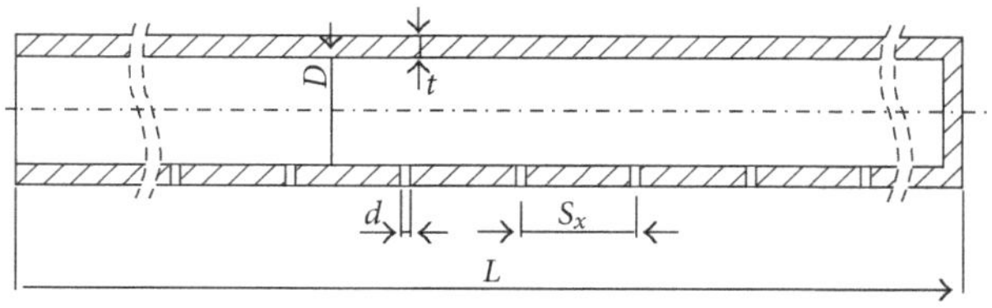

FIGURE 3: Multiple jet section geometry.

provided detailed experimental data that are, nowadays, partially available in the open literature (see [14]).

This paper deals with a simplified 1D code, developed at the University of Florence, aimed at the design of the multiple jet section of an ACC cooling tube. The main purpose of this work is to validate the in-house design tool and, by using of its predictions, to point out design practices for such kind of cooling systems.

2. Description of the Code

As stressed above, the simplified code deals with the multiple jet section of an ACC cooling tube. Such kind of geometries are depicted in Figure 3.

The aim of the 1D program is to evaluate the mass flow rate through each nozzle of the array and then to calculate the related heat transfer coefficient HTC achieved on the target surface. To do that the code, that is based on a correlative approach, implements two correlations: the first one for the nozzle C_d and the other one for the HTC evaluation. The implemented correlations are presented in the next section. Within the manifold, the total pressure losses due to friction are neglected. Then, the static pressure profile is determined by the mass flow rate bleeding operated by the holes.

3. Discharge Coefficient Correlation

The correlation for the C_d parameter considered in this study has been recently developed by DaSoghe and Andreini [9, 18]. The expression is the result of an extensive CFD analysis: it expresses the C_d of each impingement hole as a function of the ratio between the hole and the manifold mass velocity (MVR), the nozzle length-to-diameter ratio (t/d) and the local value of the β_L ratio. The discharge coefficient is evaluated using the following expression:

$$C_d = C_1 \cdot \beta_L^{\alpha(t/d)} \cdot g\left(\frac{t}{d}\right) \cdot \left(1 + C_2 \cdot C_3^{C_4 \cdot \mathrm{MVR}^{C_5}}\right), \quad (1)$$

where $\alpha(t/d)$ and $g(t/d)$ are second-order polynomial functions of the length-to-diameter ratio and the other coefficients are suitable constants.

Even developed referring to circular cross-section manifold geometries, DaSoghe and Andreini [18] proved the correlation reliability also in case of squared manifold geometries.

The ranges of applicability of the correlation are the following $\beta = 1.0$ to 1.65, $\mathrm{Re}_{\mathrm{channel}} = 0$ to $6 \cdot 10^4$, and $t/d = 0.25$ to 3. With respect to the CFD data, the correlation

TABLE 1: Geometric parameter.

Manifold diameter "D"	m
Nozzle diameter "d"	m
Manifold thickness "t"	m
Holes number "n"	—
Holes spacing "Sx"	m
Nozzle to wall distance "h"	m

TABLE 2: Thermodynamic parameter.

Cooling temperature	K
Wall temperature	K
Overall pressure ratio	—
Discharge pressure	Pa

shows a mean relative error of 1.6% and a maximum error of 4.5%, with a standard deviation of 1.2% (i.e., 95% of the C_d predicted when using the correlation leads to an error lower than 3% with respect to the CFD data set).

Finally, the correlation has been validated against the experimental data provided by Schulz and co-workers [13].

For further details about the presented correlation and the related validation study, refer to Andreini and DaSoghe papers [9, 18].

4. Heat Transfer Coefficient Correlation

The method used to calculate the HTC consists in a procedure that considers two correlations: one to evaluate the HTC peak value on the jet stagnation point, and the other to calculate the HTC value far away from the stagnation region.

The HTC peak value is evaluated using of the following expression:

$$\mathrm{Nu0} = C_6 \mathrm{Re}_d^{C_7} \mathrm{Pr}^{C_8} \left(\frac{S_x}{d}\right)^{C_9} \left(\frac{h}{d}\right)^{C_{10}}, \quad (2)$$

in which the Nu0 represents the Nusselt number peak value (i.e., the Nu at the jet stagnation point). This expression has been derived from Florschuetz and co-workers [19]. The ranges of applicability of the correlation are the following: $\mathrm{Re}_d = 2500$ to 80000, $S_x/d = 5.9$ to 31.4, and $h/d = 2.9$ to 17.8. Far away from the stagnation region, the local Nusselt number is evaluated as follow:

$$\mathrm{Nu}l = \mathrm{Nu0}\left[1 - C_{11}\left(\frac{x}{d}\right)^{C_{12}}\right] \exp\left[C_{13}\left(\frac{x}{d}\right)^{C_{14}}\right], \quad (3)$$

where Nu0 is the Nusselt number evaluated by (2) and x represents the distance between the considered point and the stagnation point.

5. 1D Procedure

The in-house code is composed of two different modules. The first one consists in a performance code: for a given ACC tube geometry and operating condition (see Tables 1 and 2),

TABLE 3: Design code: Geometric parameter.

Manifold diameter "D"	m
Manifold thickness "t"	m
Holes number "n"	—
Holes spacing "Sx"	m
Nozzle to wall distance "h"	m

TABLE 4: Validation of the 1D code: Analized geometries.

	D	d	n	t/d	Sx/d	h/d
Geom 1	12 (mm)	1 (mm)	17	2	12	7
Geom 2	12 (mm)	1 (mm)	133	2	1.5	7

the code evaluates the system performance in terms of heat transfer coefficient distribution and HTC mean value on the target surface.

By means of an iterative procedure, the code first evaluates the isentropic mass flow rate through the generic impingement hole, using the St. Venant equation:

$$\dot{m}_{\text{ideal}} = \frac{P \cdot A}{\sqrt{R \cdot T}} \cdot \sqrt{\frac{2\gamma}{\gamma - 1}\left[\left(\frac{1}{\beta_L}\right)^{1/\gamma} - \left(\frac{1}{\beta_L}\right)^{\gamma+1/\gamma}\right]}. \quad (4)$$

The actual mass flow rate through a generic hole of the array is then calculated using the following expression:

$$\dot{m}_{\text{actual}} = \dot{m}_{\text{ideal}} \cdot C_d, \quad (5)$$

where the C_d is given in (1). Once the $\dot{m}_{\text{actual}}$ is evaluated, the code calculates the jet heat transfer coefficient distribution.

The second code's module, consists in a design procedure: starting from some geometrical data and operating conditions (i.e., overall pressure ratio), the code determines the nozzles' diameter in order to obtain a desired level of heat transfer coefficient. In this case the requested geometrical parameters are reported in Table 3.

The procedure starts from the last hole of the array (the hole closest to the manifold endcap) for which the MVR is ∞. In that case the hole C_d is function of both the pressure ratio and the t/d ratio thus it can be easily estimated. Once the C_d and the isentropic mass flow rate are calculated, the code estimates the hole impinging mass flow rate and then the related mean HTC. If the calculated HTC does not match the imposed one, the code assumes a new hole diameter and then recalculates the C_d, the isentropic mass flow rate and the jet mean HTC. These steps are repeated until the calculated mean HTC matches the desired value. Once the procedure is converged for the last hole of the array, the code starts to analyse the second last hole of the array and so on.

6. Validation of the Code

In order to evaluate the accuracy and reliability of the code, the predictions of the 1-D procedure are compared with the experimental and numerical data provided by Facchini et al. [14] and Andreini and DaSoghe [9]. Two geometries have been considered (Table 4).

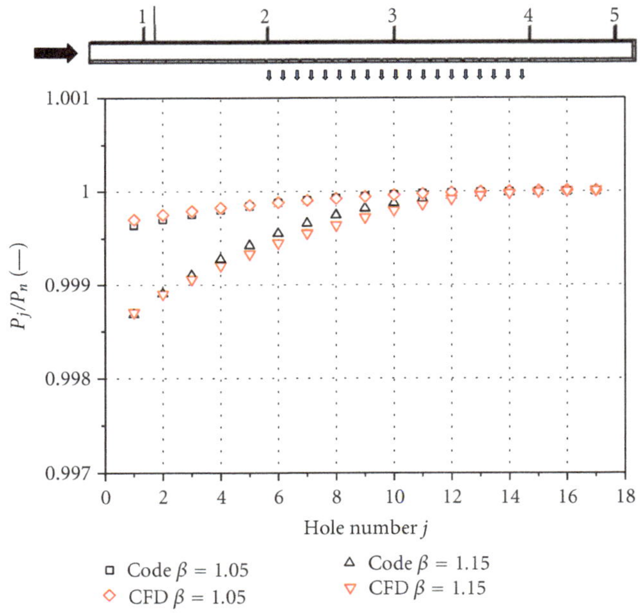

FIGURE 4: Pressure distribution for Geom 1.

The comparisons made for two operating conditions ($\beta = 1.05$ and $\beta = 1.15$), are provided in terms of pressure distribution (expressed as the ratio of the the value in correspondence of the generic hole and value at the end of the feeding pipe P_j/P_n), mass flow rate split along the manifold (expressed as the ratio of the generic hole's mass flow rate and the mass flow rate at the inlet of the feeding pipe m_j/m_{tot}) and HTC profiles.

Figures 4 and 5 show the pressure distribution and the mass flow rate split across the manifold for the geometry labelled "Geom 1".

The code predictions are compared with the CFD data provided by Andreini and DaSoghe [9]. As shown by the figures, the 1-D code well agrees with the CFD for all the tested operating conditions meaning that a reliable pressure and hole mass flow rates evaluation is provided.

The HTC profiles calculated by the in-house program for the central nozzle of the array are compared with the related experimental data of Facchini et al. [14]. The mentioned test rig consists in a real engine geometry that counts several ACC manifolds. The presence of the undercowl flow is also accounted for during the experiments. The results compared here consist in the spanwise HTC profiles (Figure 6). In the experiments, heat transfer coefficients were determined by means of a steady state technique, measuring wall temperatures on an heated surface, using TLC (thermo liquid crystal) paint. Maximum relative error referred to the heat transfer measurement is under ±12%. For further details about the experimental procedure and results, please refer to Facchini et al. [14].

In Figure 7, the HTC profile predicted by the simplified procedure in case of Geometry 1 is compared with Facchini et al.'s [14] experimental data.

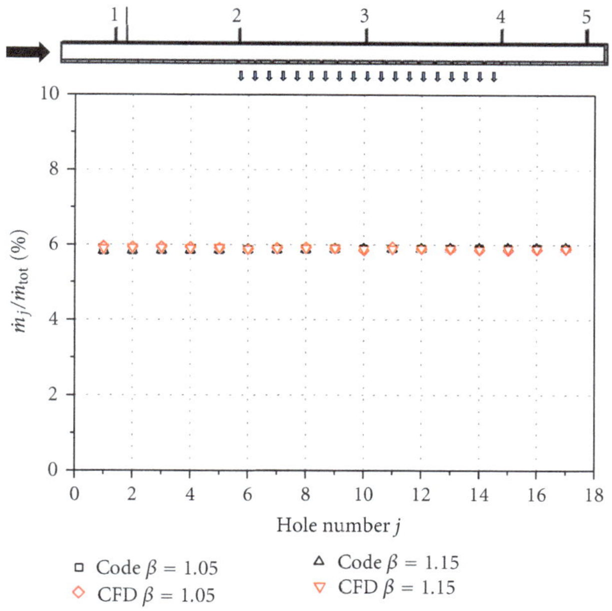

FIGURE 5: Mass flow split distribution for Geom 1 (percent).

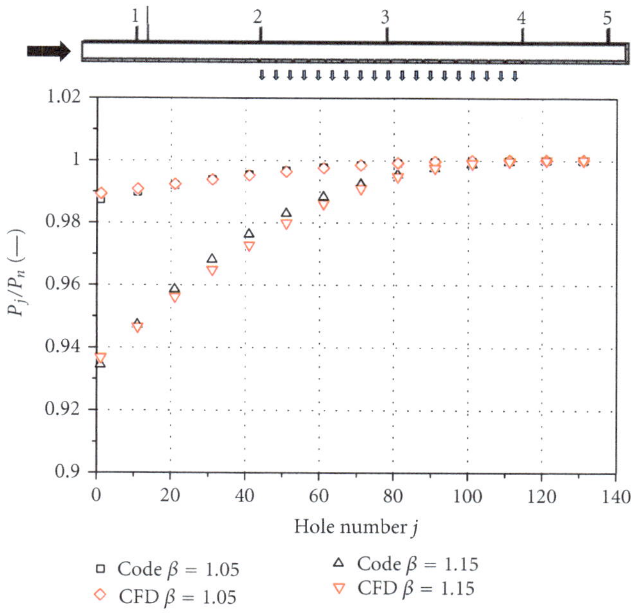

FIGURE 8: Pressure distribution for Geom 2.

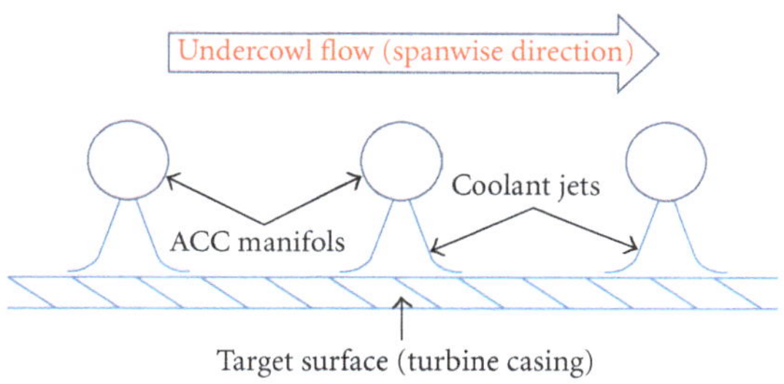

FIGURE 6: Spanwise direction.

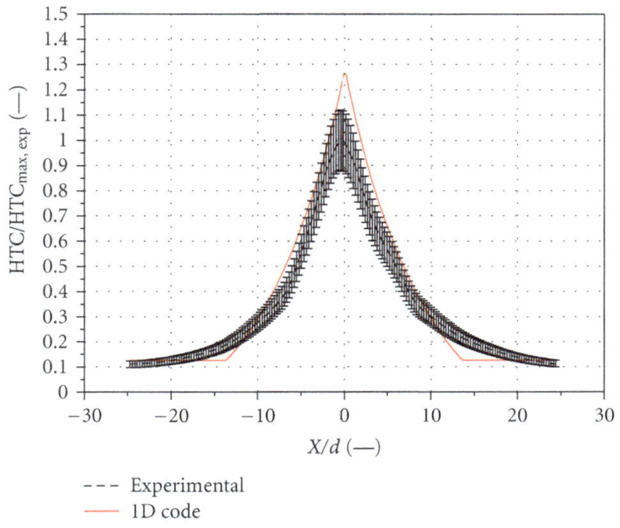

FIGURE 7: Heat transfer coefficient distribution, central hole of the array, Geom 1 β = 1.15 (experimental data from Facchini et al. [14]).

It emerges from the figures that the code is in fairly good agreement with the experiments in term of HTC profiles. The agreement becomes better when considering HTC mean values (that are the key parameter we are looking at). Similar conclusions can be drawn also for the geometry labelled "Geom 2" (Figures 8, 9, and 10).

The 1-D code can be assumed as a reliable tool for the prediction of the HTC related to the multiple jet section of an ACC cooling tube.

7. Performance Analysis of Different ACC Cooling Tube Geometries

The developed code has been used to evaluate the performance, in terms of the jet's mean HTC profile, of different ACC tube geometries.

7.1. Impact of the A_r on the Jet Mean HTC Profile. It emerges from Figure 9 that the mass flow rate split across an ACC cooling tube can be largely variable. Andreini and DaSoghe [9] point out that the mass flow rate split across a given impingement system (i.e., the local MVR) is only related to the pipe area ratio A_r defined as the ratio of the manifold inlet area and the holes cumulative exit area,

$$A_r = \frac{A_c}{n_{\text{holes}} \cdot A_j}. \quad (6)$$

Expression 1 demonstrates that, for each operating condition (i.e., for each β ratio), the discharge coefficient is roughly constant when the MVR is higher than 5. As the MVR is a function of the A_r parameter, it is possible to design the impingement system in order to assure that the mass flow rate through each hole is roughly the same. Indeed, assuming

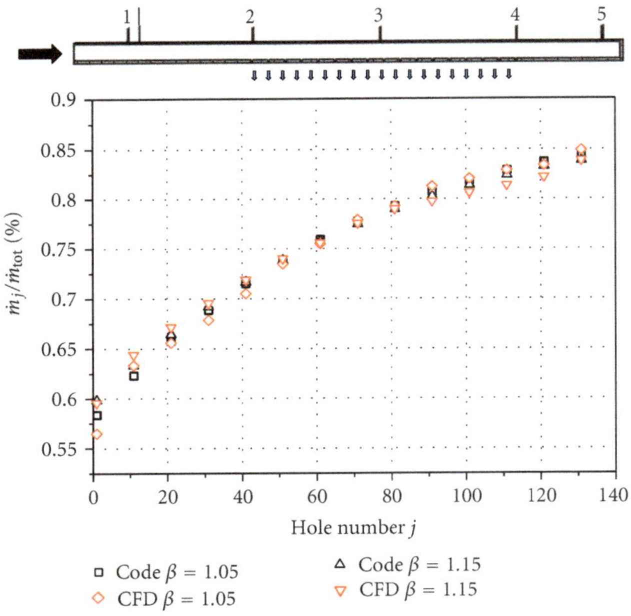

Figure 9: Mass flow split distribution for Geom 2 (percent).

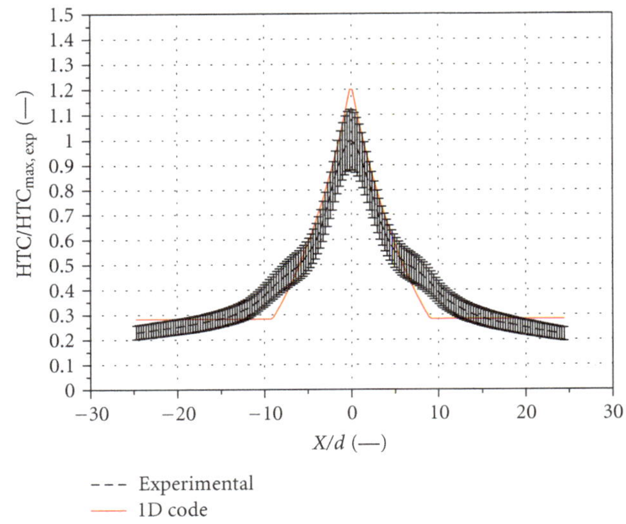

Figure 10: Heat transfer coefficient distribution, central hole of the array, Geom 1 β = 1.05 (experimental data from Facchini et al. [14]).

that the manifold inlet mass flow rate is equally distributed among the impingement holes, it follows that

$$A_c \cdot (\rho \cdot v)_c = n_{\text{holes}} \cdot A_j \cdot (\rho \cdot v)_j, \quad (7)$$

$$\text{MVR} = \frac{A_c}{n_{\text{holes}} \cdot A_j}. \quad (8)$$

The relations above are satisfied when the C_d is constant across the impingement system that is, when the MVR is higher than 5 approx. So it can be assumed that the condition:

$$\frac{A_c}{n_{\text{holes}} \cdot A_j} = A_r > 5, \quad (9)$$

defines a design rule for manifolds with equally distributed impinging mass flow rate jets.

Geometries for which the A_r is greater than 5 lead to a roughly constant impingement jet mass flow rate distribution, while, in case of geometries characterized by a low A_r, relevant differences in jet mass flow rates across the feeding pipe are expected. This last statement justifies the trends shown in Figures 5 and 9. Geom 1 is characterized by A_r = 8.4 so the mass flow split across the manifold is uniform. In case of Geom 2, for which A_r = 1.08, the impinging jet mass flow rate varies significantly from the first to the last hole of the array. It is also expected that in case of a low A_r, the impinging jet HTC could be nonuniform. Figure 11 shows the jet's mean HTC distribution obtained for Geom 2 when β = 1.05. The parameter HTC_{ref} represents the mean heat transfer coefficient evaluated for the last hole of the array of Geom 2 when β = 1.05. The figure reveals that the mean HTC varies from the inlet to the outlet of the manifold by about 25%. This variation might be higher when considering high by-pass ratio turbofans.

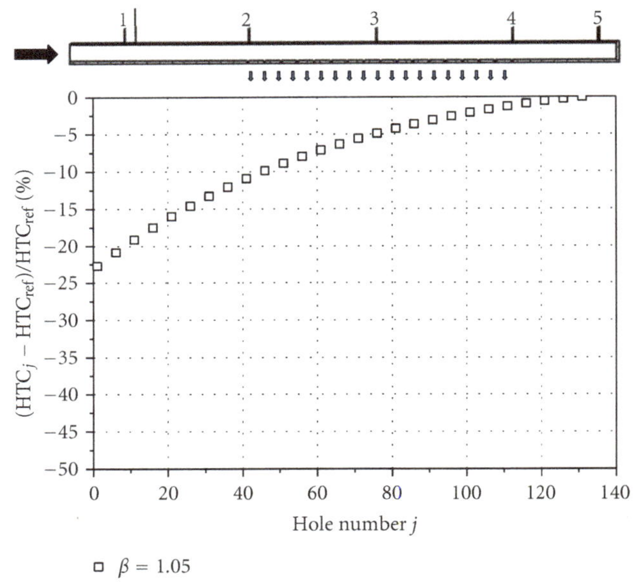

Figure 11: Mean heat transfer coefficient distribution for Geom 2 (A_r = 1.08) β = 1.05.

7.2. Impact of the Manifold Cross Section on the Jet Mean HTC Profile. Figure 12 shows the jet mean HTC distribution along the manifold for different values of the A_r parameter. The profiles are obtained assuming the same overall β ratio (β = 1.05), the same nozzle length-to-diameter ratio (t/d = 2), the same impinging holes diameter d = 1 mm, and the same nozzles number (i.e., the only parameter that changes is the manifold diameter "D").

As discussed above, the mean HTC profiles are flatter as the A_r increases. More in detail, the HTC profile's variations lie below 5% when the A_r is higher than 3. The better

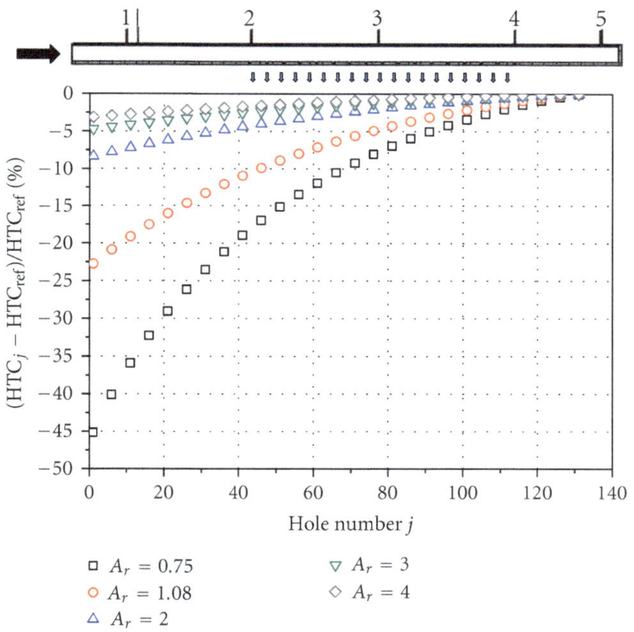

FIGURE 12: Mean heat transfer coefficient distribution varying the A_r parameter, $\beta = 1.05$.

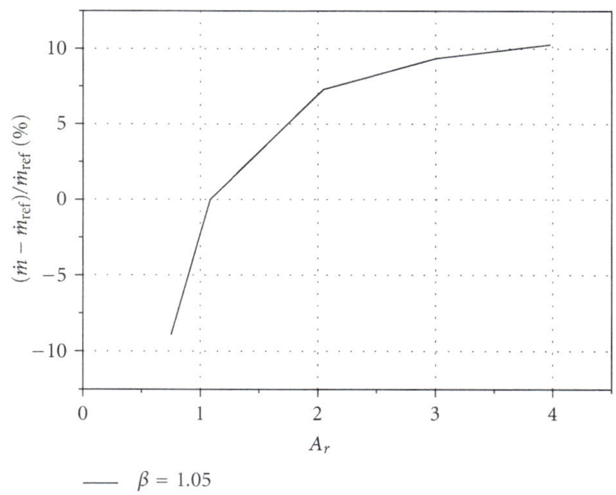

FIGURE 13: Total coolant mass flow rate varying the A_r parameter, $\beta = 1.05$.

performance obtained with the high A_r is due to a more uniform mass flow rate split along the manifold. As the mass flow through last hole of the array remains roughly the same, the total mass flow at the manifold inlet increases when the A_r is higher. Figure 13 shows the mass flow rate at the inlet of the feeding pipe for different A_r values. The parameter m_{ref} represents the total mass flow at the manifold inlet evaluated for Geom 2 when $\beta = 1.05$.

When the A_r is equal to 3 (i.e., when the changes in the HTC values along the manifold are below 5%) the mass flow rate at the manifold inlet increases, with respect to the nominal case with $A_r = 1.08$, by 9%.

The increase of the A_r by means of the manifold diameter "D" augmentation, has a beneficial effect on the jet's HTC profile uniformity. However, it has to be remarked that as the feeding pipe diameter increases, both the cooling system size and weight increase. Furthermore, the total coolant mass flow rate is higher as well. The last evidence may have a marginal effect in case of cooling air extracted from the fan flow but represents a penalty if the coolant is bled from the compressor.

7.3. Impact of the β Ratio on the Jet Mean HTC Profile. In present section, the effects of the overall β ratio are analysed. The impact of the β ratio on the jet's mean HTC profile is reported in Figure 14. To be consistent with the sensibility analysis previously conduced, the nozzles' diameter "d" have been changed in order to assure, for each case, the same mean HTC value at the last hole of the array. The manifold diameter "D" is fixed ($D = 12$ mm) and the t/d is equal to 2. Thus, as the β changes, the system A_r consequently varies.

Figure 14 shows that the increase of the system's overall pressure ratio has a beneficial effect on the HTC profile. This last statement can be motivated by observing that the β ratio affects both the isentropic mass flow rate and the impingement holes' discharge coefficient (i.e., as the β ratio increases both the C_d and the m_{ideal} result augmented). In order to obtain the same mean HTC value at the last hole of the array, the nozzles' diameter is then reduced. Thus the A_r increases significantly as the design pressure ratio is increased explaining the trend shown in Figure 14.

The reduction of the impingement holes' cross section leads, as the pressure ratio is increased, to an appreciable reduction of the coolant cumulative mass flow rate (Figure 15).

When $\beta = 1.2$, the HTC profile's variations lie below 5% and the total coolant mass flow rate is reduced by 13% with respect to the case when $\beta = 1.05$.

7.4. Impact of the Nozzles' Diameter Distribution on the Jet Mean HTC Profile. In the previous sections, for each considered geometry, the nozzles' diameter is uniform moving from the first to the last hole of the array. However in order to mitigate the changes in the nozzles' mass flow rate, the impinging manifold could be designed considering a non-uniform holes' diameter distribution. The 1-D code is then used to design manifold geometries characterized by nozzles with different diameter. Each hole's diameter is determined imposing the same mean HTC value (i.e., the designed impinging manifold guarantee a uniform HTC level moving from the first to the last nozzle of the array). When considering a manifold diameter $D = 12$ mm, $\beta = 1.05$, and $t/d = 2$, the nozzles' diameter distribution across the manifold calculated by the 1-D program is shown in Figure 16.

The first nozzle's diameter is increased by 80% with respect to the last one. For this manifold geometry, the calculated cumulative coolant mass flow rate is roughly

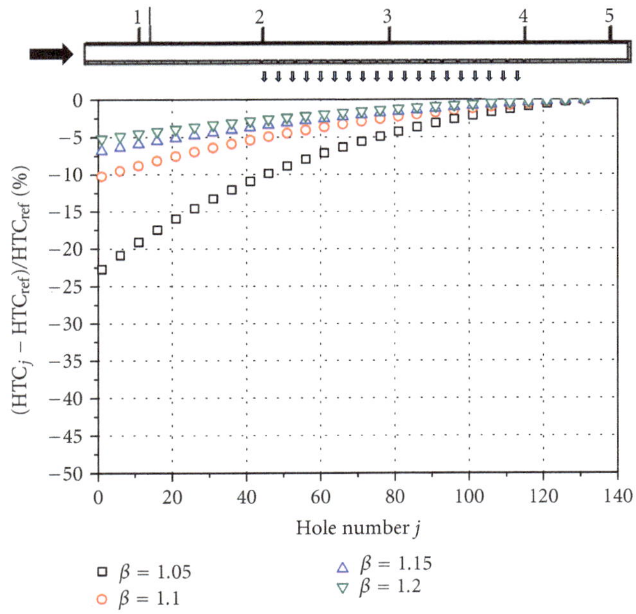

Figure 14: Mean heat transfer coefficient distribution varying the β parameter.

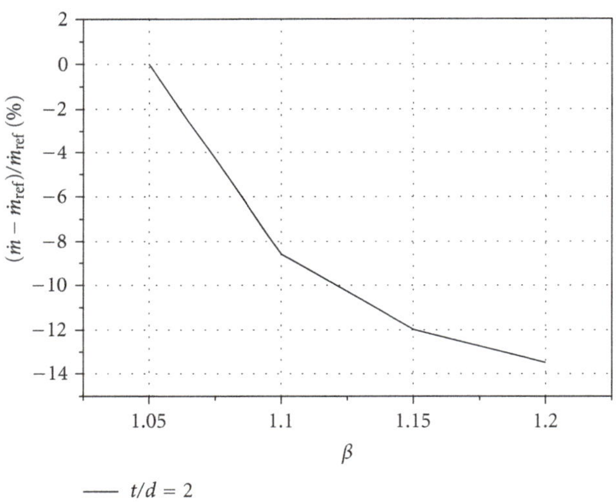

Figure 15: Total coolant mass flow rate varying the β parameter.

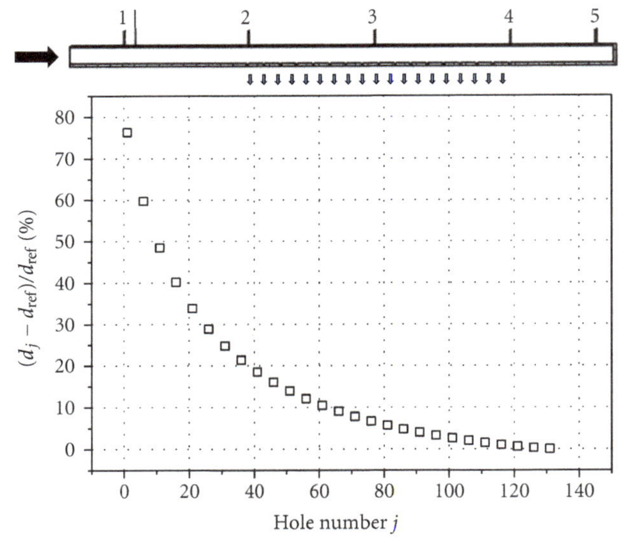

Figure 16: Nozzles' diameter distribution across the manifold $\beta = 1.05$, $D = 12$ mm, and $t/d = 2$ ($d_{ref} = 1$ mm).

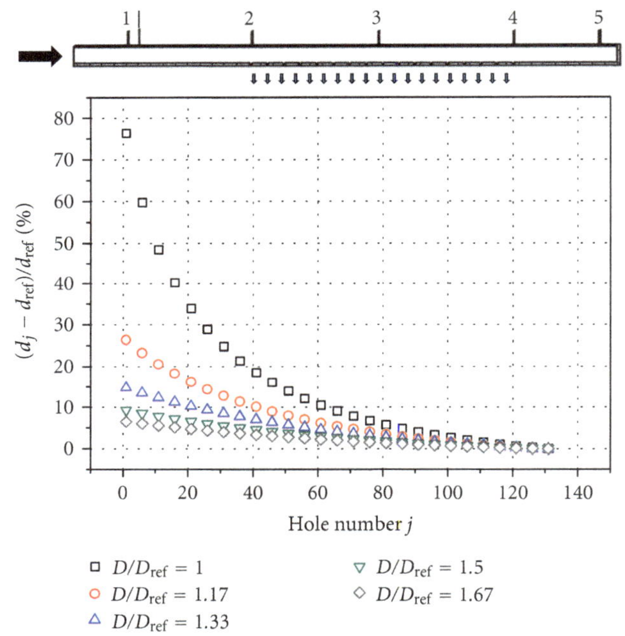

Figure 17: Nozzles' diameter distribution across the manifold $\beta = 1.05$, $t/d = 2$ varying the manifold diameter ($d_{ref} = 1$ mm and $D_{ref} = 12$ mm).

26% higher than those obtained in case of uniform holes' diameter $d = 1$ mm.

Figure 17 reports the nozzles' diameter distribution calculated in case of different manifold diameter D. As expected, the changes in the impinging holes diameter is reduced as the feeding pipe cross section is augmented. The last evidence also results in a reduction of the total coolant mass flow rate, with respect to the case for which $D/D_{ref} = 1$ (i.e., $D = 12$ mm), Figure 18.

8. Conclusions

This paper deals with a simplified 1-D code aimed at the design of the multiple jet section of an ACC cooling tube. The 1-D program is able to evaluate the mass flow rate through each nozzle of the array and then to calculate the related heat transfer coefficient (HTC) achieved on the target surface.

The code has been validated by means of both numerical and experimental data. Once validated, the program has been

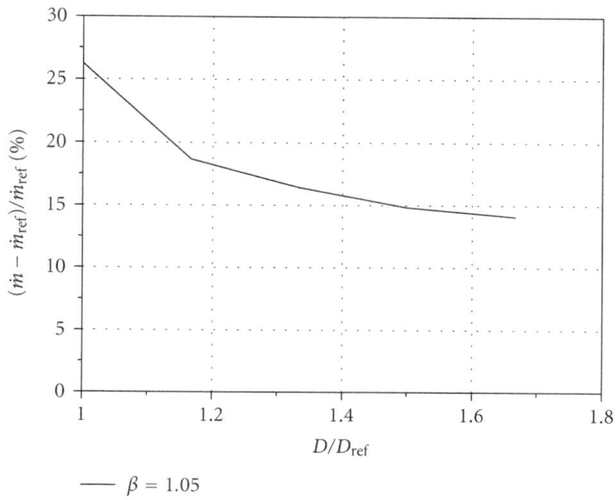

FIGURE 18: Total coolant mass flow rate varying the D/D_{ref} parameter.

used to point out some design practices for such cooling systems.

It emerges from the study that the mass flow rate split across an ACC cooling tube can be largely variable. In that way, the key parameter is the pipe area ratio A_r defined as the ratio of the manifold inlet area and the holes cumulative exit area. More in detail, geometries for which the A_r is greater than 5 lead to a roughly constant impingement jet mass flow rate distribution while, in case of geometries characterized by a low A_r, relevant differences in jet mass flow rates across the feeding pipe are expected. Variations in the mass flow rate split across an ACC cooling tube result in HTC profile nonuniformities that should be avoided or reduced for a correct system design.

The augmentation of the pipe A_r (by means of the increase of the manifold diameter) has a direct positive effect on the HTC profile uniformity. However, these solutions have few drawbacks: the increase of both the weight and size of the cooling system.

The role of the operating conditions on the manifold design has been considered. The calculations reveal that once designed considering high overall pressure ratio, the impinging system performs rather better. This last evidence has been motivated when observing that the increase of the pressure ratio leads to an augmentation of both the nozzles' discharge coefficient and the related isentropic mass flow rate. Thus, when imposing always the same mean HTC value at the last hole of the array, the increase of the β ratio involves a higher pipe A_r. The coolant total mass flow rate is also reduced as the overall pressure ratio is augmented.

Finally, manifold geometry with non-uniform nozzles' diameter have been considered. In order to assure a uniform heat load, the impinging holes' diameter changes considerably across the manifold and the cumulative coolant mass flow rate is increased as well. In case of coolant air extracted from the fan flow, this kind of arrangement could be considered to limit the ACC system size and weight.

Nomenclature

A_c: Tube internal area(m^2)
A_j: Impingement hole area (m^2)
A_r: Area ratio $A_c/(n_{holes} \cdot A_j)$ (−)
d: Cooling hole diameter (mm)
d_{ref}: Cooling hole reference diameter (1 mm) (mm)
D: Tube internal diameter (mm)
D_{ref}: Tube internal reference diameter (12 mm) (mm)
h: Holes exit to target surface distance (m)
k: Surface thermal conductivity (W/mK)
L: Tube length (mm)
$\dot{m}$: Mass flow rate (kg/s)
Ma: Mach number (−)
Nu: Nusselt number HTCd/k(−)
P: Pressure (Pa)
$\dot{Q}$: Heat flux (W/m^2)
R: Gas constant (J/kgK)
Re: Reynolds number (−)
S: Pitch (mm)
t: Nozzle length (mm)
T: Temperature (K)
x: Distance between the considered point and the stagnation point (m).

Acronyms

ACC: Active clearance control
C_d: Discharge coefficient (−)
HTC: Heat transfer coefficient $\dot{Q}/(T_{aw} - T_w)$(W/m^2K)
MV: Mass velocity ρv (kg/sm^2)
MVR: Mass velocity ratio $(\rho v)_j/(\rho v)_c$ (−).

Greek Letters

β: Global pressure ratio (P@manifold inlet/P discharge) (−)
β_L: Local pressure ratio (−)
μ: Air viscosity (kg/ms)
ρ: Air density (kg/m^3).

Subscripts

av: Averaged value
aw: Adiabatic wall
c: Main channel
d: Discharge conditions
exp: Experimental value
max: Maximum value
n: Last hole of the array
is: Isentropic
j: Jet impingement hole
ref: Reference value
s: Static
t: Total

w: Wall
x: Streamwise direction.

Acknowledgment

The authors wish to express their gratitude to F. Maiuolo and L. Tarchi for providing the experimental data.

References

[1] J. C. Han, J. S. Park, and C. K. Lei, "Heat transfer enhancement in channels with turbulence promoters," *Journal of Engineering for Gas Turbines and Power*, vol. 107, no. 3, pp. 628–635, 1985.

[2] J. F. Justak and C. Doux, "Self-acting clearance control for turbine blade outer air seals," in *Proceedings of ASME Turbo Expo (GT2009-59683)*, pp. 1229–1237, June 2009.

[3] E. Halila, D. Lenahan, and T. Thomas, *High Pressure Turbine Test Hardware*, NASA CR-167955, 1982.

[4] B. Beck and W. Fasching, *CF6 Jet Engine Performance Improvement—Low Pressure Turbine Active Clearance Control*, NASA CR-165557, 1982.

[5] F. Ahmed, B. Weigand, and K. Meier, "Heat transfer and pressure drop characteristics for a turbine casing impingement cooling system," in *Proceedings of the ASME International Heat Transfer Conference, IHTC14-22817*, vol. 5, Washington, DC, USA, August 2010.

[6] F. Ahmed, R. Tucholke, B. Weigand, and K. Meier, "Numerical investigation of heat transfer and pressure drop characteristics for different hole geometries of a turbine casing impingement cooling system," in *Proceedings of ASME Turbo Expo, GT2011-45251*, pp. 1095–1108, Vancouver, BC, Canada, 2011.

[7] A. H. Lefebvre, *Gas Turbine Combustion*, Taylor & Francis, 1998.

[8] N. Hay and D. Lampard, "The discharge coefficient of flared film cooling holes," in *Proceedings of ASME Turbo Expo, 95-GT-15*, June 1995.

[9] A. Andreini and R. DaSoghe, "Numerical characterization of aerodynamic losses of jet arrays for gas turbine applications," *Journal of Engineering for Gas Turbine and Power*, vol. 134, no. 5, article 052504, 8 pages, 2012.

[10] M. Gritsch, A. Schulz, and S. Wittig, "Method for correlating discharge coefficients of film-cooling holes," *AIAA Journal*, vol. 36, no. 6, pp. 976–980, 1998.

[11] M. Gritsch, A. Schulz, and S. Witting, "Effect of internal coolant crossflow orientation on the discharge coefficient of shaped film cooling holes," in *Proceedings of ASME Turbo Expo, 99-GT-40*, 1999.

[12] D. A. Rowbury, M. L. G. Oldfield, and G. D. Lock, "A method for correlating the influence of external crossflowon the discharge coefficients of film cooling holes," *Journal of Turbomachinery*, vol. 123, no. 2, pp. 258–265, 2001.

[13] A. Schulz, M. Gritsch, and S. Wittig, "Effect of crossflows on the discharge coefficient of film cooling holes with varying angles of inclination," in *Proceedings of ASME Turbo Expo, 2001-GT-0134*, 2001.

[14] B. Facchini, F. Maiuolo, L. Tarchi, and D. Coutandin, "Heat transfer and effectiveness evaluation of multiple impingement jet arrays for an active clearance control system," ISAIF (ISAIF10-58), 2010.

[15] H. Martin, "Heat and mass transfer between impinging gas jets and solid surfaces," *Advances in Heat Transfer*, vol. 13, pp. 1–60, 1977.

[16] J. C. Han, S. Dutta, and S. V. Ekkad, *Gas Turbine Heat Transfer and Cooling Technology*, Taylor & Francis, 1st edition, 2000.

[17] R. DaSoghe, B. Facchini, F. Maiuolo, M. Micio, and L. Tarchi, "Discharge coefficient characterization of jet array impingement holes for an active clearance control system," in *Proceedings of the 9th ETC Conference*, vol. 252, Istanbul, Turkey, March 2011.

[18] R. DaSoghe and A. Andreini, "Numerical characterization of pressure drop for turbine casing impingement cooling system," in *Proceedings of ASME Turbo Expo, GT2012-68787*, 2012.

[19] L. W. Florschuetz, C. R. Truman, and D. E. Metzger, "Streamwise flow and heat transfer distributions for jet array impingement with crossflow," *Journal of Heat Transfer*, vol. 103, no. 2, pp. 337–342, 1981.

2

Data Adaptive Spectral Analysis of Unsteady Leakage Flow in an Axial Turbine

Konstantinos G. Barmpalias,[1] Ndaona Chokani,[1] Anestis I. Kalfas,[2] and Reza S. Abhari[1]

[1] *Laboratory for Energy Conversion (LEC), Swiss Federal Institute of Technology, ETH Zurich, 8092 Zurich, Switzerland*
[2] *Department of Mechanical Engineering, Aristotle University of Thessaloniki, GR-54124 Thessaloniki, Greece*

Correspondence should be addressed to Konstantinos G. Barmpalias, konstantinos.barmpalias@power.alstom.com

Academic Editor: David Japikse

A data adaptive spectral analysis method is applied to characterize the unsteady loss generation in the leakage flow of an axial turbine. Unlike conventional spectral analysis, this method adapts a model dataset to the actual data. The method is illustrated from the analysis of the unsteady wall pressures in the labyrinth seal of an axial turbine. Spectra from the method are shown to be in good agreement with conventional spectral estimates. Furthermore, the spectra using the method are obtained with data records that are 16 times shorter than for conventional spectral analysis, indicating that the unsteady processes in turbomachines can be studied with substantially shorter measurement schedules than is presently the norm.

1. Introduction

The flows in turbomachines are inherently unsteady as it is only possible to have work exchange in the presence of a time-varying pressure field, as shown by Dean [1]:

$$\frac{Dh_t}{Dt} = \frac{1}{\rho}\frac{\partial p}{\partial t}. \tag{1}$$

The unsteadiness within turbines arises due to the relative movement of the rotor and stator blade rows. Turbulence and stochastic processes give rise to random unsteadiness, whereas periodic unsteady phenomena, including wake-to-blade interactions, potential field interactions, vortex interactions, and cavity flow interactions occur at the blade passing frequency or its harmonics. Langston [2] and Sieverding [3] referred extensively to secondary flows occurring in turbomachinery and relating them to losses. Mansour et al. [4] verified these loss generation mechanisms after having performed unsteady entropy measurements with the use of his fast response entropy probe in the harsh environment of the turbomachines. The loss mechanisms of these unsteady flow interactions are related to the efficiency of the turbomachinery. It is thus evident that an improved knowledge of the unsteady flow interactions can shed light on novel approaches to improve the efficiency of turbomachinery, as shown by Denton [5].

Making use of unsteady interactions can lead to potential loss generation reduction. Unsteady interactions of rotor inlet cavity and main flows on shrouded turbines can mitigate the adverse effects of the rotor tip passage vortex at its onset [6]. Blade-row interactions and especially the effect of wakes were exploited to control loss generation in ultrahigh-lift low pressure turbines [7]. Mathematical modeling of unsteadiness in turbomachinery flows can be used in order to improve design as well as producing integrated blading and flow path design. Chaluvadi et al. [8] developed a model to quantify vortex transport inside the downstream bladerow breaking down the unsteady loss generated upstream. Behr et al. [9] showed that clocking may impact on loss generation through axial and radial redistribution of low energy fluid while Barmpalias et al. [10] associated the stretching and tilting of the vortex with secondary loss generation.

Spectral analysis is a tool that can be used to quantify unsteady signals. Ullum et al. [11] reported on the onset and identification of the frequency of rotating stall based on the power spectrum analysis of velocity and pressure data in a centrifugal pump. In a completely different area, Chen Li and

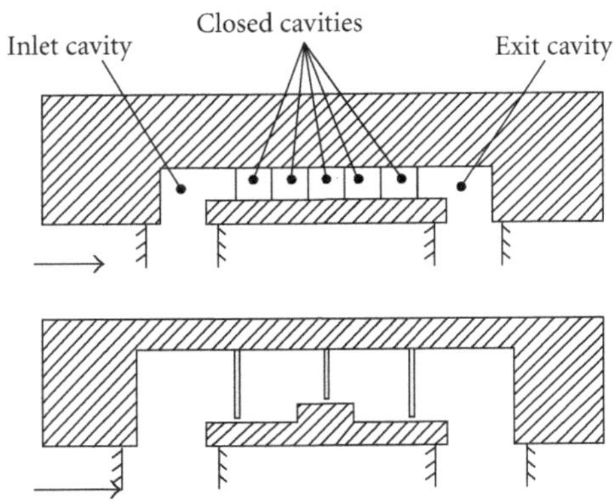

FIGURE 1: Examples of rotor tip labyrinth seal configurations (from Pfau [14]).

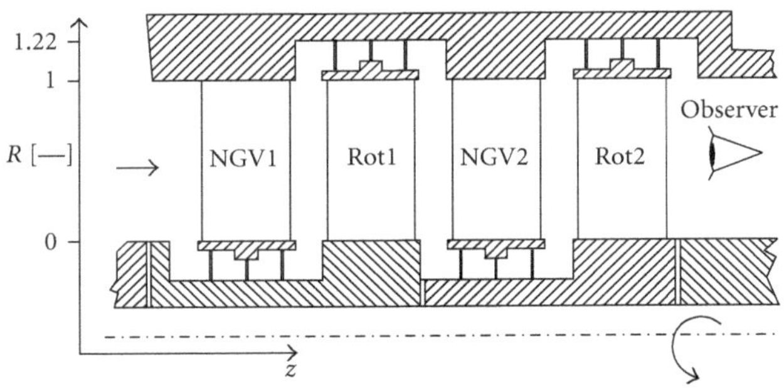

FIGURE 2: Meridional cut of the test section.

Li Ti-Pei [12] studied quasiperiodic oscillations in the X-ray emissions from a star using power spectrum density derived from an autoregressive model instead of the classical Fourier transformation so as to overcome the shortcomings of the latter. The work presented here is focused on the cavity flows that are associated with the rotor tip labyrinth seal, as seen in Figure 1. The rotor tip gap is required to provide a relative motion between the rotor blade and the casing. However, the associated unsteady cavity flow has a profound effect on the efficiency of the turbine. Pfau et al. [13] made highly resolved, three-dimensional flow measurements of the inlet cavity of a turbine rotor tip labyrinth seal in the two-stage axial research turbine. Additional measurements showed that this labyrinth seal configuration results in a turbine efficiency reduction of about 1.6%. Pfau [14] showed that the gap size has a major effect on turbine losses. Unsteady wall pressure measurements, as shown in Figure 2, comparing 0.3% and 1.0% gap cases indicate that there is a pronounced change in the cavity pressure distribution.

However, the changes in the pressure distribution could not be clearly discerned from the spectra. This difficulty arises due to the ensemble averaging procedures that are needed to improve the signal-to-noise ratio in Fourier-based power spectra. The objective of the current work is to demonstrate an alternative approach to obtaining spectral estimates. This alternate approach, the maximum entropy method of spectral analysis, has superior resolving power compared with more conventional spectral analysis.

TABLE 1: Main characteristics of the test turbine.

Pressure ratio	1.32	Mass flow	9.86 kg/sec
Max. power	400 kW	Turbine speed	2700 rpm
T_{inlet}	40°C	p_{exit}	Ambient
Mach	0.1–0.4	Reynolds number	10^5
Number of blades on rotor/stator	42	Tip diameter	800 mm
Blade passing frequency	1890 Hz	Blade aspect ratio	1.8

The maximum entropy method (MEM) was first introduced by Burg [15], with his interest being focused on the analysis of geophysical data series. This method was subsequently used in related studies [16, 17]. Burg commented on the better resolution and the more realistic power estimates of the MEM compared to the conventional methods of spectral estimation. Later Ulrych and Clayton [18] reviewed the principles of maximum entropy spectral analysis and the closely related topic of autoregressive time series modeling. Barrodale and Erickson [19, 20] developed an algorithm to solve the underlying least-squares linear prediction problem in maximum entropy spectral analysis. At the same time, Theodoridis and Cooper [21] applied the maximum entropy spectral analysis technique to signals with spectral peaks of finite width and compared their results to that of the conventional Fourier method. They reported that the spectra were smoother and better resolved. Well-defined spectral peaks could be obtained using fewer points. Maximum entropy methods for spectral estimates now have broader acceptance and use in the field of vibration data analysis [22] and even in the textile industry [13]. Morgenstern and Chokani [23] applied MEM for purposes of spectral analysis of pressure signals in hypersonic flows past open cavities and noted that extremely short data lengths are required to yield the spectra.

In general, the method is suited to data with poor signal-to-noise ratios and also for extremely short data lengths. In the next section we briefly review conventional methods of spectral analysis and highlight their shortcomings. The formulation of the maximum entropy method of spectral analysis is then outlined, and the results of its application to the unsteady wall pressure measurements of Pfau [14] are presented. Finally, we conclude with a summary of the work.

1.1. The Research Facility. The measurements were performed in the "LISA" two-stage axial research turbine at the Laboratory for Energy Conversion (LEC), ETH Zurich. The turbine inlet temperature is kept constant at 400 K with an accuracy of 0.9 K. A DC generator maintains a constant operating speed of 2700 ± 0.5 RPM (±0.02%). A more detailed description of the test facility is available in Schlienger et al. [24]. The main characteristics of the turbine are summarized in Table 1.

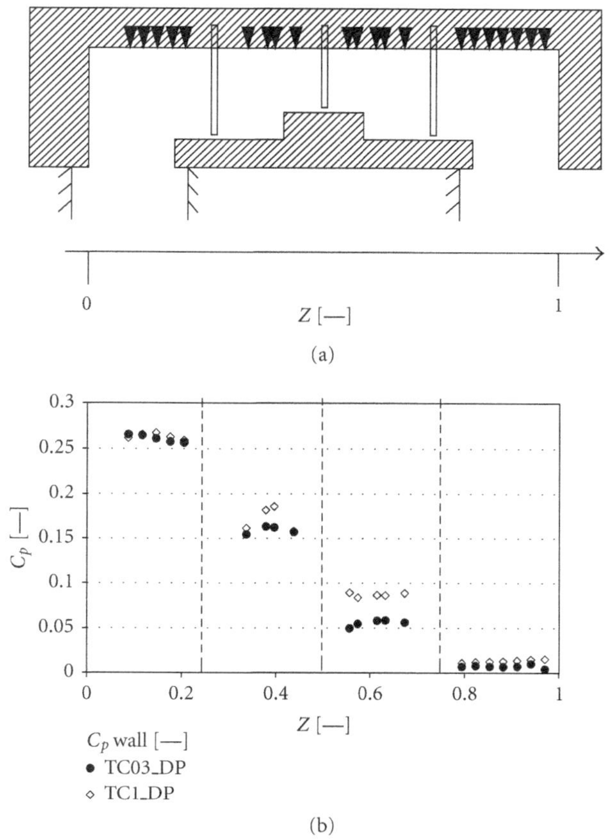

FIGURE 3: Axial locations of pressure taps in labyrinth groove (a) and pitch-wise averaged wall pressures Cp for 0.3% gap case (TC03_DP) and 1% gap case (TC1_DP) (b) (from Pfau [14]).

The constant annulus of the turbine and the four blade rows are depicted in Figure 2. The geometry under investigation is representative of a steam turbine. The unsteady wall measurements performed during this campaign were made at the rotor inlet cavity downstream of the 2nd stator.

1.2. Unsteady Wall Pressure Measurements. The unsteady pressure measurements presented here were acquired at the inlet cavity of the 2nd rotor shown in Figure 3. Six pressure holes in an axial direction, distanced 2.5 mm apart, were used. The inlet cavity itself has a 25 mm axial wall. The 1st measurement hole is located 15% downstream from the upstream radial wall. The potential to circumferentially clock the cavity ring against the stator permits multimeasurements in the circumferential direction. One stator pitch is thus covered by 23 circumferential measurements.

2. Conventional Spectral Analysis

The most widely used method of spectral estimation is that based on the use of the fast Fourier transformation. In this approach, if we consider a uniformly sampled time series x_n where $n = 1, 2, \ldots, N$, then the estimate of the power spectral density is given as

$$\mathrm{PSD}_m = \frac{1}{N \cdot \Delta t}\left(X_m X_m^*\right). \tag{2}$$

The index m denotes the discrete frequency, f_m, at which the power spectral density is estimated. The FFT approach for spectral estimates is widely used as computationally efficient methods for the determination of (2) have been developed. However, although the FFT technique is widely used, there are several aspects of the technique that limit its reliability. Firstly, the technique assumes that the signal of the sampled time series is a linear series of sine waves. Secondly, the spectral resolution is inversely proportional to the duration of the signal; therefore, for small frequency resolution a long signal is required. Thirdly, the signal is assumed to be stationary—that is the processes generating the signal are assumed to be the same from the beginning through to the end of the analysis period. Furthermore, a window function is required to minimize the leakage in the spectral estimates; this window function distorts the estimates in adjacent frequencies. Finally, when the signal-to-noise ratio of the signal is large, the spectral estimates are unreliable. Alternatively, the Wavelet Transform (WT) may be employed so as to overcome the FFT shortcomings. This would allow for the analysis of signals that have highly concentrated time localized high-frequency components superimposed on longer lived low-frequency components. This feature would allow for feature detection apart from signal noise removal and data compression [25]. The WT has a time window interval width that is inversely proportional to its wave width for all frequencies.

3. Data Adaptive Spectral Analysis

In an attempt to overcome the aforementioned limitations of FFT conventional spectral analysis method, Burg [15] suggested the maximum entropy method for spectral analysis. Burg's interest was in the analysis of geophysical data series, and in that application it was found that this new method gave better spectral resolution and more realistic estimates of power. The development of the MEM as suggested by Burg is as follows. The entropy of a Gaussian process is given as:

$$H = \frac{1}{4f_N}\int_{-f_N}^{f_N} \log \mathrm{PSD}(f)\,df. \tag{3}$$

In terms of the auto covariance of the data time series, (3) can be rewritten as

$$H = \frac{1}{4f_N}\int_{-f_N}^{f_N} \log\left[\sum_{k=-\infty}^{\infty} \rho(k) e^{-2\pi i f_k \Delta t}\,df\right]. \tag{4}$$

If H is maximized by the use of Lagrange multipliers, subject to the constraint that the spectral estimate is consistent with the known autocovariances, $\rho(k)$, $-p < k < p$, the solution to the variational problem is

$$P_{\mathrm{MEM}}(f) = \frac{P_p \Delta t}{\left|1 - \sum_{k=1}^{p} a_k e^{-2\pi i f_k \Delta t}\right|^2}. \tag{5}$$

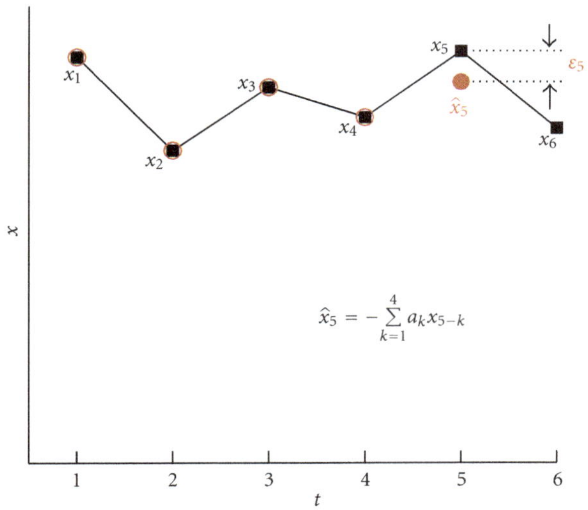

FIGURE 4: The Maximum entropy method of spectral analysis from the perspective of time series modeling. In this example, black squares denote measured data; unfilled red circles show past measured values that are used to estimate a modeled data point (filled red symbol).

Equation (5) is the expression for the spectral estimates obtained from MEM and is equivalent to the estimate derived using the FFT method, shown by (2). The maximizing of (4) results in the rather unfortunately named—in the sense that it is confusing from the perspective of examining loss generation in turbomachines—*maximum entropy method for spectral analysis.*

Van den Bos [26] investigated the duality between the *maximum entropy method for spectral analysis* and autoregressive representation. If we briefly review this duality, the interpretation of (5) is then clearer. Consider an autoregressive model that is consistent with the measured data:

$$\hat{x}_n = -\sum_{k=1}^{p} a_k x_{n-k}. \tag{6}$$

Note that (6) represents a pth-order model, in which the modeled data $\hat{x}_n$ are determined from a finite number, p, of previous measured data points.

Thus, for example, if we have a 4th order model, a linear combination of 4 past values is used to predict each modeled data point, as in Figure 4. Although we lose data points, in the sense that if N measured data points are available, then only (N-p) modeled data points can be determined, and the modeled data are solely based on known, measured data. The total squared error between the modeled and measured data is given as

$$\begin{aligned} E &= \sum_{n=p}^{N-1} (\hat{x}_n - x_n)^2 \\ &= \sum_{n=p}^{N-1} \left(-\sum_{k-1}^{p} a_k x_{n-k} - x_n \right)^2. \end{aligned} \tag{7}$$

The minimization of (7) with respect to the coefficients a_1, $a_2, \ldots, a_p$ then yields a spectrum of (N-p)-length time series of error, $\varepsilon_n = \hat{x}_n - x_n$, that is equivalent to that of "white" noise. It is therefore evident from (3) that the uncertainty ε_n is maximized, and therefore this maximization yields the most random (i.e., maximum entropy) error between the modeled and measured data. On the other hand, for this maximization (therefore known coefficients $a_1, a_2, \ldots, a_p$) (6) represents a time series model that is *adapted* to the measured data. Thus we prefer to refer to Burg's approach as a *data adaptive spectral analysis method.*

The basic working equation for the spectral estimates determined using the *data adaptive spectral analysis method* is given by (5). Barmpalias et al. [10] present an algorithm to determine the coefficients $a_1, a_2, \ldots, a_p$ and the model constant, $P_p = P_{p-1}(1 - a_p^2)$, from the measured data.

It is pertinent to highlight two salient differences between the spectral estimates from conventional spectral analysis, as shown in (2), and from *data adaptive spectral analysis*, as in (5). Firstly in (5) no windowing is used, unlike in (2). Thus (5) is based on actual measured data, whereas (2) is based on measured data that are modified by the windowing procedure. Secondly, in (5) the frequency of the spectral estimate, f, is an independent variable. Thus the spectral resolution in (5) is independent of the sampling frequency and the user chooses the desired spectral resolution. By contrast, in conventional spectral analysis the frequency resolution is the inverse of the duration of the length of the measured data, and therefore in (2) high frequency resolution can only be accomplished by having a long measured signal. Burg [15] has noted that (5) is well suited for short data lengths.

The determination of the order, p, in the time series model, (6), is of central importance to the *data adaptive spectral analysis method.* Stated simply, if p is too small, then a smooth spectrum is obtained; on the other hand, if p is too large, then there are spurious peaks in the spectrum. The most commonly used approaches to determine the optimum p involve the final prediction error, Akaike [25]. That is given by

$$\mathrm{FPE}_p = \frac{N + p + 1}{N - p - 1} P_p \tag{8}$$

and/or the Akaike [22] information criterion, that is,

$$\mathrm{AIC}_p = N \log_{10} P_p + 2p. \tag{9}$$

The optimum p is determined as the first occurrence of $\mathrm{FPE}_{p+1} > \mathrm{FPE}_p$ for successively increasing p. This criterion was not found to be reliable in the present work. Instead, as the primary focus in this work is periodic unsteady phenomena and cavity flow interactions that occur at the blade passing frequency, the optimum p is determined as that value for which the FPE first exhibits a local maximum in the region of interest as p is increased. This range is bounded by the occurrence of a smooth spectrum for relatively small p and by spurious peaks in the spectrum when a certain value of p is exceeded. For this application the optimum p was found to be within the range of 200 to 400. As seen in

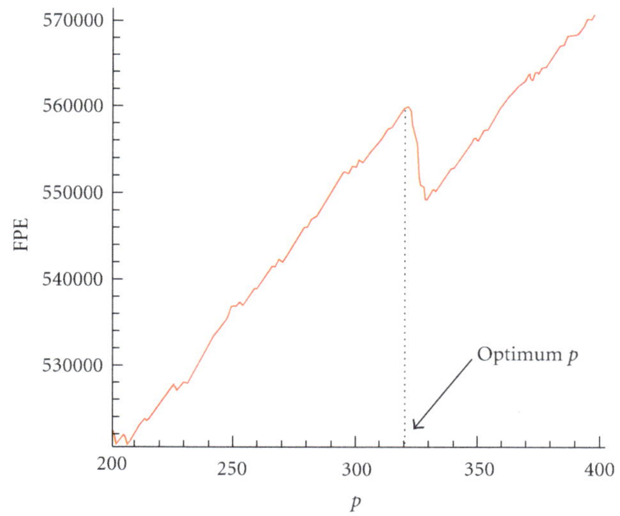

FIGURE 5: Determining the optimum p as the first occurrence of a local maximum in FPE as p increases.

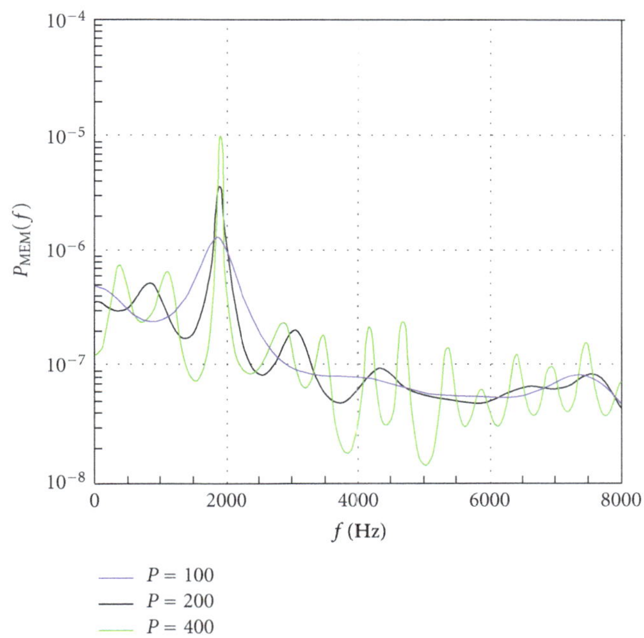

FIGURE 6: The effect of the time series model order, p, see (6), on the spectra predicted using the data adaptive spectral analysis method.

Figure 5, FPE has a local maximum at $p = 320$, which is then taken as the optimum p.

4. Results and Discussion

An analysis of the unsteady wall pressure measurements of Pfau [14] is presented to illustrate the *data adaptive spectral analysis method*. The labyrinth seal configuration consists of 4 cavities with a gap size of 0.1 mm, which corresponds to 1% of blade height. The data are sampled at a rate of 200 kHz.

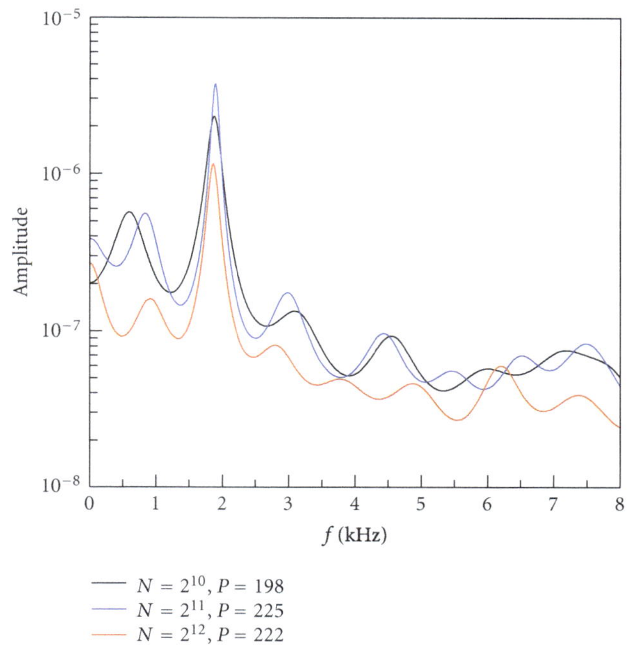

FIGURE 7: The effect of the length of measured data, N, on the spectra predicted using the data adaptive spectral analysis method.

For the present work the data acquired in cavity 1 at the midcircumference station are presented.

Figure 6 examines the effect of the time series model order, p, as shown in (6) on the spectra predicted using the *data adaptive spectral analysis method*. Three different model orders, 100, 200, and 400, are examined. For all three cases, the spectral resolution is 12.2 Hz and the spectra show a dominant peak at the blade passing frequency, 1840 Hz. The optimum value of p is 200, and the corresponding spectrum shows additional peaks at the harmonic frequency and non-linear interaction frequencies (i.e., subharmonic frequency and difference interaction (fundamental and subharmonic) frequency). For $p = 100$ (< 200), it can be seen that the spectrum is relatively smooth compared to the spectrum of the optimum p, where, for example, the harmonic frequency is not detected. For $p = 400$ (> 200), on the other hand, there are many spurious peaks in the spectrum. These results clarify the importance of determining the order of the time series model in the *data adaptive spectral analysis method*.

The effect of the length of the measured data record, N, on the spectra predicted using the *data adaptive spectral analysis method* is examined in Figure 7. Three different data record lengths, 2^{10}, 2^{11}, and 2^{12}, are examined. For each data record length the spectral resolution is specified as 12.2 Hz, and the optimum order of time series model, p, is determined and found to be 198, 225, and 222, respectively, for the three data record lengths. It can be seen that for all three cases, the spectra show essentially the same information. This result shows the ability of the *data adaptive spectral analysis method* to extract the same spectral information for both long and short data records. It should be noted that there are small differences in the spectral

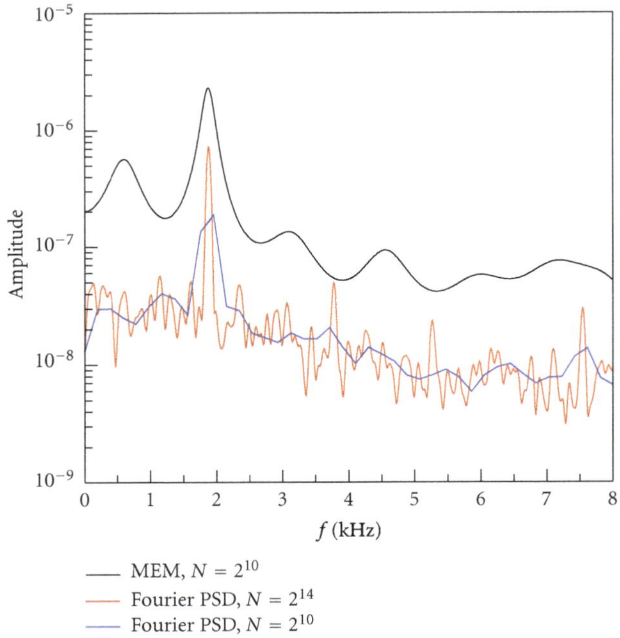

FIGURE 8: Comparison of data adaptive and conventional spectral estimates of power spectral densities.

amplitudes. These differences arise since the model constant, $P_p = P_{p-1}(1 - a_p^2)$, in (5) is related to the coefficients of the time series model. However, the normalized spectra are indistinguishable, as noted by Burg [15].

Figure 8 compares the power spectral estimates of the data adaptive and conventional methods of spectral analysis. For sake of clarity, the conventional spectral estimates are offset by one decade from the data adaptive spectral estimate. In all three cases, the peak at the blade passing frequency is identified as the most dominant feature of the spectra. In the case where the data record lengths are the same, $N = 2^{10}$, it is evident that the conventional spectral estimate has a relatively poor spectral resolution, 195.2 Hz. Furthermore, in the conventional spectral estimate, no peaks other than that at the blade passing frequency are clearly identified. In order to improve the spectral resolution in the conventional spectral estimate to the spectral resolution, 12.2 Hz, of the data adaptive spectral estimate, the data record length must be increased by a factor 16 to $N = 2^{14}$. The resultant spectrum, as seen in Figure 8, captures additional peaks in the spectra. However, it is evident that the conventional spectral estimate is noisier. On the other hand, the data adaptive spectral estimate, which is obtained with a shorter data record, yields a smoother spectrum. It is therefore evident that using the *data adaptive spectral analysis method*, the measurement schedule could be reduced by up to factor 16 in terms of data acquisition, while still being able to extract available information from the spectra.

5. Concluding Remarks

A *data adaptive spectral analysis method*, which has previously been used in the analysis of geophysical data series, has been introduced in the context of examining the periodic unsteadiness within turbines. This *data adaptive spectral analysis method* employs a time series model of measured data. The spectrum of the model's error is the most random (i.e., maximum entropy) that is consistent with measured data. No windows are employed in the determination of the spectral estimates. Spectral estimates are obtained with a data record that is 16 times shorter than that required for Fourier power spectrum estimates. The use of shorter data records has the potential to reduce measurement schedules for measurement test campaigns.

The application of this *data adaptive spectral analysis method* (maximum entropy method) as employed within the context of this study can be summarized in the following steps for a given number of data points, N.

(1) Calculate the final prediction error, FPE, using (8) as the order of p is increased.

(2) Determine the optimum order of p as the first occurrence of a maximum in FPE, as in Figure 5.

(3) Calculate the spectrum using (5).

Nomenclature

a_k: Coefficients of time series model
ε_n: Model error, $\hat{x}_n - x_n$
E: Total error
F: Frequency
f_{bp}: Blade passing frequency
f_m: Discrete frequency, $(m\text{-}1)\,\Delta f$
f_N: Nyquist frequency
Δf: Spectral resolution
H: Information entropy of a Gaussian process
M: Index for spectral estimate
N: Index for data time series
P: Order of time series model, or pressure
P_p: Output power of p-length time series model
P_{MEM}: MEM power spectral estimate
R: Blade span
T: Time
Δt: Sampling rate
x_n: Measured data time series
$\hat{x}_n$: Modeled data time series
X_m: Fourier transformation of x_n
X_m^*: Complex conjugate of X_m
Z: Axial direction
T_{inlet}: Temperature at stage inlet
p_{exit}: Pressure at turbine exit.

Abbreviations

FFT: Fast Fourier transformation
FPE: Final prediction error

MEM: Maximum entropy method for spectral analysis
PSD_m: Power spectral density estimate.

Acknowledgment

The authors are grateful to Axel S. Pfau for making available the unsteady wall pressure data used here.

References

[1] R. C. Dean, "On the necessity of unsteady flow in fluid machines," *ASME Journal of Basic Engineering*, vol. 81, pp. 24–28, 1959.

[2] L. S. Langston, "Secondary flows in axial turbines—a review," *Annals of the New York Academy of Sciences*, vol. 934, pp. 11–26, 2001.

[3] C. H. Sieverding, "Recent progress in the understanding of basic aspects of secondary flows in turbine blade passages," *Journal of Engineering for Gas Turbines and Power*, vol. 107, no. 2, pp. 248–257, 1985.

[4] M. Mansour, N. Chokani, A. I. Kalfas, and R. S. Abhari, "Time-resolved entropy measurements using a fast response entropy probe," *Measurement Science and Technology*, vol. 19, no. 11, Article ID 115401, 2008.

[5] J. D. Denton, "Loss mechanisms in turbomachines," *Journal of Turbomachinery*, vol. 115, no. 4, pp. 621–656, 1993.

[6] K. G. Barmpalias, A. I. Kalfas, and R. S. Abhari, "Design considerations for axial steam turbine rotor inlet cavity volume and length scale," in *ASME Turbo Expo*, Vancouver, Canada, June 2011.

[7] H. P. Hodson and R. J. Howell, "Bladerow interactions, transition, and high-lift aerofoils in low-pressure turbines," *Annual Review of Fluid Mechanics*, vol. 37, pp. 71–98, 2005.

[8] V. S. P. Chaluvadi, A. I. Kalfas, H. P. Hodson, H. Ohyama, and E. Watanabe, "Blade row interaction in a high-pressure steam turbine," *Journal of Turbomachinery*, vol. 125, no. 1, pp. 14–24, 2003.

[9] T. Behr, L. Porreca, T. Mokulys, A. I. Kalfas, and R. S. Abhari, "Multistage aspects and unsteady effects of stator and rotor clocking in an axial turbine with low aspect ratio blading," in *ASME Turbo Expo*, pp. 1211–1222, June 2004.

[10] K. G. Barmpalias, A. I. Kalfas, N. Chokani, and R. S. Abhari, "The dynamics of the vorticity field in a low solidity axial turbine," in *ASME Turbo Expo*, June 2008.

[11] U. Ullum, J. Wright, O. Dayi et al., "Prediction of rotating stall within an impeller of a centrifugal pump based on spectral analysis of pressure and velocity data," *Journal of Physics: Conference Series*, vol. 52, no. 1, pp. 36–45, 2006.

[12] L. Chen and T. P. Li, "Autoregressive spectral estimation for quasi-periodic oscillations," *Chinese Journal of Astronomy and Astrophysics*, vol. 5, no. 5, pp. 495–507, 2005.

[13] A. Pfau, J. Schlienger, D. Rusch, A. I. Kalfas, and R. S. Abhari, "Unsteady flow interactions within the inlet cavity of a turbine rotor tip labyrinth seal," *Journal of Turbomachinery*, vol. 127, no. 4, pp. 679–688, 2005.

[14] A. Pfau, *Loss mechanisms in labyrinth seals of shrouded axial turbines ETH Dissertation No. 15226*, Swiss Federal Institute of Technology, ETH, Zürich, Switzerland, 2003.

[15] J. P. Burg, *Maximum entropy spectral analysis*, Ph.D. thesis, Department of Geophysics, Stanford University, 1975.

[16] R. O. Vicente and R. G. Currie, "Maximum entropy spectrum of long-period polar motion," *Royal Astrological Society Geophysical Journal*, vol. 46, pp. 67–73, 1976.

[17] H. R. Radoski, E. J. Zawalick, and P. F. Fougere, "The superiority of maximum entropy power spectrum techniques applied to geomagnetic micropulsations," *Physics of the Earth and Planetary Interiors*, vol. 12, no. 2-3, pp. 208–216, 1976.

[18] T. J. Ulrych and R. W. Clayton, "Time series modelling and maximum entropy," *Physics of the Earth and Planetary Interiors*, vol. 12, no. 2-3, pp. 188–200, 1976.

[19] I. Barrodale and R. E. Erickson, "Algorithms for least-squares linear prediction and maximum entropy spectral analysis—part I: theory," *Geophysics*, vol. 45, no. 3, pp. 420–432, 1980.

[20] I. Barrodale and R. E. Erickson, "Algorithms for least-squares linear prediction and maximum entropy spectral analysis—part II: fortran program," *Geophysics*, vol. 45, no. 3, pp. 433–446, 1980.

[21] S. Theodoridis and D. C. Cooper, "Application of the maximum entropy spectrum analysis technique to signals with spectral peaks of finite width," *Signal Processing*, vol. 3, no. 2, pp. 109–122, 1981.

[22] T. M. Romberg, A. G. Cassar, and R. W. Harris, "A comparison of traditional Fourier and maximum entropy spectral methods for vibration analysis," *Journal of Vibration, Acoustics, Stress, and Reliability in Design*, vol. 106, no. 1, pp. 36–39, 1984.

[23] A. Morgenstern and N. Chokani, "Hypersonic flow past open cavities," *AIAA Journal*, vol. 32, no. 12, pp. 2387–2393, 1994.

[24] J. Schlienger, A. Pfau, A. I. Kalfas, and R. S. Abhari, "Effects of labyrinth seal variation on multistage axial turbine flow," in *ASME Turbo Expo*, pp. 173–185, June 2003.

[25] H. Akaike, "Fitting autoregressive models for prediction," *Annals of the Institute of Statistical Mathematics*, vol. 21, pp. 243–247, 1969.

[26] A. van den Bos, "Alternative interpretation of maximum entropy spectral analysis," *IEEE Transactions on Information Theory*, vol. 17, no. 4, pp. 493–494, 1971.

Numerical Research on Effects of Splitter Blades to the Influence of Pump as Turbine

Yang Sun-Sheng,[1] Kong Fan-Yu,[1] Fu Jian-Hui,[2] and Xue Ling[1]

[1] *Research Center of Fluid Machinery Engineering and Technology, Jiangsu University, Zhenjiang, Jiangsu 212013, China*
[2] *Zhejiang Jiali Technology Co., Ltd, Hangzhou, Zhejiang 311241, China*

Correspondence should be addressed to Yang Sun-Sheng, yangsunsheng@126.com

Academic Editor: Luis Gato

Centrifugal pumps can be operated in reverse as small hydropower recovery turbines and are cheaper than bespoke turbines due to their ease of manufacture. Splitter blades technique is one of the techniques used in flow field optimization and performance enhancement of rotating machinery. To understand the effects of splitter blades to the steady and unsteady influence of PAT, numerical research was performed. 3D Navier-Stokes solver CFX was used in the performance prediction and analysis of PAT's performance. Results show that splitter blades have a positive impact on PAT's performance. With the increase of splitter blades, its required pressure head is dropped and its efficiency is increased. Unsteady pressure field analysis and comparison show that the unsteady pressure field within PAT is improved when splitter blades are added to impeller flow passage. To verify the accuracy of numerical prediction methods, an open PAT test rig was built at Jiangsu University. The PAT was manufactured and tested. Comparison between experimental and numerical results shows that the discrepancy between numerical and experimental results is acceptable. CFD can be used in the performance prediction and optimization of PAT.

1. Introduction

Small hydroelectric power stations became attractive for generating electrical energy after the oil price crisis of the seventies. However cost per kW energy produced by these stations is higher than the hydroelectric power plants with large capacity. We considered the idea of using pumps as hydraulic turbines an attractive and important alternative. Pumps are relatively simple machine, easy to maintain, and readily available in most developing countries. From the economical point of view, it is often stated that pumps working as turbines in the range of 1 to 500 kW allow capital payback periods of two years or less which is considerably less than that of a conventional turbine [1].

Current research on PAT mainly focuses on the development of accurate prediction models for the turbine operation of different designs of centrifugal pumps [2–4]. However, the accuracy of these prediction models requires continuous verification and development. The optimization of PAT means generating more energy from current small hydropower resources. So some papers have been concentrated on the optimization of PAT [5–8]. Splitter blades technique is one of the techniques frequently used in turbomachinery design and application [9–11]. Thus, we consider the idea of adding splitter blades to PAT's impeller flow passage.

Because of the impulse of the blades that rotates with respect to the stator, the transmission of fluid machinery is unsteady [12]. The interaction of rotor and stator could result in the generation of pressure pulsations [13], component fatigue [14], and excessive noise and vibration [15]. Adding of splitter blades to impeller flow passage will cause an increase of impeller blade number and a change of the three-dimensional unsteady pressure flow field within PAT. Therefore, investigation into splitter blades to the influence of unsteady pressure field can be useful.

CFD has been widely used in the flow field visualization, optimization, and unsteady flow field analysis of rotating machinery [16, 17]. Yet, investigation into unsteady pressure field of PAT using CFD is few [18]. In this paper, the all geometry of PAT including volute, impeller, and chambers

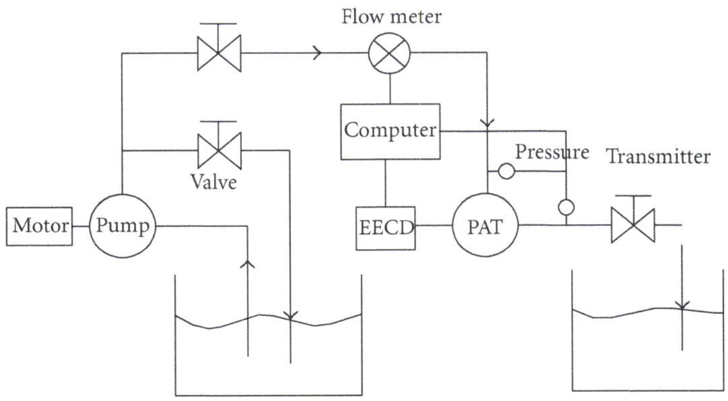

FIGURE 1: An open PAT test rig.

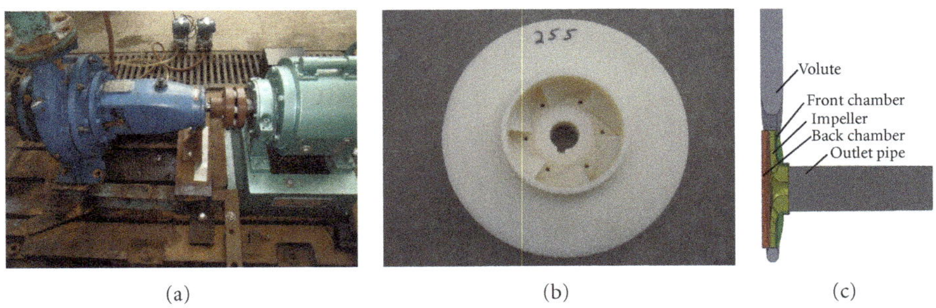

FIGURE 2: (a) Test PAT. (b) Impeller. (c) 3D model.

TABLE 1: Main geometric parameters of the designed PAT.

b_2	14.38
D_1	102
D_2	255
D_3	266
D_4	65
D_h	30
z	6/6+6
e	4
L_{sb}	$0.8\,L_{mb}$
β_1	19.5
β_2	20
L	15
θ	0.6
ϕ	131.5

has been simulated, using commercial flow solver. The effects of splitter blades to the PAT's influence of steady performance and unsteady flow field are investigated.

A complete micro-hydropower test rig was established at Jiangsu University and was used for experimental verification of numerical results. The simulated PAT was made and tested using this test ring. All required parameters were measured for obtaining complete characteristic curves of the PAT. The accuracy of CFD results for steady state simulation was verified.

2. Experimental Setup

A complete laboratory model of PAT an open test rig, as shown in Figure 1, was set up at Jiangsu University. High-pressure fluid required for PAT energy recovery is supplied by a feed pump. An electric eddy current dynamometer (EECD) is installed to measure and consume energy generated by PAT and to regulate PAT's rotational speed. Its accuracy of torque measurement is ±0.4% N·m. The discharge is measured by a turbine flow meter with an uncertainty of ±0.5%. PAT inlet and outlet pressure are measured by pressure transmitter with an uncertainty of ±0.4%. After measuring all parameters, PAT required pressure head, shaft power, and efficiency were obtained.

3. PAT Main Geometric Parameters

The focus of the investigation is a single-stage centrifugal PAT with a rotational speed of 1500 rpm. Table 1 lists the main geometric parameters of the designed PAT. Figure 2(b) shows the tested PAT without splitter blades.

4. Numerical Investigation

ANSYS-CFX is a commercial 3D Navier-stokes CFD code that utilizes a finite-element-based finite-volume method to discrete the transport equations [19]. It is a fully-implicit solver; thus it creates no time step limitation and is considered easy to implement. It is also a coupled solver

(a) (b)

Figure 3: (a) Mesh of impeller. (b) Mesh assembly.

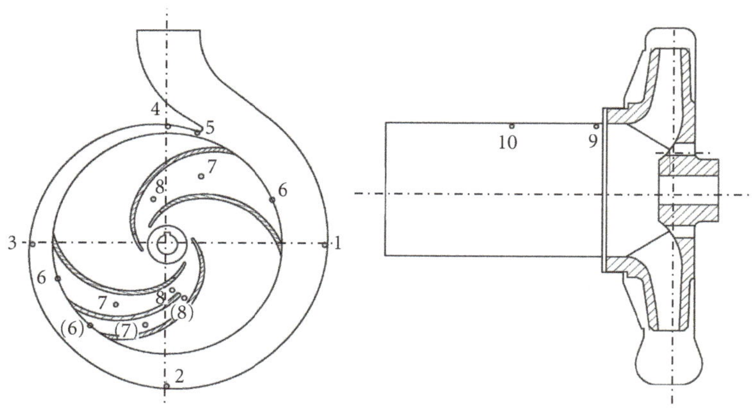

Figure 4: Location of pressure monitoring points.

Table 2: BEPs of experimental and numerical results.

Items	Q (m^3/h)	η (%)	P (kW)	H (m)
Exp	96.99	62.53	6.78	41.04
Without splitter	95.00	65.77	6.92	40.69
With splitter	90.00	69.19	6.24	36.80

meaning that the momentum and continuity equations are solved simultaneously. This approach reduces the number of iterations required to obtain convergence and no pressure correction term is required to retain mass conversion, which leads into a more robust and accurate solver.

The fluid flow in a pump front and back chamber is usually neglected during CFD simulation. This simplification, which only has a minor effect on pump performance, is reasonable [20]. Unfortunately, the same is not true when considering in turbine mode. Due to the importance of PAT inlet flow, relatively large error has been reported in turbine mode [20]. In order to get a more accurate result, all fluid flow domain within the PAT control volume was modeled. The fluid flow domain was split into five component parts: volute, impeller passage, front and back chambers, and outlet pipe as indicated in Figure 2(c). This separation allowed each mesh to be generated individually and tailored to the flow requirements in that particular component. To get a relatively stable inlet and outlet flow, four times of the pipe diameter was extended in the PAT inlet and outlet section.

4.1. Mesh Generation. ICEM-CFD was used to generate structured hexahedral grid for each component part [19]. A grid-independent test of PAT's performance was performed; it was found that when mesh numbers were around 1 million, the variation of efficiency was within 0.5%. The final number of mesh elements of volute, six blades impeller, front and back chambers, outlet pipe, and total number was 479710, 325164, 134956, 129700, 174720, and 1244250, respectively. For impeller with splitter blades, the mesh number was almost the same.

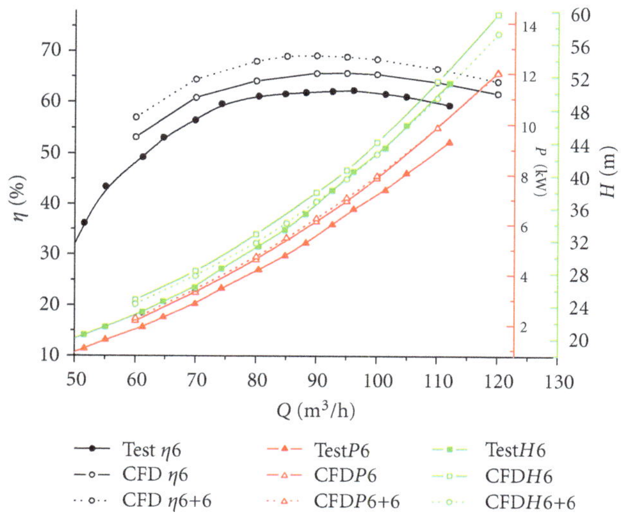

FIGURE 5: Comparison between experimental and numerical results.

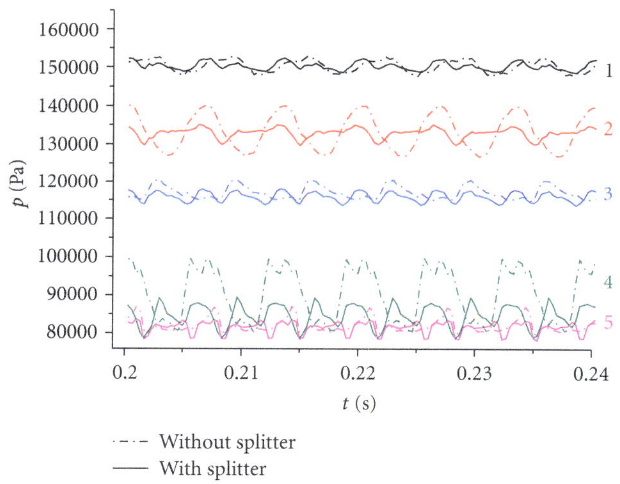

FIGURE 6: Time domain of pressure pulsation at volute monitoring points.

Care was taken in the PAT is outlet pipe mesh generation in order that mesh between the interface of outlet pipe and front chamber leakage is almost the same. The y^+ near the boundary wall was around 40. Due to the complexity of generating a structured mesh based on geometry, great efforts had been taken in the mesh generation of volute. Figure 3 gives a general view of the generated meshes.

4.2. Solution Parameters. The turbulence model selected was k-ε model. Standard wall function was used in the near wall treatment. The advection scheme was set to high resolution. The convergence criterion was 10^{-6}. The fluid selected was water at 25°C. All the wall surface roughness within the control volume was set to 50 μm. The inlet and outlet boundary conditions were set to static pressure inlet and mass flow rate outlet for both steady and unsteady simulation [19]. By changing the mass flow rate, performance curves of the PAT were acquired.

As the motion of the impeller blades relative to the stationary volute was central to the investigation, the analysis must involve multiple frames of reference. The volute and outlet pipe were set in stationary frame of reference and the impeller was set in rotary frame of reference. The interfaces between two stationary components, rotary and stationary components, were set to general grid and rotor stator interface, respectively.

Steady-state simulation was served as initial condition for unsteady simulation. The total time was set to 0.44 s, and the time period set for each time step is 3.33×10^{-4} s. The time period of one cycle is 0.04 s, so 11 rounds were simulated for the unsteady simulation. To get a relatively stable flow field, the final 6 rounds were selected for analysis. Figure 4 shows the position of monitoring points. The interfaces between two stationary components, rotary and stationary components, were set to general grid and transient rotor stator interface, respectively.

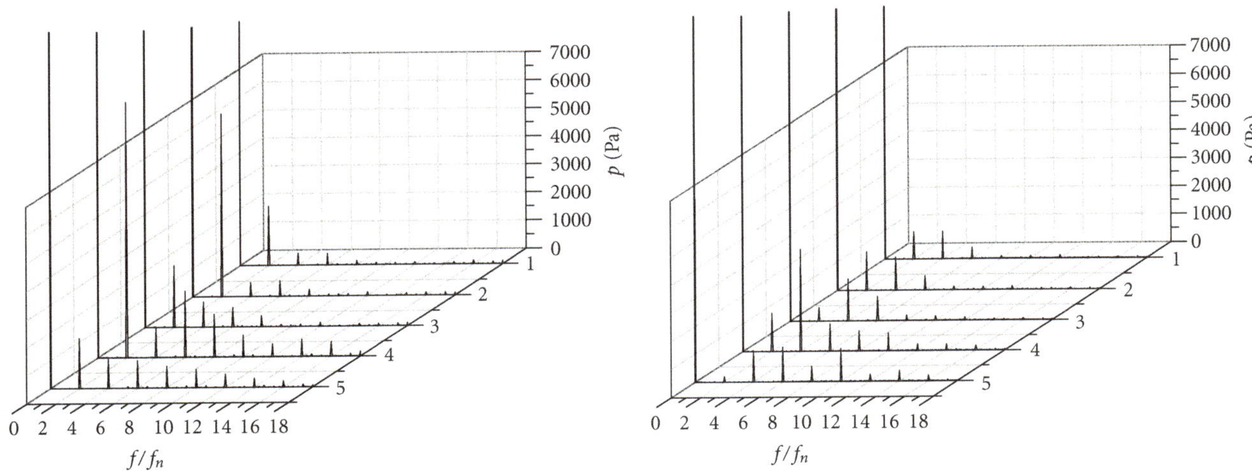

(a) Without splitter blades (b) With splitter blades

FIGURE 7: Frequency domain of pressure pulsation at volute monitoring points.

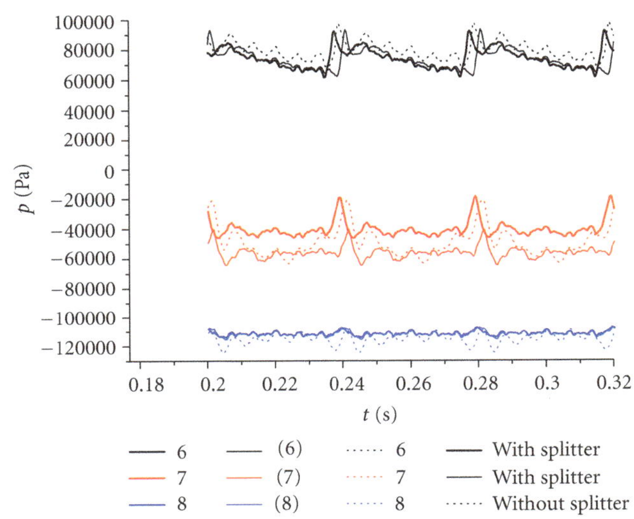

FIGURE 8: Time domain of pressure pulsation at impeller monitoring points.

5. Results Analysis

5.1. Performance Comparisons of Experimental and Numerical Results. An impeller with six blades was manufactured and tested. Numerical simulation of impeller with and without splitter blades was performed. Comparison of experimental and numerical results is presented in Figure 5. Table 2 lists its experimental and numerical results at the best efficiency points (BEPs). As is shown in Figure 5, the tendency of PAT's numerical predicted performance curves is in agreement with that of experimental. Numerical predicted efficiency, pressure head, and shaft power are higher than those of experimental. This overpredict of efficiency, pressure head, and shaft power may be attributed to the neglection of leakage loss through balancing holes and mechanical loss caused by mechanical seal and bearings. The comparison between experimental and steady-state numerical results indicates that the grid discretization and turbulence model are good for the PAT numerical simulation.

Numerical results comparison of PAT's impeller with and without splitter blades show that PAT's efficiency over the full operating range is increased when splitter blades are added to impeller flow passage. The required pressure head of PAT with splitter blades is decreased compared with that of without splitter blades. The variation of shaft power with and without splitter blades is minor.

5.2. Unsteady Pressure Field Comparison. The interaction of rotating impeller and stationery volute could cause strong pressure pulsation. To analyze PAT's unsteady pressure field and compare the pressure difference caused by the increase of splitter blade to PAT's impeller, unsteady pressure field analysis and comparison at a flow rate of $90\,m^3/h$ are presented.

TABLE 3: Amplitude of frequency domain and MPF at volute monitoring points (see Figures 7(a) and 7(b)).

Items	Frequency	1	2	3	4	5
Without splitter	50 Hz	2101.52	6497.02	2194.44	9119.44	1797.85
	100 Hz	425.34	491.59	911.75	1111.94	1051.60
	150 Hz	420.67	560.64	699.09	2390.07	987.26
	MPF	1.44	3.73	1.66	5.24	2.03
With splitter	50 Hz	956.43	1361.59	492.59	1379.13	190.40
	100 Hz	963.56	1160.57	1495.59	3628.93	1120.18
	150 Hz	400.03	498.83	870.31	984.72	1249.82
	MPF	1.02	1.55	1.25	3.09	1.48

TABLE 4: Amplitude of frequency domain and MPF at impeller monitoring points (see Figures 9(a), 9(b), and 9(c)).

Items	Frequency	6/(6)	7/(7)	8/(8)
Without splitter (a)	8.33 Hz	6018.02	11049.18	2256.43
	16.67 Hz	1649.96	2377.97	281.91
	33.33 Hz	2255.83	4226.27	1600.46
	41.67 Hz	2872.05	3407.42	1617.94
	MPF	8.05	11.47	4.34
With splitter (b)	8.33 Hz	7445.42	3633.10	164.20
	16.77 Hz	2417.61	2764.29	645.44
	33.33 Hz	2466.57	3954.63	1000.15
	41.67 Hz	2674.59	2972.82	458.17
	MPF	8.55	7.58	2.02
With splitter (c)	8.33 Hz	7195.57	1958.94	424.43
	16.77 Hz	2589.07	909.47	756.64
	33.33 Hz	2413.62	3585.52	1085.72
	41.67 Hz	2313.00	2206.07	617.10
	MPF	8.59	6.75	2.30

5.2.1. Volute. The absolute pressure pulsation comparisons with and without splitter blades within volute are plotted in Figure 6. As is shown in Figure 6, the absolute pressure gradually decreases along the volute flow channel. The maximum pressure pulsation among the five monitoring points is point 4, which is about 20° before the volute tongue. When splitter blades are added to the impeller, there is a significant drop of pressure fluctuation of the five monitoring points, especially at pressure monitoring points of 2 and 4.

Figure 7 presents the frequency domain of PAT's absolute pressure pulsation after Fast Fourier Transform (FFT) with and without splitter blades. Table 3 lists the amplitude value at different frequencies. The maximum pressure fluctuation (MPF) is calculated from

$$\text{MPF} = \frac{p_{\max} - p_{\min}}{\rho g H} \times 100, \qquad (1)$$

where $p_{\max}$ and $p_{\min}$ are the maximum and minimal pressure; ρ is the density of fluid; H is PAT's required pressure head.

Figure 7(a) shows that the main frequency of PAT's without splitter blades is 50 Hz. PAT's rotational speed is 1500 rpm. The impeller rotational frequency f_n is 25 Hz. So PAT's main frequency of impeller without splitter blades is 2 times of the impeller rotational frequency. Also the high-frequency component of the pressure pulsation gradually increases as the flow channel is near the volute cutwater. Figure 7(b) indicates that when splitter blades are added to the impeller, its main frequency changes to 100 Hz. Also the value of pressure amplitude is greatly dropped.

TABLE 5: Amplitude of frequency domain and MPF at outlet pipe monitoring points (see Figure 11).

Points	Frequency	9	10
Without splitter	50 Hz	3646.66	3112.41
	100 Hz	523.45	705.60
	150 Hz	385.90	383.59
	MPF	2.48	2.25
With splitter	50 Hz	745.41	1416.07
	100 Hz	754.98	524.37
	150 Hz	544.05	510.22
	MPF Hz	1.02	1.30

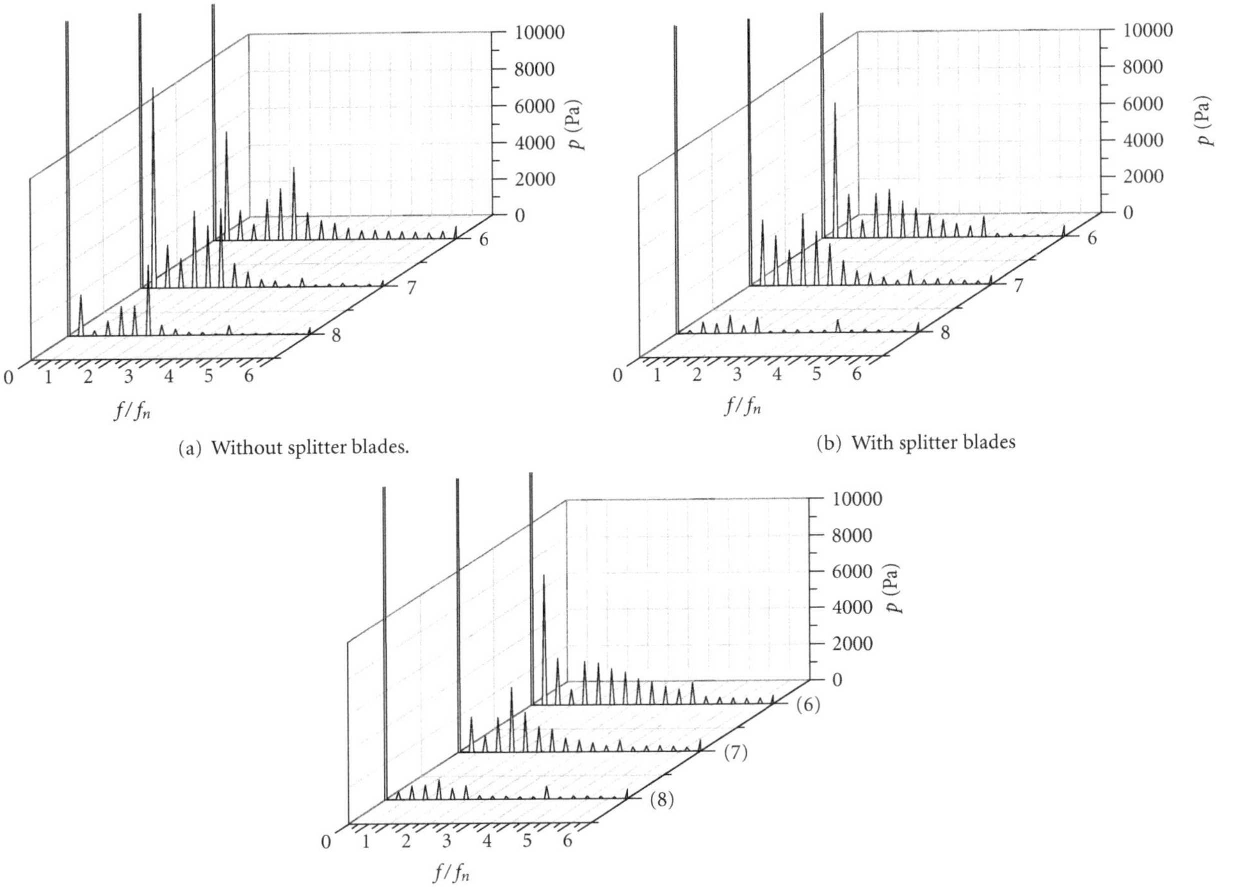

(a) Without splitter blades.

(b) With splitter blades

(c) With splitter blades

FIGURE 9: Frequency domain of pressure pulsation at impeller monitoring points.

Of the five investigated monitoring points, the minimum and maximum decreases of MPF are 24.70% and 58.44%, respectively.

5.2.2. Impeller. The absolute pressure comparisons within impeller are plotted in Figure 8. Figure 9 presents the frequency domain of the monitoring points after FFT. Table 4 lists the amplitude value at different frequencies.

As is shown in Figure 8, there is only one peak value within one impeller rotational period. This is different from that of volute. The absolute pressure decreases dramatically along the impeller flow passage. This great drop of absolute pressure is due to the energy transfer of high-pressure energy to PAT's shaft power.

Figure 9 and Table 4 indicate that the main frequency of pressure fluctuation within impeller is 8.33 Hz. So the blade passing frequency is 4.17 Hz. Thus, we could see that the main frequency of pressure fluctuation within impeller is 2 times of the blade passing frequency.

Figures 8, 9 and Table 4 demonstrate that the maximum amplitude value of pressure fluctuation among the three monitoring points for impeller without splitter blades is point 7. When splitter blades are added to the impeller, the amplitude value of pressure fluctuation at point 7 is decreased. And this decrease of pressure fluctuation also happens at pressure monitoring point 8. The pressure fluctuation at impeller inlet area is almost the same for impeller with and without splitter blades. Also, it is observed that the amplitude of pressure fluctuation within two flow passages caused by the increase of splitter blades to impeller passage is almost the same.

5.2.3. Outlet Pipe. The variations of absolute pressure at monitoring points 9 and 10 within outlet pipe are plotted in Figure 10. Figure 11 shows the frequency domain of the pressure pulsation after FFT. Table 5 lists the amplitude value of pressure pulsation at different frequencies.

Figure 10 indicates that there are six peak values within one impeller rotational period. This is the same as that of volute. Figure 11 and Table 5 indicate that the main frequency of pressure fluctuation within outlet pipe is 50 Hz. So PAT's main frequency is 2 times of the impeller rotational frequency. Also seen from Figures 10, 11 and Table 5, when splitter blades are added to the impeller, the amplitude of pressure fluctuation is decreased. And the MPF at monitoring points 9 and 10 is decreased by 58.87% and 42.22%, respectively.

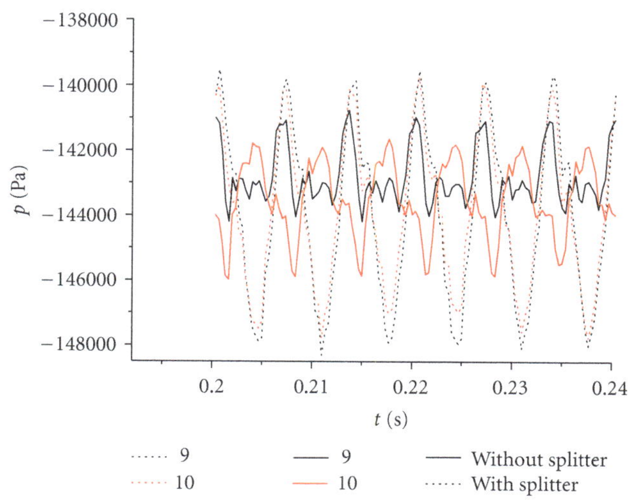

FIGURE 10: Time domain of pressure pulsation at outlet pipe monitoring points.

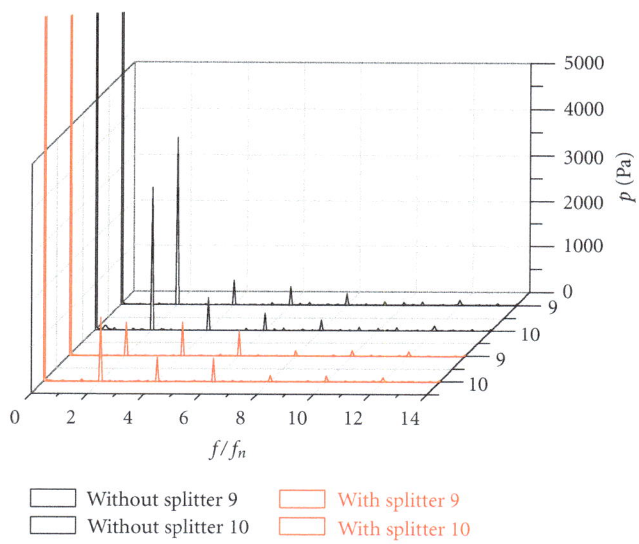

FIGURE 11: Frequency domain of pressure pulsation at outlet pipe monitoring points.

6. Conclusions

Numerical investigation into splitter blade to the performance influence of pump as turbine was carried out.

An open PAT test rig was built. Experimental and numerical research on a single-stage PAT of impeller without splitter blades was carried out. In the process of numerical simulation, all domains except the leakage through the balancing holes and mechanical seals are included. Comparison between experimental and numerical results shows that CFD results are in good agreement with those of experimental and can be used in the performance prediction and optimization of PAT. CFD could be used in the optimization and analysis of pump as turbine.

Numerical predicted performance comparison of impeller with and without splitter blades shows that PAT's efficiency is increased and its required pressure head is decreased when splitter blades are added to impeller flow passage.

Investigation into unsteady pressure field within volute shows that the absolute pressure gradually decreases along the volute flow channel. The maximum amplitude value of pressure pulsation within volute is located before the volute tongue. When splitter blades are added to the impeller, there is a significant decrease of pressure fluctuation.

Due to the energy transfer of high-pressure energy to PAT's shaft power, the absolute pressure within PAT's impeller decreases dramatically along impeller flow passage. For impeller without splitter blades, the maximum value of pressure pulsation within impeller happens in the middle of impeller. When splitter blades are added, its amplitude of pressure pulsation is greatly decreased. Also it is observed that the pressure pulsation within the two flow passages caused by the increase of splitter blade to impeller flow channel is almost the same.

Analysis of pressure pulsation within PAT's outlet pipe shows that the absolute pressure gradually decreases along the outlet pipe. When splitter blades are added to the impeller, the amplitude of pressure fluctuation is also decreased.

Nomenclature

b_2: Impeller inlet width, mm
D_1: Impeller outlet diameter, mm
D_2: Impeller inlet diameter, mm
D_3: Volute base circle diameter, mm
D_4: Volute inlet diameter, mm
D_h: Impeller hub diameter, mm
e: Impeller blade thickness, mm
f_n: Impeller rotational frequency, Hz
H: Head, m
L: Length of impeller wear ring, mm
L_{mb}: Length of main blade, mm
L_{sb}: Length of splitter blade, mm
n: Rotational speed, rpm
P: Power, W, kW
Q: Discharge, m^3/h
y^+: y plus
z: Number of blades.

Greek Symbols

η: Efficiency
θ: Splitter blade offset angle
ϕ: Blade wrap angle.

Acknowledgments

Special thanks are given to the national project entitled *The Exploitation and Utilization of Hydraulic Power Recovery Turbines (2009GJC10007)* and Doctoral innovation foundation of Jiangsu University entitled *Research on High Pressure Fluid Power Recovery Device of Pump as Turbine (CX10B_012X).*

References

[1] J. M. Chapallaz, P. Eichenberger, and G. Fischer, *Manual on Pumps Used as Turbines*, Vieweg, Braunschweig, Germany, 1992.

[2] M. Amelio and S. Barbarelli, "A one-dimensional numerical model for calculating the efficiency of pumps as turbines for implementation in micro-hydro power plants," in *Proceedings of the 7th Biennial ASME conference Engineering systems Design and Analysis (ESDA '04)*, pp. 65–72, Manchester, UK, July 2004.

[3] P. Singh and F. Nestmann, "An optimization routine on a prediction and selection model for the turbine operation of centrifugal pumps," *Experimental Thermal and Fluid Science*, vol. 34, no. 2, pp. 152–164, 2010.

[4] S. Derakhshan and A. Nourbakhsh, "Experimental study of characteristic curves of centrifugal pumps working as turbines in different specific speeds," *Experimental Thermal and Fluid Science*, vol. 32, no. 3, pp. 800–807, 2008.

[5] S. Derakhshan, B. Mohammadi, and A. Nourbakhsh, "Efficiency improvement of centrifugal reverse pumps," *Journal of Fluids Engineering, Transactions of the ASME*, vol. 131, no. 2, Article ID 021103, 9 pages, 2009.

[6] S. Derakhshan, B. Mohammadi, and A. Nourbakhsh, "The comparison of incomplete sensitivities and Genetic algorithms applications in 3D radial turbomachinery blade optimization," *Computers and Fluids*, vol. 39, no. 10, pp. 2022–2029, 2010.

[7] P. Singh, *Optimization of internal hydraulics and of system design for pumps as turbines with field implementation and evaluation*, Ph.D. Thesis, Karlsruhe University, 2005.

[8] P. Singh and F. Nestmann, "Internal hydraulic analysis of impeller rounding in centrifugal pumps as turbines," *Experimental Thermal and Fluid Science*, vol. 35, no. 1, pp. 121–134, 2011.

[9] M. Harano, K. Tani, and S. Nomoto, "Practical application of high-performance francis-turbine runner fitted with splitter blades at ontake and shinkurobegawa no. 3 power stations of the Kansai electric power CO., Inc," *Hitachi Review*, vol. 55, no. 3, pp. 109–113, 2006.

[10] G. Mustafa, P. Yasar, and S. Yakup, "Energy saving in a deep well pump with splitter blade," *Energy Conversion and Management*, vol. 47, no. 5, pp. 638–651, 2006.

[11] J. P. Solano, V. Pinilla, G. Paniagua, S. Lavagnoli, and T. Yasa, "Aero-thermal investigation of a multi-splitter axial turbine," *International Journal of Heat and Fluid Flow*, vol. 32, no. 5, pp. 1036–1046, 2011.

[12] R. Barrio, J. Parrondo, and E. Blanco, "Numerical analysis of the unsteady flow in the near-tongue region in a volute-type centrifugal pump for different operating points," *Computers and Fluids*, vol. 39, no. 5, pp. 859–870, 2010.

[13] K. A. Kaupert and T. Staubli, "The unsteady pressure field in a high specific speed centrifugal pump impeller—part I: influence of the volute," *Journal of Fluids Engineering, Transactions of the ASME*, vol. 121, no. 3, pp. 621–626, 1999.

[14] R. Spence and T. Purdom, "Numerical prediction of transient loadings on multistage pump impellers," in *Advances of CFD in Fluid Machinery Design*, IMechE, London, UK, 2003.

[15] L. K. Stanmore, "Field problems relating to high-energy centrifugal pumps operating at part-load," in *IMechE Conference C343/88 Pumps in Part Load*, 1988.

[16] H. Keck and M. Sick, "Thirty years of numerical flow simulation in hydraulic turbomachines," *Acta Mechanica*, vol. 201, no. 1–4, pp. 211–229, 2008.

[17] S. Yang and F. Kong, "Numerical simulation and comparison of pump and pump as turbine," in *ASME, Fluids Engineering Summer Meeting*, pp. 1–10, Montreal, Canada, 2010.

[18] S. Carlos, M. Jesus, and M. Katia, "Numerical modelling and flow analysis of a centrifugal pump running as a turbine: unsteady flow structures and its effects on the global performance," *International Journal for Numerical Methods in Fluids*, vol. 65, no. 5, pp. 542–562, 2011.

[19] Help Navigator, ANSYS CFX, Release 12.0 CFX-Solver modeling Guide.

[20] S. Derakhshan and A. Nourbakhsh, "Theoretical, numerical and experimental investigation of centrifugal pumps in reverse operation," *Experimental Thermal and Fluid Science*, vol. 32, no. 8, pp. 1620–1627, 2008.

4

Technology Review of Modern Gas Turbine Inlet Filtration Systems

Melissa Wilcox,[1] Rainer Kurz,[2] and Klaus Brun[1]

[1] *Machinery, Southwest Research Institute, San Antonio, Tx, USA*
[2] *Systems Analysis, Solar Turbines Incorporated, San Diego, CA, USA*

Correspondence should be addressed to Melissa Wilcox, melissa.wilcox@swri.org

Academic Editor: Mauro Venturini

An inlet air filtration system is essential for the successful operation of a gas turbine. The filtration system protects the gas turbine from harmful debris in the ambient air, which can lead to issues such as FOD, erosion, fouling, and corrosion. These issues if not addressed will result in a shorter operational life and reduced performance of the gas turbine. Modern day filtration systems are comprised of multiple filtration stages. Each stage is selected based on the local operating environment and the performance goals for the gas turbine. Selection of these systems can be a challenging task. This paper provides a review of the considerations for selecting an inlet filtration system by covering (1) the characteristics of filters and filter systems, (2) a review of the many types of filters, (3) a detailed look at the different environments where the gas turbine can operate, (4) a process for evaluating the site where the gas turbine will be or is installed, and (5) a method to compare various filter system options with life cycle cost analysis.

1. Introduction

Gas turbines ingest a large amount of ambient air during operation. Because of this, the quality of the air entering the turbine is a significant factor in the performance and life of the gas turbine. A filtration system is used to control the quality of the air by removing harmful contaminants that are present. The selection of the filtration system can be a daunting task, because there are many factors to consider. The system should be selected based on the operational philosophy and goals for the turbine, the contaminants present in the ambient air, and expected changes in the contaminants in the future due to temporary emission sources or seasonal changes. This paper outlines the primary considerations for selecting and installing a gas turbine inlet filtration system. First, the consequences that can occur due to improper inlet filtration are reviewed, then the different characteristics are discussed, after this the components of a filtration system and considerations for the operating environment are outlined, and lastly, a procedure for quantitatively comparing inlet filtration system options is provided.

2. Consequences of Poor Inlet Filtration

When the quality of the air entering the gas turbine is not well controlled, there are several consequences which can occur. Some of the most common degradation mechanisms are reviewed below including erosion, fouling, and corrosion.

2.1. Erosion. Erosion occurs when solid or liquid particles approximately 10 μm and larger impact rotating or stationary surfaces in the gas turbine. The particles will impact the surface and remove tiny particles of metal which eventually lead to changes in the geometry of the surface. This change in geometry causes deviations in the air flow path, roughening of smooth surfaces, alteration of clearances, and reduction of cross-sectional areas of metal components possibly in high stressed regions. Erosion is a non-reversible process; therefore, the gas turbine components must be replaced in order to regain their original condition. However, particles 10 μm and larger are easily removed by commercial filters [1–3].

2.2. Fouling. Fouling of compressor blades is an important mechanism leading to performance deterioration in gas turbines over time. Fouling is caused by the adherence of particles to airfoils and annulus surfaces. Particles that cause fouling are typically smaller than 2 to 10 μm. Smoke, oil mists, carbon, and sea salts are common examples. Fouling can be controlled by an appropriate air filtration system and often reversed to some degree by detergent washing of components. The adherence is impacted by oil or water mists. The result is a build-up of material that causes increased surface roughness and to some degree changes the shape of the airfoil (if the material build up forms thicker layers of deposits). Fouling in turn causes a decrease in the performance of the gas turbine.

Commercial filters can remove the majority of particles that cause fouling. But there are several submicron particles that are difficult to remove from the flow stream. The build-up of particles not removed by the inlet filtration system is removed with the use of compressor washing. This process recovers a larger portion of the compressor performance but cannot bring the gas turbine back to its original condition [1–5].

2.3. Corrosion. When chemically reactive particles adhere to surfaces in the gas turbine, corrosion can occur. Corrosion that occurs in the compressor section is referred to as "cold corrosion" and is due to wet deposits of salts, acid, and aggressive gases such as chlorine and sulfides. Corrosion in the combustor and turbine sections is called "hot corrosion." It is also referred to as a high temperature corrosion. Hot corrosion requires the interaction of the metal surface with another chemical substance at elevated temperatures. Hot corrosion is a form of accelerated oxidation that is produced by the chemical reaction between a component and molten salts deposited on its surface. Hot corrosion comprises a complex series of chemical reactions, making corrosion rates very difficult to predict. It is the accelerated oxidation of alloys caused by the deposit of salts (e.g., Na_2SO_4). Type I or high temperature hot corrosion occurs at a temperature range of 1346 to 1742°F (730 to 950°C). Type II or low temperature hot corrosion occurs at a temperature range of 1022 to 1346°F (550 to 730°C). Some of the more common forms of hot corrosion are sulfidation, nitridation, chlorination, carburization, and vanadium, potassium, and lead hot corrosion. Sulfidation Hot Corrosion requires the interaction of the metal surface with sodium sulfate or potassium sulfate, salts that can form in gas turbines from the reaction of sulfur oxides, water, and sodium chloride (table salt) or potassium chloride, respectively. It is usually divided into Type I and Type II hot corrosion, and Type I hot corrosion takes place above the melting temperature of sodium sulfate (1623°F (884°C)), while Type II occurs below this temperature. Hot corrosion is caused by the diffusion of sulfur from the molten sodium sulfate into the metal substrate which prevents the formation of the protective oxidation film and results in rapid removal of surface metal. One should note that for hot corrosion to occur both sulfur and salt (e.g., sodium chloride or potassium chloride or chloride) have to be present in the very hot gas stream in and downstream of the combustor. Sulfur and salt can come from the inlet air, from the fuel, or water (if water is injected). The potassium hot corrosion mechanism is similar to sulfidation but is less frequently observed in gas turbines, unless the fuel contains significant quantities of potassium.

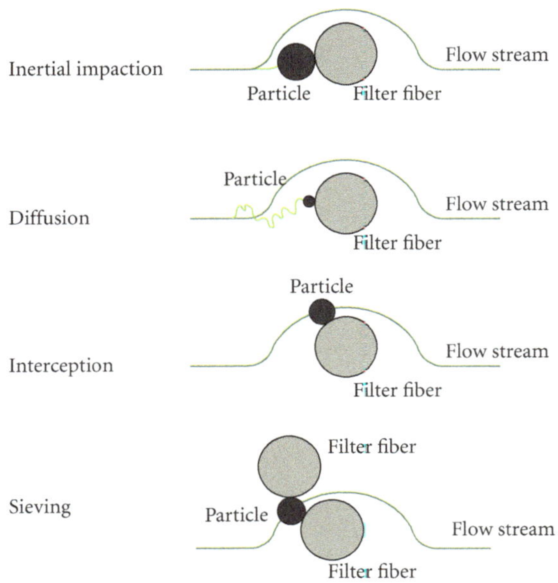

Figure 1: Common filtration mechanism [29].

Corrosion is a nonreversible degradation mechanism. Therefore, corroded components must be replaced in order to regain the original gas turbine performance. Corrosion also initiates or advances other damage mechanisms in the gas turbine. For example, corrosion can intrude into cracks or other material defects and accelerates crack propagation [1–3].

3. Filtration Characteristics

3.1. Filtration Mechanisms. Filters in the filtration system use many different mechanisms to remove contaminants from the air. The filter media, fiber size, packing density of the media, particle size, and electrostatic charge influence how the filter removes contaminants. Each filter typically has various different mechanisms working together to remove the contaminants. Four filtration mechanisms are shown in Figure 1.

The first filtration mechanism is inertial impaction. This type of filtration is applicable to particles larger than 1 μm in diameter. The inertia of the large heavy particles in the flow stream causes the particles to continue on a straight path as the flow stream moves around a filter fiber. The particulate then impacts and is attached to the filter media and held in place as shown in the top picture of Figure 1. This type of filtration mechanism is effective in high-velocity filtration systems.

The next filtration mechanism, diffusion, is effective for very small particles typically less than 0.5 μm in size. Effectiveness increases with lower flow velocities. Small particles interact with nearby particles and gas molecules. Especially in turbulent flow, the path of small particles

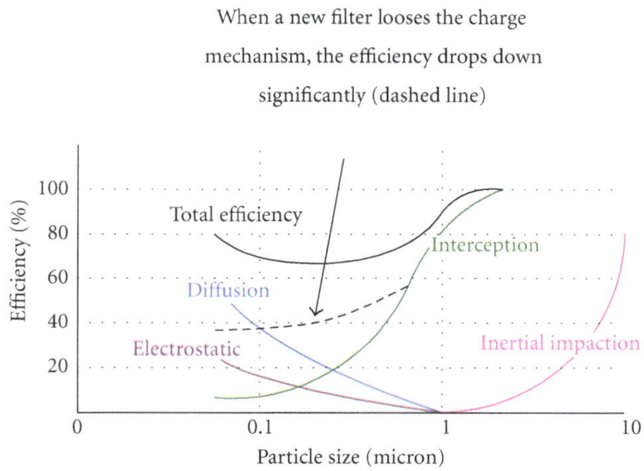

FIGURE 2: Combination of filtration mechanisms to obtain filter efficiency at various particle sizes [29].

fluctuates randomly about the main stream flow. As these particles diffuse in the flow stream, they collide with the fiber and are captured. The smaller a particle and the lower the flow rate through the filter media leads to a higher probability that the particle will be captured.

The next two filtration mechanisms are the most well known: interception and sieving. Interception occurs with medium-sized particles that are not large enough to leave the flow path due to inertia or not small enough to diffuse. The particles will follow the flow stream where they will touch a fiber in the filter media and be trapped and held. Sieving is the situation where the space between the filter fibers is smaller than the particle itself, which causes the particle to be captured and contained.

Another mechanism not shown in Figure 1 is electrostatic charge. This type of filtration is effective for particles in the 0.01 to 1 μm size range (Figure 2). The filter works through the attraction of particles to a charged filter. In gas turbine applications, this charge is applied to the filter before installation as a result of the manufacturing process. Filters always lose their electrostatic charge over time because the particles captured on their surface occupy charged surface area, therefore neutralizing their electrostatic charge. As the charge is lost, the filter efficiency for small particles will decrease. On the other hand, as the filter is loaded, the filtration efficiency increases, thus counteracting the effect of lost charge to some extent. This will offset some of the loss of filtration efficiency due to the lost charge. Figure 2 shows a comparison of a filter's total efficiency based on the various filtration mechanisms that are applied. The figure shows the difference between the filter's efficiency curve before and after the charge is lost. The performance of the filter should be based on the discharged condition [6, 7].

3.2. Filter Efficiency and Classification. Filter efficiency is a broad term. In general, the filter efficiency is the ratio of the weight, volume, area, or number of particles captured in the filter to the weight, volume, area, or number of the particles entering the filter, respectively. A general efficiency calculation is shown in (1), where W is the variable for which efficiency is being calculated. The efficiency can be expressed in several ways: maximum, minimum, or average lifetime value. Many filters have poor performance against small particles at the beginning of their lives, but as the filter media becomes loaded with particles, it is able to catch smaller particles. In this case, the average efficiency would actually be higher than the initial efficiency. Some of the filters will never reach the quoted maximum efficiency before they are replaced

$$\eta = \frac{W_{\text{entering}} - W_{\text{leaving}}}{W_{\text{entering}}} * 100\%. \qquad (1)$$

Filter efficiency is a trade-off against the pressure loss across the filter. Normally, the filtration system pressure loss will increase with an increase in filtration efficiency. As filters become more efficient, less dust penetrates through them. Also, the air flow path is more constricted with higher efficiency filters. This leads to higher pressure loss. Filter engineers must determine the acceptable pressure loss and efficiency for their application. Studies have shown that a higher pressure loss due to using a high efficiency filter has a lower effect on gas turbine power degradation than poor inlet air quality.

The efficiency of a filter cannot be stated as a general characteristic. The filter efficiencies vary with particle size, typically being lower for small particles and higher for large particles. They also vary with operational velocity. Filters designed for medium and low velocities will have a poor performance at higher velocities and vice versa. Therefore, a particle size range and flow velocity must be associated with the stated efficiency. For example, a filter may have 95 percent filtration efficiency for particles greater than 5 μm at a volumetric flowrate of 3000 cfm (5097 m^3/h), but the efficiency could be reduced to less than 70 percent for particles less than 5 μm or at a volumetric flowrate of 4000 cfm (6796 m^3/h).

Filters are rated for performance based on standards established in the United States of America and Europe. These filter ratings are based on the results of standard performance tests. In the United States, ASHRAE standard 52.2:2007 outlines the requirements for performance tests and the methodology to calculate the efficiencies. In this standard, the efficiencies are determined for various ranges of particles sizes. The filter is given a Minimum Efficiency Reporting Value (MERV) rating based on its performance on the particle size ranges (particle count efficiency) and the weight arrestance (weight efficiency). The weight arrestance is a comparison of the weight of the dust penetrating the filter to the dust feed into the flow stream. In this standard, a filter with a MERV of 10 will have 50 to 65 percent minimum efficiency for particles 1 to 3 μm in size and greater than 85 percent for particles 3 to 10 μm in size [8, 9].

The European standards used to determine performance are EN 779:2002 and EN 1822:2009. EN 779:2002 is used to rate coarse and fine efficiency filters. EN 1822:2009 presents a methodology for determining the performance of high efficiencies filters: Efficient Particulate Air filters (EPA), High

Table 1: Summary of filter classification for ASHRAE 52.2:2007 [10], EN 779:2002 [11], and EN 1822:2009 [12–16].

ASHRAE 52.2:2007 [10]					EN 779:2002 [11]		EN 1822:2009 [12–16]	
ASHRAE filter class	Average particles size efficiencies in X-Y micron (%)			EN filter class				
	E_1	E_2	E_3		Average separation efficiency (A_m)	Average separation efficiency (E_m)	Total filtration separation efficiency (%)	Local filtration separation efficiency (%)
MERV	0.3–1.0	1.0–3.0	3.0–10.0					
1			<20	G1	$50 \le A_m < 65$			
2			<20					
3			<20	G2	$65 \le A_m < 80$			
4			<20					
5			20–35	G3	$80 \le A_m < 90$			
6			35–50					
7			50–70	G4	$90 \le A_m$			
8			>70					
9		<50	>85	F5		$40 \le E_m < 60$		
10		50–65	>85					
11		65–80	>85	F6		$60 \le E_m < 80$		
12		>80	>90					
13	<75	>90	>90	F7		$80 \le E_m < 90$		
14	75–85	>90	>90	F8		$90 \le E_m < 95$		
15	85–95	>90	>90	F9		$95 \le E_m$		
				E10			85	
16	>95	>95	>95	E11			95	
				E12			99.5	
17				H13			99.95	99.75
18								
19				H14			99.995	99.975
20								
				U15			99.9995	99.9975
				U16			99.99995	99.99975
				U17			99.999995	99.9999

Note: Correlations between ASHRAE and EN standard are approximate.

Efficiency Particulate Air filter (HEPA), and Ultra Low Particulate Air filter (ULPA). In EN 779:2002, the performance is found with average separation efficiency, which is an average of the removal efficiency of 0.3 μm particles at four test flowrates (particle count efficiency) for fine filters and with an average arrestance (weight efficiency) for coarse particle filters. This standard rates the filters with a letter and number designation: G1 to G4 (coarse filters) and F5 to F9 (fine filters). Filter performance is determined by the Most Penetrating Particle Size efficiency (MPPS) in EN 1822:2009. The MPPS is defined as the particle size, which has the minimum filtration efficiency or maximum penetration during the filter testing. The particle sizes tested range from 0.15 to 0.3 μm. The filter efficiency is calculated based on particle count. These filters are given a rating of E10 to E12 for EPA-type filters, H13 to H14 for HEPA-type filters, and U15 to U17 for ULPA filters. Table 1 gives a general overview of the efficiencies for each filter rating and a comparison of the filter ratings between American and European standards [10–16].

If an engine ingests 220 lb/year (100 kg/year) of contaminants if there were no filtration system in a typical offshore application, an F5 filter would reduce this to about 46 lb/year (21 kg/year), an F6 filter to 13 lb/year (6 kg/year), an F7/E10 filter system to 0.44 lb/year (0.20 kg/year), and an F7/F9/E10 system to as little as 0.11 lb/year (0.05 kg/year). This indicates two conclusions: while large particles have an impact on fouling degradation, a significant amount is due to the finer particles. The overall contaminant ingestion can be influenced by several orders of magnitude by using an appropriate air filtration system. Also, with filtration systems of this type, there are virtually no particles larger than a few microns entering the engine [17].

3.3. Filter Pressure Loss. As mentioned above, a higher pressure loss occurs with a more efficient filter due to air flow restrictions. Pressure loss has a direct impact on the gas turbine performance, as it causes a reduction in compressor inlet pressure. For the compressor to overcome the inlet system losses, the gas turbine will consume more fuel, and it also has a reduced power output. As the pressure loss increases the power decreases and the heat rate increases linearly. A 0.2 inH2O (50 Pa) reduction of pressure loss can result in a 0.1 percent improvement in power output. Typical pressure losses on inlet filtration systems can range from 2 to 6 inH2O (500 to 1500 Pa) [18].

The filter's performance needs to be assessed for the full pressure loss range over its life, not just when it is new. The pressure loss will increase over the lifetime of the filter. Therefore, one can expect a lower gas turbine performance over the life of the filter, or filters have to be changed or cleaned periodically in order to maintain a low pressure loss. The change of pressure loss over time is highly dependent upon the filter selection and the type and amount of contaminants experienced.

3.4. Filter Loading (Surface or Depth). During operation as the filter collects particles, it is slowly loaded until it reaches a "full" state. This state is usually defined as the filter reaching a specified pressure loss, or when the end of maintenance interval. Filters are loaded in two different ways: surface and depth loading.

Depth loading is the type of filtration where the particles are captured inside of the filter material. To regain the original pressure loss or condition, the filter must be replaced.

The other type of filter is a surface-loaded filter. With this type of loading, the particles collect on the surface of the filter. Few of the particles may penetrate the fiber material, but not enough to call for a replacement of the filter. Surface loaded filters are most commonly used in, but not restricted to, self-cleaning systems, because the dust can easily be removed with pulses of air once the filter differential pressure reaches a certain level. Once the filter is cleaned, the pressure loss across the filter will be close to its original condition. The surface loaded filter's efficiency actually increases as the surface is loaded with dust, because a dust cake develops on the surface of the media, creating an additional filtration layer, and also decreases the amount of available flow area in the filter media [19, 20].

3.5. Face Velocity. Filtration systems are distinctively classified as high, medium, or low velocity systems. The velocity of the filtration system is defined as the actual volumetric air flow divided by the total filter face area. Low velocity systems have air flow at less than 500 ft/min (2.54 m/s) at the filter face. Medium velocities are in the range of 610 to 680 ft/min (3.1 to 3.45 m/s). High velocity systems have air flows at the filter face in excess of 780 ft/min (4 m/s).

3.5.1. High Velocity Systems. Historically, high velocity systems are used on marine vessels and offshore platforms where space and weight are premiums. However, today, low, medium, and high velocity systems are found on marine and offshore applications. High velocity systems have the advantages of reduced size (cross-sectional area), weight, and initial cost. Filter efficiencies for small particles are significantly lower than those of lower velocity systems, and dust holding capacities are lower.

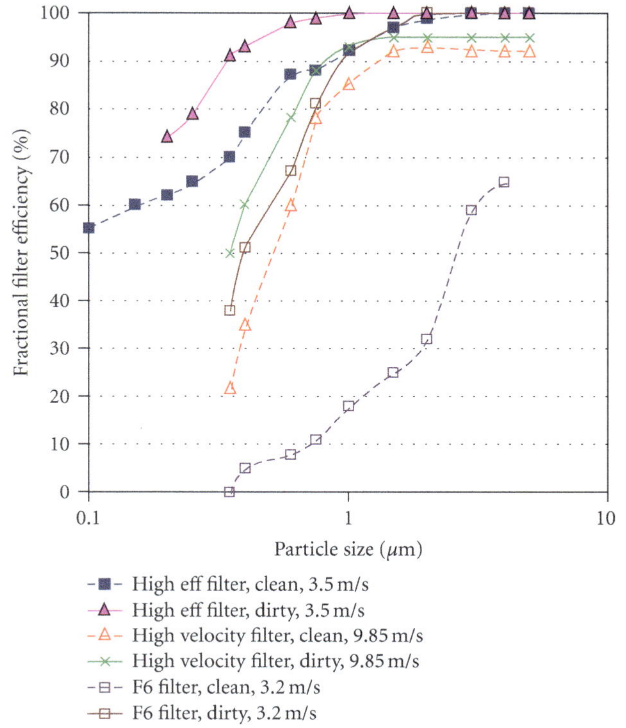

FIGURE 3: Comparison of fractional efficiency for filter elements from different suppliers and different face velocities in new and dirty conditions [25].

High velocity systems typically use vane separators upstream, and often also downstream, of the filter media to remove water from the air stream. For the vanes to work effectively, higher flow velocities are necessary. Ultimately, this type of system requires more filter replacements when compared to the lower velocity system of similar performance [21, 22].

3.5.2. Low Velocity Systems. Low velocity systems are the standard on land-based applications; however, high velocity systems are also used in some coastal applications. The low velocity systems are characterized by large inlet surface areas, large filter housings, and usually multiple stages of filters. The two or three stage filters provide an advantage over high velocity systems, because they have a high efficiency filter stage as the final stage to remove many small particles (especially salt) below 1 μm. Recently developed filter media can also keep water from penetrating the media, and thus entering the gas turbine. The lower velocity also provides a lower pressure loss or higher filtration efficiency. Using prefilters to remove the majority of the particles, the life of the high efficiency filters is extended. Overall, low velocity systems can be more effective at reducing the mass of contaminants that enter a system, thus extending the water wash intervals for the engine (Figure 3).

3.6. Water and Salt Effects. Many environments where gas turbines operate will have wet ambient conditions. This could be in a tropical environment where it rains a significant

TABLE 2: Different types of moisture experience in inlet filtration systems [2, 23].

Description	Liquid Size (μm)
Humidity	vapor form
Smog (more smoke than humidity)	0.01 to 2
Cooling tower aerosols	1 to 50
Water mist	1 to 50
Clouds and fog	2 to 150
Water spray (ship wake, ocean spray)	10 to 500
Drizzle	50 to 400
Rain	400 to 1000

amount of time or coastal location with ocean or lake mist. Table 2 is a list of the different types of moisture that can be experienced together with their particle size. The difference between filter operation in wet and dry conditions can be significant. In some cases, the pressure loss across a filter can increase significantly even with a little moisture. This is true for cellulose fiber filters which swell when they are wet. These filters will also retain the moisture that can lead to long periods of time when the pressure loss across the filter is elevated.

Salt can have a direct effect on the life of a gas turbine if not removed properly. It is often carried into the engine dissolved in water spray. Salt can lead to fouling and corrosion. Gas turbine manufacturers usually recommend stringent criteria on the amount of salt which can be allowed to enter the gas turbine (less than 0.01 ppm). In coastal environments, the airborne salt can easily range from 0.05 to 0.5 ppm on a typical day. If the filtration system is not equipped to handle the salt, it can enter the compressor and the hot section of the gas turbine. Salt is present in the air, either as salt dust or dissolved in seawater, and contains sodium chloride, magnesium chloride, and calcium sulfate. Salt may also come from localized sources such as a dry salt bed [2, 23]. The salt on compressor blades must be removed through water washing methods or direct scrubbing of the blades.

4. Components of a Filtration System

In order to protect the gas turbine from the variety of contaminants present in the ambient air, several filtration devices are used. Each of the devices used in modern filtration systems are discussed below.

4.1. Weather Protection and Trash Screens. Weather louvers or hoods and trash screens are the most simplistic of the filtration mechanisms, but they are important in order to reduce the amount of moisture and solid contaminants, which enter the main filtration system. These are not classified as filters, but they are part of the filtration system and provide assistance in removal of large objects or contaminants carried in the flow stream.

Weather hoods are sheet metal coverings on the entrance of the filtration system (see Figure 4). The opening of

FIGURE 4: Weather Hood on inlet filtration system [34].

the hood is pointed downward so the ambient air must turn upwards to flow into the inlet filtration system. The turning of the air is effective at minimizing rain and snow penetration. Weather hoods and louvers are used on the majority of inlet filtration systems, and they are essential for systems in areas with large amounts of rainfall or snow. Weather hoods or another comparable weather protection system are strongly recommended for all systems with high efficiency filter.

After the weather hood is a series of turning vanes called weather louvers, which redirect the air so that it must turn. The weather louvers are also effective at minimizing water and snow penetration. After the weather hood or louver is a trash or insect screen. Trash screens capture large pieces of paper, cardboard, bags, and other objects. The screens also deflect birds, leaves, and insects. Screens that are installed specifically for preventing insects entering the filtration system are referred to as insect screens. These screens will have a finer grid than trash screens. Weather hoods, louvers, trash screens, and insect screens are used on the majority of filtration systems due to their inexpensive cost and construction, and negligible pressure loss [9].

4.2. Anti-Icing Protection. Anti-icing protection is used in climates with freezing weather. Freezing climates with rain or snow can cause icing of inlet components, which can result in physical damage to inlet ducts or to the gas turbine compressor. This ice can also affect the performance of the gas turbine. If ice forms on filter elements, then ice on those filters will be blocking the flow path, which will cause the velocity at the other filters to increase. This causes a decrease in filtration efficiency. Also, the filter elements with ice can be damaged. Figure 5 shows an example of ice formation on filters due to cooling tower drift.

Heaters, compressor bleed air, or self-cleaning filters are often used in the inlet system in frigid environments to prevent the build-up of ice on the inlet bell mouth or filter elements. It should be noted that any location in the inlet system (even past the filter system) that creates a pressure drop can potentially have ice formation. In some plant operations, the Inlet Guide Vanes (IGVs) are used for flow control at part load operations. If the IGVs are partially closed, then under the right weather conditions, ice build can

Figure 5: Pulse cartridge filters with frost build up due to cooling tower drift [29].

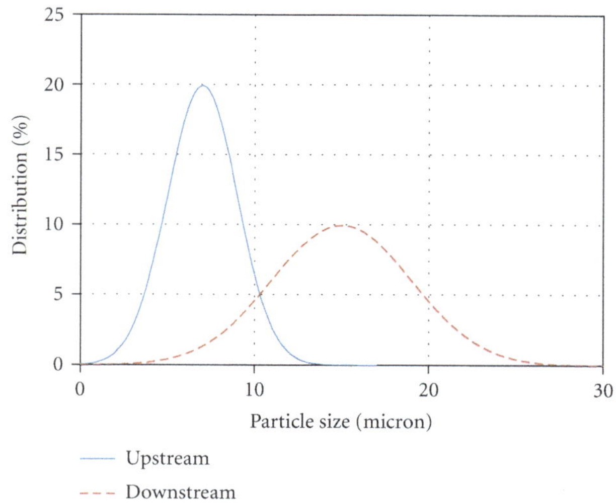

Figure 6: Coalescer droplet formation distribution [29].

occur at this location. In order to prevent ice build-up in this situation, it may be necessary to limit the closer of the IGVs or have provisions to heat the inlet air to avoid ice formation.

4.3. Inertial Separators. Inertial separation takes advantage of the physical principles of momentum, gravity, centrifugal forces, and impingement, and the physical difference between phases to cause particles to be moved out of the gas stream in such a way that they can be carried off or drained. The higher momentum of the dust or water particles contained in the air stream causes them to travel forward, while the air can be diverted to side ports and exit by a different path than the dust. There are many types of inertial separators, but the ones commonly used with gas turbine inlet filtration are vane and cyclone separators [7, 9].

4.4. Moisture Coalescers. In environments with high concentration of liquid moisture in the air, coalescers are required in order to remove the liquid moisture. The coalescer works by catching the small water droplets in its fibers. As the particles are captured, they combine with other particles to make larger water droplets. Coalescers are designed to allow the droplets to either drain down the filter or be released back into the flow stream. If the larger drops are released, then they are captured downstream by a separator. Figure 6 shows an example of how the droplet size distribution changes across the coalescer which releases the droplets [7, 9].

4.5. Prefilters. The air has a mixture of large and small particles. If a one-stage high efficiency filter is used, the build-up of large and small solid particles can quickly lead to increased pressure loss and filter loading. Prefilters are used to increase the life of the downstream high efficiency filter by capturing the larger solid particles. Therefore, the high efficiency filter only has to remove the smaller particles from the air stream which increases the filter life. Prefilters normally capture solid particles greater than 10 μm, but some prefilters will also capture the solid particles in the 2 to 5 μm size range. These filters usually consist of large diameter synthetic fiber in a disposable frame structure. Bag filters are also commonly used for prefilters. These offer higher surface area that reduces the pressure loss across the filter [7, 9]. In many installations, the prefilters can be exchanged without having to shut the engine down.

4.6. High-Efficiency Filters. As discussed above, there are filters for removing larger solid particles, which prevent erosion and FOD. Smaller particles which lead to corrosion, fouling, and cooling passage plugging, are removed with high efficiency filters. These types of filters have average separations greater than 80 percent. Three common types of high efficiency filters are EPA, HEPA, and ULPA. EPA and HEPA filters are defined as having a minimum efficiency of 85 percent and 99.95 percent, respectively, for all particles greater than or equal to 0.3 μm. ULPA filters have a minimum efficiency of 99.9995 percent for particles the same size or larger than 0.12 μm [11–16]. Often, these names are used loosely with discussion of high efficiency filtration. However, the majority of the high efficiency filters used in gas turbine inlet filtration do not meet these requirements.

The high-efficiency filters used with gas turbines have pleated media that increase the surface area. In order to achieve the high filtration efficiency, the flow through the filter fiber is highly restricted which creates a high pressure loss, unless the face velocity is kept low. The pleats help reduce this pressure loss. Initial pressure loss on high efficiency filters can be up to 1 inH2O (250 Pa) with a final pressure loss in the range of 2.5 inH2O (625 Pa) for rectangular filters and 4 inH2O (2000 Pa) for cartridge filters. The life of the filters is highly influenced by other forms of filtration upstream. If there are stages of filtration to remove larger solid articles and liquid moisture, then these filters will have a longer life. Minimal filtration before high efficiency filters will lead to more frequent replacement or cleaning. High efficiency filters are rated under various standards. The majority of filters used in gas turbines are not classified as EPA, HEPA, or ULPA. The filters used in gas turbines are rated with ASHRAE 52.2:2007 and EN 779:2002.

There are many different constructions of high efficiency type filters: rectangular, cylindrical/cartridge, and bag filters.

FIGURE 7: Rectangular high-efficiency filters [35, 36].

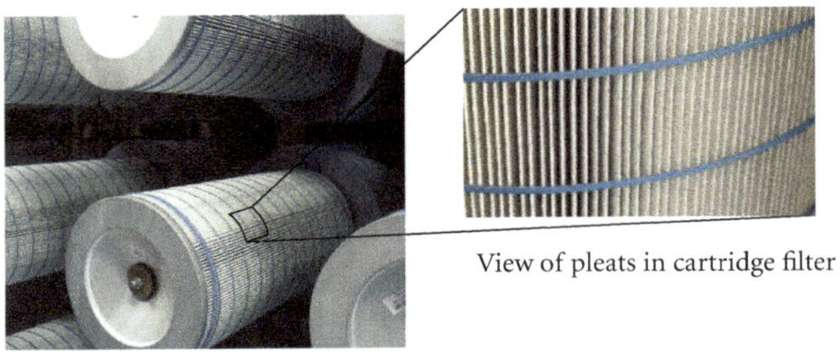

FIGURE 8: High-efficiency cartridge filters [29].

The rectangular high efficiency filters are constructed by folding a continuous sheet of media into closely spaced pleats in a rectangular rigid frame. Rectangular filters are depth loaded; therefore, once they reach the maximum allowable pressure loss, they should be replaced. Two examples of rectangular high efficiency filters are shown in Figure 7. High efficiency filters can also be made from media that do not allow water to seep through the filter media.

Cartridge filters are also made up of closely spaced pleats, but they are in a circular fashion (Figure 8). Air flows radially into the cartridge. They are installed in a horizontal or vertical fashion (hanging downward). These types of filters can be depth or surface loaded. The surface loaded filters are commonly used with a self-cleaning system, but not all of them are designed for self-cleaning. Cartridge filters used in self-cleaning systems require a more robust structural design in order to protect the filter fiber media during the reverse air pulses. The more common structural support is a wire cage around the pleated media on the inside and outside of the filter. The filters shown in Figure 8 are *not* designed for a self-cleaning system since there are no structural supports on the outside of the filter. Self-cleaning filtration systems are discussed in the next section [7, 9].

4.7. Self-Cleaning Filters. All of the filters with fiber-type media previously discussed are required to be replaced once they reach the end of their usable life. In some environments, the amount of contaminants can be excessive to the point where the filters previously discussed would have to be replaced frequently to meet the filtration demand. A prime example of one of these environments is a desert with sand

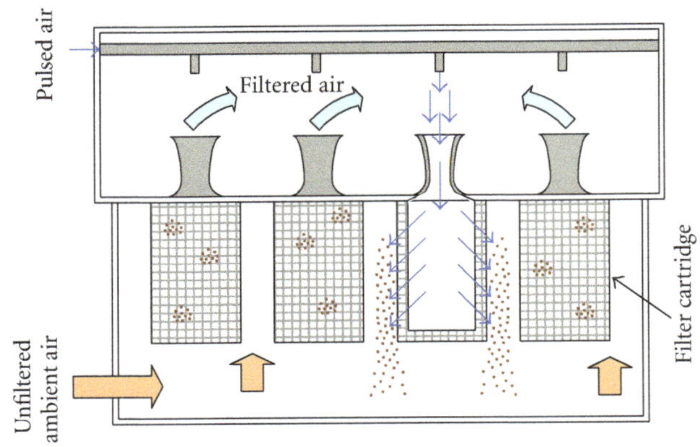

FIGURE 9: Example of operation of an updraft self-cleaning filters [29].

storms. In the 1970s, the self-cleaning filtration system was developed for the Middle East where gas turbines are subject to frequent sand storms. Since then, this system has been continually developed and utilized for gas turbine inlet air filtration.

The self-cleaning system operates primarily with surface-loaded high-efficiency cartridge filters. The surface loading allows for easy removal of the dust, which has accumulated with reverse pulses of air (Figure 9). The pressure loss across each filter is continuously monitored. Once the pressure loss reaches a certain level, the filter is cleaned with air pulses. The pressure of the air pulses ranges from 80 to 100 psig (5.5 to 6.9 barg). The reverse jet of compressed air (or pulse) occurs for a length of time between 100 and 200 ms. To avoid disturbing the flow and to limit the need for compressed air, the system typically only pulses 10 percent of the elements at a given time. With this type of cleaning, the filter can be brought back to near the original condition [19, 24].

4.8. Staged Filtration. Any gas turbine application typically needs more than one type of filter, and there are no "universal filters" that will serve all needs. Therefore, two-stage or three-stage filtration systems are used. In these designs, a prefilter or weather louver can be used first to remove erosive contaminants, rain, and snow. The second may be a low-to-medium-performance filter selected for the type of finer-sized particles present or a coalescer to remove liquids. The third filter is usually a high-performance filter to remove smaller particles less than 2 μm in size from the air. Figure 10 shows a generalized view of a filtration arrangement. This arrangement is not correct for all cases due to the fact that the filter stages are highly influenced by the environment they are operating in.

5. Operating Environment

The selection of the inlet filtration system should be primarily dependent on the environment where it operates. This includes the contaminants in the ambient air from surrounding vegetation, weather events, local emissions, temporary emissions, and seasonal changes. Several different

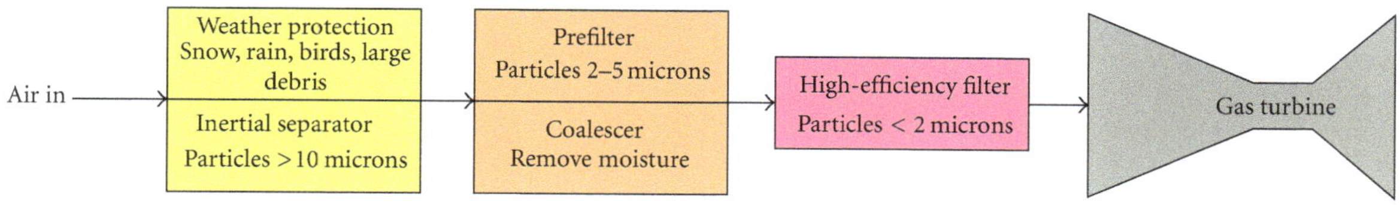

FIGURE 10: Multistage filtration system.

common environments with their typical contaminants are reviewed below.

5.1. Coastal, Marine, or Offshore. Gas turbines operating near or on the ocean are classified as being in a coastal, marine, or offshore environment. The gas turbine is in a coastal environment when the gas turbine is installed on land and within 10 miles (16 km) of the ocean. At distances from approximately 8 to 12 miles (13 to 19 km) from the shoreline, the salt concentration in the air drops to natural background levels for an environment far away from the ocean [23]. The offshore and marine environments are defined as being in the middle of the ocean. The gas turbine is considered in an offshore environment when it is located at least 100 ft (30.5 m) off of the ocean surface. Gas turbines located below 100 ft (30.5 m) are considered to be in a marine environment.

The primary contaminant that is a concern in the coastal, marine, and offshore environments is salt. Salt as previously discussed can lead to fouling and corrosion. Salt is prevalent in these environments due to the sea water. In coastal environments, it is present as dry contaminants (areas with lower humidity), sticky contaminants (humidity between 40 and 70 percent), or as liquid aerosols (humidity greater than 70 percent) [2]. In the offshore environment, salt is usually present in the sticky particle or liquid state. The marine environment (closest to the ocean surface) has salt present in the liquid aerosol state. In all these environments, the amount of salt depends on the wind speed and direction and the elevation of the gas turbine.

Coastal environments also have land-based contaminants that must be removed from the air. These will be discussed in more detail below. Offshore environments have industrial contaminants such as exhaust fumes, by-products of maintenance (such as dust from grit blasting), and unburned hydrocarbons from flares. Many of these particles are on the submicron size; therefore, high efficiency filtration is often employed. The marine environment most often does not have as many additional contaminants to remove from the air. However, when a ship is near a coast, land-based contaminants may be present. In addition, icing in colder environments is often a concern. Icing can also be an issue in offshore and coastal environments.

The filtration system for a coastal environment is similar to that of a land-based environment, which will be discussed later. However, in coastal environment it is important to have mist eliminators for water and salt water removal and high efficiency filtration for salt removal. The filtration system in an offshore environment is similar to a coastal filtration system but may have increased air velocity due to size and weight limitations. The filtration system on a marine vessel is most commonly comprised of a vane-coalescer-vane system. This system has two vane axial separators with a coalescer in between them. This system is a high velocity system that is designed for removing salt water. It has limited solid particle removal capability [2, 23, 25, 26].

5.2. Land-Based Environment. The land-based environment is very diverse. It can be classified in many different ways depending on weather patterns, vegetation, and local emission sources. Several land-based environments are described below.

5.2.1. Desert. The desert is classified as an area with a dry and hot climate. Large amount of dust is present and there is little vegetation. Sand storms are common and can quickly load filters to their maximum dust holding capacity. The main regions of the world which can be characterized by desert like environments are across the Sahara desert in Africa, the Middle East, and parts of Asia. However, small localized areas with high dust concentrations do exist. These can include gas turbines installed near quarries, dried lakebeds, loess, industrial areas, dirt tracks, dry agricultural land, and construction sites. There are three typical conditions that exist in the desert: clean air, dust haze, and sand storms. Dust is the main contaminant in the desert for these conditions. This can be sand or other fined grained material such as desert pavement. Desert pavement is the layer of large stones left on the floor of the desert. While these stones are not harmful in their solid state, they can easily be broken by human or animal traffic and crumbled into fine particles. These particles can range from large (500 μm) to very fine (submicron size). Due to the lack of vegetation and protection of the ground dust from the wind, more dust can be lofted into the air than in other environments. This leads to a high concentration of dust.

The filtration systems in deserts are usually solely designed for dust removal. However, some desert locations experience periods of dense fog and high humidity. This is especially true for deserts near a coastal region. The moisture can collect on the surface of cartridge filters on self-cleaning systems and cause the dirt to form a cake on the filter. This cake of dust can significantly reduce the effectiveness of filtration and pulse cleaning. If fog and high humidity are present at the desert-type site, then this should be considered for the filtration system.

Dust loads in the desert can range from mild (low wind) to fairly high (dust storms). Conventional non-self-cleaning filtration systems can quickly become loaded and require frequent filter change outs. Also, high pressure losses can trigger a shutdown if they become excessive. In order to avoid the constant maintenance and labor required for changing filters out, a self-cleaning system is needed. Filtration systems without self-cleaning filters have proven to be more expensive due to the labor cost and maintenance required with filter replacements [26, 27].

5.2.2. Arctic. The arctic environment is characterized by freezing weather (below 32°F (0°C)) for an extended period of time. The location will not necessarily be classified as arctic for the entire year. It will have other land-based contaminants, which must be considered. However, the arctic seasons of the year will influence the design of the inlet filtration system.

Ice build-up is the primary concern in this environment during the cold months. Ice can form from the ingestion of snow or freezing rain and also due to the depression or cool humid air in the inlet system. Placement of the inlet of the filtration system, weather hoods with large openings (referred to as snow hoods), and self-cleaning filters is often adequate to protect against the ingestion of snow and freezing rain. To prevent the formation of ice from the depression of cool humid air requires an anti-icing system such as recirculated exhaust air or a compressor bleed.

In addition to ice, warm season contaminants must be considered for the design of the inlet filtration system. These contaminants can be similar to any of the other land-based environments discussed in this section [26].

5.2.3. Tropical. Tropical areas are characterized by hot climate, high humidity, monsoons, high winds, and insect swarms. Due to the extensive vegetation, there is not much erosion concern. It is considered a low-dust environment. The area has little seasonal variation with the exceptions of periods of intense rainfall. Typhoons, dust, insects, and the remoteness of systems in the tropics should be considered when choosing the correct system.

The main contaminants in this area are water (from rain), insects, and salt (if the location is near a shoreline). Dust is minimal, since the overgrown vegetation protects the ground dust from winds. Of course, there are always exceptions to this. If the gas turbine is installed in a construction site, then the dust levels will be higher than normal. Also, unpaved roads can contribute to the dust in the environment. Pollen can be an issue. Salt will be present in aerosol form due to the high humidity and moisture present.

The filtration systems for tropical environments are specifically built to handle large amounts of rain. Weather hoods are used as a primary defense. Extended area insect screens are used for blocking insects. These screens have a lower air velocity (in the range of 260 ft/min (1.3 m/s)), which allows the insects to move away from the screens. This prevents obstruction of the inlet air flow. This is followed by a mix of prefilters, coalescers, and vane separators. The water removal system must be designed in order to handle the highest expected water ingestion and prevent corrosion. If this is not done, then water will be able to travel farther downstream in the inlet filtration system. Any prefilters or high efficiency filters used should be selected to prevent water travel through the filter. If water is allowed to penetrate the filter, then it can absorb the capture soluble contaminants and transport them through the filter into the gas turbine. This can have detrimental effects if salt is being removed from the air stream. These filters should also be selected for the expected contaminants such as pollen and road dust [26, 28].

5.2.4. Rural. The rural countryside is a diverse environment. Depending upon where the gas turbine is located in this environment, it can be subjected to hot, dry climate, rain, snow, and fog throughout the year. The majority of the year there is a nonerosive environment with low dust concentrations in the range of 0.02 to 0.1 ppm (0.01 to 0.05 grains per 1000 ft^3 (28.3 m^3)). The area can be near a local forest or be near agricultural activities.

The contaminants in this environment vary depending on the season. Throughout the year, insects and airborne particulate will need to be filtered. If the gas turbine is installed near an agricultural area, then during plowing and harvesting season, the concentration of dust will increase. During plowing, insecticides and fertilizers will be airborne. At harvest, the particles or grains from cutting plants down will be lofted into the air. The particles that travel to the gas turbine are relatively small (less than 10 μm), unless strong winds are present to carry large particles. Gas turbines near forests may not have as high dust concentration. The foliage of the forest will protect the ground dust from being lifted by the wind. With the change in season, snow, rain, fog, pollen, airborne seeds, and insects will be present. This climate has one of the most diverse filtration requirements as compared to other environments.

These systems are typically comprised of three stages: weather hood, prefilter, and high-efficiency filter. The weather hood protects the filters farther downstream from rain and snow. They also minimize the amount of dust entering the filtration system. Insect screens are used, especially if insects are present in swarms during parts of the year. The prefilter is used to remove any erosive dust present in the air. The prefilter also protects the high efficiency filter from being overloaded too quickly. The high efficiency filter removes the smaller particles. If the gas turbine is installed near an agricultural area, the filter engineer may consider a self-cleaning system. This type of system would be beneficial during plowing or harvest season when the air has a high erosive dust concentration. A self-cleaning system can also be beneficial in an area with a dry, cold climate during the winter season. It can effectively prevent ice from forming on the filter elements and influencing the gas turbine operation [26].

5.2.5. Large City. Large cities can experience all the types of gas turbine degradation: corrosion, erosion, and fouling. Contaminants from many different sources ensure the requirement of a multistaged filtration system.

All different types of weather can occur throughout the year in a large city. The amount of contaminants varies throughout the season as discussed above for the rural countryside. One example is salt or grit that is laid down on icy roads during the winter. The city also has smog and pollution. These can also be seen in the countryside due to high winds, but are much more concentrated in the city. Some other considerations for large cities are noise issues and vandals.

The system has a multistage approach with specific filters installed for the local contaminants. Weather hoods are used the majority of the time due to the changing weather conditions with seasons. This protects the system from rain, snow, and windy conditions. The filtration system is composed of a prefilter and a high-efficiency filter. The prefilter removes the larger erosive particles. The high-efficiency filter is typically of the non-self-cleaning type with rectangular filters or cartridges filters. The self-cleaning systems are not used due to the sticky aerosols present in the air. If freezing conditions are expected, then an anti-icing system is included. Urban/industrial areas typically do not have airborne particulate concentrations that warrant the use of self-cleaning filtration systems, but self-cleaning systems are used successfully in these areas, when these are in regions of heavy snow and minimal sticky contaminants [26].

5.2.6. Industrial Area. Many gas turbines are installed in heavy industrial areas. These locations can be in any of the environments discussed above, but they have additional concerns. There are several emission sources in an industrial location, which contribute to the contaminants that must filtered out.

The most prevalent contaminant in industrial areas is contaminants from exhaust stacks. These can be in the form of particles, gases, and aerosols. Many of the particles emitted by the exhaust stack are in the submicron size range. These size particles are difficult to filter and can collect on compressor blades and cause fouling. The gases emitted in the exhaust can contain corrosive chemicals. For example, exhaust gases from fossil fuel plants has SO_x, which contains sulfur. Sulfur is one of the corrosive components that can lead to hot corrosion in the turbine section. Gas cannot be removed by mechanical filtration. Aerosols also present a challenge. These are typically on the submicron size and difficult to filter. Many of these aerosols are sticky, and when they are not removed by the filters, they stick to compressor blades, nozzles, and other surfaces. One example of this already mentioned in this guideline is the compressor blade fouling due to oil vapors.

Industrial locations can also experience contaminants that are not typically seen, unless near a localized source. Some examples of these are dust from mining operations, sawmills, foundries, and other industrial facilities. Also, if the gas turbine is near a petrochemical plant, the air may be contaminated with specific chemicals. These chemicals could be harmless, but they also could have corrosive properties.

One commonality between all industrial locations is that the inlet of the filtration system is subjected to the local plant emissions. This condition typically requires a more robust high-efficiency filtration system to remove fine particles that are entrained in the air. One way to reduce the amount of emissions that are ingested into the inlet is to direct the inlet air flow away from these emission sources. Several recommendations in regards to the inlet placement and site layout are discussed in a later section. Even so, there are still some emissions that are ingested by the turbine. Additional filter elements should be included in the filtration system to address these emission particles. For example, if the industrial location is near an open coal storage site, then the gas turbine should have prefilters and high-efficiency filters to remove the coal dust that is in the air.

One contaminant that is often in the air at industrial locations is sticky aerosols. These aerosols can be from oil vapors from lubrication systems or unburned hydrocarbons emitted from exhaust stacks. These aerosols are very difficult to remove from the air and often lead to blade fouling. High-efficiency filters should be used to minimize the aerosol's effect on the gas turbine, but a compressor washing scheme is needed to keep the compressor blades clean and to minimize the effects of fouling on gas turbine performance [26].

5.3. Temporary and Seasonal Contaminant Sources. In many of the applications discussed above, temporary or seasonal conditions are mentioned. As gas turbines become more advanced and more sensitive to the inlet air quality, it becomes more important to address these conditions.

In order to address seasonal changes, the expected conditions must first be defined. During the design phase, the air quality at the site where the gas turbine is going to be installed should be monitored for at least 1 year. This will provide the filter engineer with information about which contaminants they can expect in each season. Also, the filter engineer should map out any potential construction, agricultural, or dust-generating projects that will occur in the first 5 to 10 years of the life of the gas turbine. Combining the expected contaminants will allow the filter engineer to develop a more holistic approach to their inlet filtration.

Currently, the majority of the filtration systems installed have a fixed filtration system. The number of stages, types of filters, and level of filtration remain constant throughout the operation. If the future site for the gas turbine is expected to have high variability in the type of contaminants experienced (temporary or seasonal), the filter engineer may consider a filtration system which allows the use of many different filters. This would then allow the filtration system to be adapted to the current conditions.

5.4. Site Layout. The layout of the site where the gas turbine is installed can have a significant effect on the type and amount of contaminants that need to be removed from the inlet air. This has been mentioned in several of the environmental-type discussions above but is summarized here for completeness. Listed below are general recommendations. Gas turbine manufacturers may have their own set of guidelines for placing the gas turbine [29].

(i) When installing other combustion-type equipment, such as a diesel engine, near the gas turbine, the exhaust of the equipment should be directed away from the gas turbine inlet. This reduces the possibility of the exhaust gas entering the gas turbine inlet system. This exhaust can contain unburned hydrocarbons or corrosive gases.

(ii) Cooling towers can be a major source of aerosol drift. Cooling towers are open to the atmosphere and, therefore, release aerosols into the air due to agitation from cross winds and the flow of the water down the tower. The water in the cooling tower also contains water treatment chemicals that could be detrimental to the gas turbine. The drift of aerosols from a cooling tower is confined within a few hundred feet. If possible, the gas turbine inlet should be positioned away from cooling towers and placed upstream of the prevailing wind direction to minimize the aerosol drift. CFD can be a useful tool to model how the wind will carry aerosols over to the gas turbine inlet. This will help the filter engineer to properly place the gas turbine to minimize cooling tower drift effects.

(iii) Pressure relief valves are installed on many gas lines and equipment to protect the equipment in case of an over pressurization event. The vents to these relief devices should be directed away from the gas turbine inlet. Release of any hydrocarbon could result in high concentration ingestion at the filtration system. The filters at the inlet to the gas turbine do not remove gas phase contaminants.

(iv) Piping connections on gas, fluid, or steam lines will generally leak after some time. The leaks at these connections can impact the filtration system. Piping should be routed away from the inlet in order to prevent this influence.

(v) Lube oil vents should be directed away from the inlet to prevent oil vapor ingestion.

(vi) The exhaust of the gas turbine should be directed away from the inlet of the gas turbine. Carbon smoke and hydrocarbon fumes released at the exhaust could lead to accelerated fouling of the compressor blades.

(vii) The gas turbine inlet system should not be directed toward or installed near any exhaust stacks. These exhaust stacks release chemical exhaust and unburned hydrocarbons, which can lead to compressor fouling and corrosion.

(viii) Avoid placing the inlet near gravel or dirt roads. The dust thrown into the air from vehicle traffic and wind can be carried into the inlet of the gas turbine.

  (a) If the gas turbine is operated during construction activities, consider adding more robust filters to remove the excess dirt that will be ingested.

(ix) Direct the inlet away from any open storage of coal, salt, or other grainy particles. The wind can carry the smaller grains from the storage area into the inlet of the gas turbine.

5.5. Site Evaluation. As discussed previously, there are several different types of environments where a gas turbine can operate. Also, there are many possible local, seasonal, and temporary contaminants that can be present. Therefore, each gas turbine installation site has a unique make-up of contaminants. When selecting the inlet filtration system, this make-up should be determined. This includes determining what contaminants and how much are present at the site. Once this information is known, the types of filters needed and filtration efficiency required can be established. Below is a list of items that should be considered when evaluating the site where the gas turbine will be installed [9, 29]:

(i) environment where the gas turbine will be installed: Coastal, marine, offshore, desert, arctic, tropical, industrial area, rural countryside, or large city,

(ii) contaminants present in that environment,

(iii) local contaminants (mining operating, foundries, agricultural activities, inland salt lakes, etc.)

(iv) temporary contaminants (construction activity, dirt roads, etc.),

(v) future emission sources (new industrial facility or residential development),

(vi) site layout (vents and exhaust, cooling tower drift, open storage of grainy particles, etc.),

(vii) weather patterns.

6. Life Cycle Cost Analysis

When selecting a filtration system, the filter engineer is burdened with deciding the level of quality they want their system to achieve. This includes the efficiency of filtration, the particle size to be filtered, the amount of maintenance that will be needed to maintain the filtration system, what rate of degradation is acceptable for the gas turbine, the required availability and reliability of the gas turbine, what type of washing scheme will be used (online, offline, or a combination of both), and cost of the filtration system. The cost impact of each of the items mentioned can be quantified. A Life Cycle Cost (LCC) analysis provides a convenient means to compare different filtration system options quantitatively.

6.1. Life Cycle Cost Basics. This section covers the inputs that should be considered for the LCC analysis for a filtration system. It also provides methods to calculate the cost impact for each input. This type of analysis focuses on the overall or lifetime cost of a system. It is a tool that estimates the total cost to purchase, install, operate, maintain, and dispose of equipment. This analysis can assist in determining the best design options, which will minimize the overall cost of a system.

It is important to include initial cost in the analysis, but it is just as important to include operation and maintenances cost. The operating and maintenance cost over the life of a piece of equipment can have a more significant effect, especially if a poorly designed system is chosen. An LCC analysis can help to determine which system configuration can minimize lifetime costs. Some of the costs that are typically considered are shown below. Examples of how this would apply to filtration systems are provided in parentheses:

(i) initial cost (filters, filtration system, spares filters, instrumentation),

(ii) installing and commissioning costs (labor, cost of installation equipment (such as cranes), shipping costs),

(iii) energy costs (pulse system for self-cleaning filters),

(iv) operating costs (labor, inspections),

(v) maintenance (replacing filters, repairing system, labor for maintenance),

(vi) downtime (replace filters, complete offline water washes, anything outside of normal shutdowns for other maintenance),

(vii) gas turbine effects (degradation, performance loss),

(viii) decommissioning and disposal (disposal of filters).

In an LCC analysis, estimates are provided for each cost component of the system. An inflation rate can be applied to the costs which will occur later in the life of a system (such as 10 years from the installed date). Once these costs are established, they are brought back to present value using (2). The Net Present Value (NPV) term represents the value of the cost in present terms. A is the value of the cost in the year it occurs. The term i is the discount rate and n is the year the cost occurs in. If there is a price increase (inflation) or decrease, then this can be accounted for by using (3). The term e is the increase or decrease in price:

$$\mathrm{NPV} = A(1+i)^{-n}, \tag{2}$$

$$\mathrm{NPV} = A(1+(i-e))^{-n}. \tag{3}$$

Projected costs over the lifetime of the system cannot be combined directly when calculating the LCC, because the funds spent at different times have different values to the investor. The discount rate, i, is used to bring the costs to present terms, where they can be directly added together, and is defined as the rate of return that is used to compare expenditures at different points in times. For example, the investor would be equally satisfied to have one amount received earlier and the other amount received later.

If a cost occurs yearly, the NPV of the total recurring costs can be calculated with (4). If inflation or price escalation is considered in the analysis, the NPV of the total recurring cost can be calculated with (5):

$$\mathrm{NPV} = \frac{A}{i}\left(1-[1+i]^{-n}\right), \tag{4}$$

$$\mathrm{NPV} = A\left(\frac{1+e}{1-e}\right)\left(1-\left[\frac{1+e}{1+i}\right]^{n}\right). \tag{5}$$

The NPVs must be determined for each cost. Then the cost will be added together to obtain the total NPV or LCC cost [30].

6.2. Considerations for an Inlet Filtration System. In an LCC analysis for a gas turbine inlet filtration, there are six main parameters: purchase price/initial cost, maintenance cost, availability/reliability of the gas turbine, gas turbine degradation and compressor washing, pressure loss, and failures of the filtration system or gas turbine due to inlet air quality [29].

6.2.1. Purchase Price/Initial Cost. The purchase price occurs in the first year of the LCC analysis. It is the cost to purchase and install the inlet filtration system. An estimate for this value can be obtained from the filter vendor or gas turbine manufacturer.

6.2.2. Maintenance Cost. The maintenance cost includes the cost of filter replacement and disposal and any maintenance to auxiliary systems for the inlet filtration system. It is a recurring cost that should be included in each year that the cost is acquired. This cost can be calculated based on estimated filter change out frequencies, cost of filters from vendors, labor cost for maintenance, and cost of downtime to replace filters.

6.2.3. Availability/Reliability of Gas Turbine. The availability/reliability of a gas turbine impacts the cost due to the lost production as a result of the nonavailability of the gas turbine. Filter exchanges requiring the shutdown of the engine, as well as on-crank water washing negatively, impacts the availability of the engine. On the other hand, if the engine is not used 100 percent of the time, for example, because it is a standby or peaking unit, the cost of degradation has to be adjusted accordingly.

6.2.4. Gas Turbine Degradation and Compressor Washing. Gas turbine degradation is perhaps the most important cost in the analysis. This is often the cost which drives the analysis to favor one filtration system option over another. The cost of gas turbine degradation is calculated based on the reduced power output and increased heat rate due to inlet air quality. The rate of degradation due to inlet air quality is difficult to calculate and is best found from past operating history. There are several degradation models discussed in the literature, which can provide estimates of the expected degradation rate. A few examples are the models presented by Zaba and Lombardi [31], Kurz and Brun [32], and Meher-Homji et al. [33].

Once the degradation rate is calculated the lost profit due to reduced gas turbine output can be calculated. If the gas turbine is operating at full load, then it is expected that the fuel cost will decrease due to the lower power output. For part load operations, it is expected that the fuel cost will go up since the engine will be operated at the desired power output. The change in fuel cost should be calculated based on the change in heat rate and operational philosophy and be included in the analysis. This cost should be included in each year of the analysis.

Compressor washing is often performed in a gas turbine in order to minimize the effects fouling on the performance of the gas turbine. The use of compressor washing may reduce the rate of degradation in the gas turbine. However, the most effective type of washing is on-crank washing, which requires that the engine is shut down. This results in a lower availability of the engine, and, associated with this, may cause the cost of lost production.

6.2.5. Pressure Loss. The pressure loss across the inlet filtration system can also have a significant effect on the cost of the inlet filtration system. An increase in the pressure loss across the filtration system leads to reduced power output from the gas turbine and an increased heat rate. The cost of these effects should be included yearly in the LCC analysis.

6.2.6. Failure/Event Cost. The last cost is any cost associated with a failure or event that occurs due to the inlet filtration system or inlet air quality. This could be a failure of a filter material, which requires shutdown for replacement or a failure of a gas turbine blade which occurred due to corrosion from poor inlet air quality. These costs are often included based on past experience with the gas turbine or other filtration systems.

7. Conclusions

In summary, the selection and operation of an inlet filtration system is highly dependent on the environment where the gas turbine is operating. The contaminant present in the ambient air will dictate the type filters that are used. It is important to quantify what type and size of contaminants are present in order to correctly select the filters to be used. Temporary and seasonal variations must also be considered for the inlet filtration system. A life cycle cost analysis provides a convenient method to quantify and compare various filtration system options such that the optimal system can be selected.

Nomenclature

e: Escalation rate
i: Discount rate
n: Year cost occurs
A: Cost in present value
EPA: Efficient particulate air filter
HEPA: High efficiency particulate air filter
IGV: Inlet guide vanes
MERV: Minimum efficiency reporting value
MPPS: Most penetrating particle size
NPV: Net present value
ULPA: Ultra low particulate air filter
W: Weight, volume, area, or particle number
η: Efficiency.

References

[1] R. Kurz and K. Brun, "Gas turbine tutorial—maintenance and operating practices effects on degradation and life," in *Proceedings of the 36th Turbomachinery Symposium*, 2007.

[2] P. T. McGuigan, "Salt in the marine environment and the creation of a standard input for gas turbine air intake filtration systems," in *Proceedings of the ASME Turbo Expo Power for Land, Sea, and Air*, Vienna, Austria, 2004, GT2004-53113.

[3] S. Howes, "Selecting gas-turbine inlet air systems for new, retrofit applications," *Combined Cycle Journal, Second Quarter*, 2004.

[4] E. Syverud, O. Brekke, and L. E. Bakken, "Axial compressor deterioration caused by saltwater ingestion," *Journal of Turbomachinery*, vol. 129, no. 1, pp. 119–126, 2007.

[5] T. Z. Baden, "Losses in gas turbines due to deposits on the blading," *Brown Boveri Review*, vol. 67, no. 12, pp. 715–722, 1980.

[6] *HEPA Filtration Facts*, Donaldson Filtration Solutions, 2009.

[7] *Principles of Air Filtration*, Mueller Environmental Design, 2009.

[8] M. Owens, *Engineering Bulletin—Compressor Fouling Benhmark*, AAF International, 2009.

[9] R. L. Loud and A. A. Slaterpryce, "Gas Turbine Inlet Air Treatment," GE Power Generation, GER-3419A, 1991.

[10] ASHRAE 52.2, *Method of Testing General Ventilation Air-Cleaning Devices for Removal Efficiency by Particle Size*, American Society of Heating, Refrigeration and Air-Conditioning Engineers, 2007.

[11] DIN EN 779, *Particulate Air Filters for General Ventilation—Determination of the Filtration Performance*, European Committee for Standardization, 2002.

[12] DIN EN–1 1822, *High Efficiency Air Filters—Part 1: Classification, Performance Testing, Marking*, European Committee for Standardization, 2009.

[13] DIN EN–2 1822, *High Efficiency Air Filters—Part 2: Aerosol Production, Measuring Equipment, Particle Counting Statistics*, European Committee for Standardization, 2009.

[14] DIN EN–3 1822, *High Efficiency Air Filters—Part 3: Testing Flat Sheet Filter Media*, European Committee for Standardization, 2009.

[15] DIN EN–4 1822, *High Efficiency Air Filters—Part 4: Determining Leakage of Filter Element (Scan Method)*, European Committee for Standardization, 2009.

[16] DIN EN–5 1822, *High Efficiency Air Filters—Part 5: Determining The Efficiency of Filter Element*, European Committee for Standardization, 2009.

[17] R. Kurz and K. Brun, "Fouling mechanisms in axial compressors," in *Proceedings of the ASME Turbo Expo, Power for Land, Sea, and Air*, Vancouver, Canada, 2011, GT2011-45012.

[18] *Gas Turbine World 2009 GTW Handbook*, vol. 27, 2009.

[19] T. J. Retka and G. S. Wylie, "Field experience with pulse-jet self-cleaning air filtration on gas turbines in an arctic environment," *Journal of Engineering for Gas Turbines and Power*, vol. 109, no. 1, pp. 79–84, 1987.

[20] A. Klink and T. Schroth, "New solutions for improved intake air filtration of gas turbines and turbocompressors," in *Proceedings of the ASME Turbo Expo Power for Land, Sea, and Air*, Birmingham, UK, 1996.

[21] R. K. Mudge and S. D. Hiner, "Gas turbine intake systems—high velocity filtration for marine gas turbine installation," in *Proceedings of the ASME Turbo Expo, Power for Land, Sea, and Air*, New Orleans, La, USA, 2001, 2001-GT-0584.

[22] A. D. Oswald and S. D. Hiner, "More efficient applications for naval gas turbines—addressing the mismatch between available technology and the requirements of modern naval gas turbine inlets," in *Proceedings of the ASME Turbo Expo Power for Land, Sea and Air*, Barcelona, Spain, 2006, GT2006-90305.

[23] J. P. Stalder and J. Sire, "Salt percolation through gas turbine air filtration systems and its contribution to total contaminant level," in *Proceedings of the International Joint Power Generation Conference*, pp. 445–456, New Orleans, La, USA, June 2001, JPGC2001/PWR-19148.

[24] R. G. Neaman and A. W. Anderson, "Development and operating experience of automatic pulse-jet self-cleaning air filters for combustion gas turbines," in *Proceedings of the Gas Turbine Conference and Products Show*, New Orleans, La, USA, 1980, 80-GT-83.

[25] O. Brekke, L. Bakken, and E. Syverud, "Filtration of gas turbine intake air in offshore installations: the gap between test standards and actual operating conditions," in *Proceedings of the ASME Turbo Expo, Power for Land, Sea, and Air*, Orlando, Fla, USA, 2009, GT2009-59202.

[26] D. G. Hill, "Gas turbine intake systems in unusual environments," in *Proceedings of the Gas Turbine Conference and Products Show*, Washington, DC, USA, 1973, 73-GT-38.

[27] C. Brake, "Identifying areas prone to dusty winds for gas turbine inlet specification," in *Proceedings of the Turbo Expo, Power for Land, Sea, and Air*, Montreal, Canada, 2007, GT2007-27820.

[28] R. E. Cleaver, "Gas turbine filtration in tropical environments," in *Proceedings of the Turbomachinery Maintenance Congress*, 1990.

[29] M. Wilcox, R. Baldwin, A. Garcia-Hernandez, and K. Brun, "Guideline for gas turbine inlet air filtration systems," in *Proceedings of the Gas Machinery Research Council*, 2010.

[30] Europump and Hydraulic Institute, *Pump Life Cycle Costs: A Guide to LCC Analysis for Pumping Systems*, 1st edition, 2001.

[31] T. Zaba and P. J. Lombardi, "Experience in the operation of air filters in gas turbine installation," *Brown Boveri Review*, vol. 72, no. 4, pp. 165–171, 1985.

[32] R. Kurz and K. Brun, "Degradation in gas turbine systems," *Journal of Engineering for Gas Turbines and Power*, vol. 123, no. 1, pp. 70–77, 2001.

[33] C. B. Meher-Homji, M. Chaker, and A. F. Bromley, "The fouling of axial flow compressors—causes, effects, susceptibility and sensitivity," in *Proceedings of the ASME Turbo ExpoPower for Land, Sea, and Air*, Orlando, Fla, USA, 2009, GT2009-59239.

[34] *Offshore Filter Systems*, Camfil Farr Brochure, 2009.

[35] *Cam GT for Turbomachinery*, Camfil Farr Product Sheet, 2009.

[36] *Industrial Filtration*, Burgess-Manning Bulletin BM-2-301A, 2009.

5

Gas Turbine Health State Determination: Methodology Approach and Field Application

Michele Pinelli, Pier Ruggero Spina, and Mauro Venturini

Dipartimento di Ingegneria, Università degli Studi di Ferrara, Via G. Saragat, 1-44122 Ferrara, Italy

Correspondence should be addressed to Mauro Venturini, mauro.venturini@unife.it

Academic Editor: Rainer Kurz

A reduction of gas turbine maintenance costs, together with the increase in machine availability and the reduction of management costs, is usually expected when gas turbine preventive maintenance is performed in parallel to on-condition maintenance. However, on-condition maintenance requires up-to-date knowledge of the machine health state. The gas turbine health state can be determined by means of Gas Path Analysis (GPA) techniques, which allow the calculation of machine health state indices, starting from measurements taken on the machine. Since the GPA technique makes use of field measurements, the reliability of the diagnostic process also depends on measurement reliability. In this paper, a comprehensive approach for both the measurement validation and health state determination of gas turbines is discussed, and its application to a 5 MW gas turbine working in a natural gas compression plant is presented.

1. Introduction

Maintaining high levels of availability and reliability is an essential objective for all production units, especially for those that are subject to high costs due to loss of production. Nonscheduled stops due to unforeseen faults cause relevant costs related to the reduction or the interruption of the process and to the consequent repairing actions. For this reason, in strategic applications, stand-by machines are usually required to ensure the desired level of availability.

In the last decades, gas turbines have been more and more used either for power generation or as mechanical drive (e.g., in natural gas compression plants), thanks to their favorable characteristics with respect to other technologies, such as low emissions and high availability and reliability. In particular, the latter issues represent winning features of gas-turbine-based power plants. Hence, in order to utilize these systems as effectively as possible, the management of machine maintenance must be optimized.

The optimization of maintenance management, which should lead to cost saving and increase in machine availability, can be performed by supporting gas turbine preventive maintenance (which comes from manufacturer experience in terms of component life and performance degradation versus working hours and is performed according to *a priori* schedules, regardless of the effective gas turbine health state) with on-condition maintenance, which consists of "ad hoc" actions descending from gas turbine actual operating state [1–7]. Therefore, On-condition maintenance requires up-to-date knowledge of the machine health state in real time.

One of the most successful approaches for gas turbine health state determination consists of the application of Gas Path Analysis (GPA) techniques, which use a physical thermodynamic model of the machine. Linear GPA techniques have been developed and applied since the seventies [8], while nonlinear GPA techniques have been applied since the nineties [9–12].

As an alternative to physics-based methodologies, diagnostic information can also be obtained through black-box models [13, 14], which usually prove successful and robust [15].

In case the machine under investigation is often operated under transient conditions, steady-state models (e.g., GPA models) cannot be used, and dynamic models have instead to be applied. In this field, the authors developed several simulation models, both through a physics-based approach [16–21] and through a black-box approach [22, 23].

In this paper, a nonlinear GPA technique, together with its potential and limitations, is presented. The GPA-based diagnostic process uses gas turbine field measurements to determine, by means of a gas turbine thermodynamic cycle model (cycle program (CP)), the actual values of the parameters that are indices of the gas turbine health state (Health Indices (HIs)), such as efficiencies, characteristic flow passage areas and pressure drops along the gas path [9, 12, 24–27]. One of the main effects of deterioration and fault is the modification of compressor and turbine performance maps. Since detailed information about actual modification of component maps is usually unavailable, the effects of deterioration and fault are simulated by scaling the map itself, that is, by calculating the map scaling factors through the inverse solution of the CP, in order to reproduce the measurements taken on the gas turbine [9, 12].

Recently, the authors developed a stage-by-stage model to investigate the effects of compressor and turbine stage deterioration [28]. This model uses generalized stage performance curves matched by means of a stage-stacking procedure [29] and was also successfully applied for wet compression modeling [30, 31]. Moreover, the analysis of stage performance deterioration was carried out by means of CFD simulations [32–35]. The CFD approach allows to overcome the lack of experimental data necessary to correlate measurements, HI variations, and causes of degradation. Some experimental findings were reported by the authors in [36–38].

In this paper, a GPA tool, developed by the authors [12] and validated through testbench measurements [39], is used. By comparing the actual and the expected values of the HIs, it is possible to determine (i) how far the actual machine operating condition is from the expected one, (ii) which components are degraded, and (iii) the causes of malfunctioning.

One of the most critical problems that has to be faced when GPA techniques are applied is the reliability of the information that can be obtained, which depends on several factors [40, 41].

(i) Capability of the CP to accurately reproduce the actual gas turbine thermodynamic cycle [42].

(ii) Accuracy of field measurements. To minimize measurement error effects, it is usually advisable to support GPA techniques by means of methodologies for measurement validation [43–48]. In this way, it is possible (i) to determine whether a measurement set is reliable and, if it is recognized, that an outlier is present, and (ii) to adapt the technique for the health state determination, for example, by excluding such a measurement set from the diagnostic process.

(iii) Limited availability of measured quantities on the gas turbine, which causes problems to correctly assess the actual health state. In fact, for example, a single failure can lead to the same effects (same measurement variations) as those that can be induced by a series of concurrent failures. Furthermore, some typologies of failures, as clearance increase or combustor malfunctioning, are usually detectable with difficulty [12]. So, only an adequate number of measured quantities can help to distinguish among different failures.

(iv) Some of the HIs to be estimated have to be kept constant during the calculations. In fact, since the number of the available measured quantities is usually lower than the number of HIs that have to be determined, some of them have to be considered constant. This causes an estimation error on the HIs considered as problem variables, when variations due to aging or deterioration occur on the HIs which were considered as fixed HIs [40, 41].

Thus, methodologies for the improvement in HI determination accuracy are required [43, 44, 49–52]. In particular, since the methodology makes use of field measurements, the reliability of the diagnostic process also depends on measurement reliability. For this reason, two techniques for measurement validation are presented in the paper: the first one is based on the use of acceptability bands [43, 53], while the other uses statistical-based methods for outlier identification [44]. Analytical redundancy techniques for sensor fault detection and isolation can also be used [54]. In any case, if there is insufficient data or measurement accuracy is low, a "reality check" should be always made to verify plausibility of the information obtained through the GPA-based tool.

In this paper, a comprehensive methodology developed by the authors for both measurement validation and health state determination of gas turbines is presented. The methodology is applied to a 5 MW gas turbine working in a natural gas compression plant. Finally, this paper illustrates the main features of a software, which was implemented in the considered compression plant to automate the presented methodology.

2. Methodology for Measurement Validation and Health State Determination of Gas Turbines

The methodology for measurement validation and health state determination of gas turbines requires the availability of a CP, which should reproduce the particular gas turbine under consideration as accurately as possible. Two situations can occur:

(1) The cycle deck developed by the manufacturer is available. This CP reproduces a gas turbine type, which presents average characteristics among gas turbine units of the same model.

(2) A generalized CP is available. In this case, the program has to be tuned to reproduce the machine type under investigation, for instance, by using the performance curves supplied by the manufacturer to the user.

In any case, both the cycle deck and the generalized CP have to be tuned in order to represent the particular gas turbine

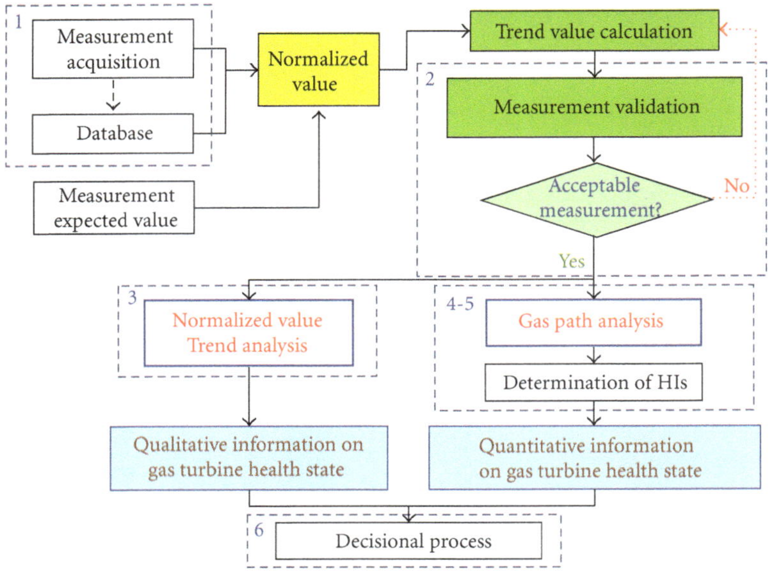

Figure 1: Comprehensive methodology for measurement validation and health state determination of gas turbines.

unit under consideration. CP tuning procedure is reported in detail in [55].

A low number of measured quantities is usually available on gas turbines in field operation. This fact limits the number of HIs that can be determined through the GPA method. Therefore, the optimal set of HIs, that can be determined by the set of available measurements, has to be identified. This requires an *a priori* optimized selection of the HIs which have to be considered as problem variables and of the ones to be kept constant during the calculation [41, 43, 52].

Once a CP is tuned on the particular gas turbine and the optimal set of HIs is identified, the main steps of the comprehensive methodology for measurement validation and health state determination are sketched in Figure 1 and reported below.

(1) Acquisition of field measurements and storage on a historic database, to perform off-line data processing.

(2) Measurement validation. This allows the identification of measurements that have a level of uncertainty higher than a fixed threshold. Thus, the measurement set can be excluded from the gas turbine diagnostic process, to avoid an incorrect evaluation of the machine health state.

(3) Analysis of the normalized measurement trend (trend analysis). This analysis is required for measurement validation and also provides useful information for the determination of gas turbine health state, so improving the results obtainable by using the GPA technique alone [43, 53].

(4) Use of acceptable data to perform gas turbine health state determination, which consists in the determination of gas turbine HIs by using a GPA-based technique.

(5) Improvement of the diagnostic process through the determination of a higher number of gas turbine HIs by considering more than one operating point (multipoint analysis). In fact, the multiple operating point analysis allows the determination of a number of HIs higher than available measurements, since it compensates for the lack of measurements with the measurements taken at different operating points [24, 56].

(6) Decisional process, which consists of the identification of unacceptable measurements, the planning of the machine stop for maintenance, on the basis of the values of machine HIs, the possibility of on-line actions, such as compressor washing, the adaptation of the gas turbine control logic to its actual health state, and, in the worst case, the immediate stop of the machine.

A software that automates the described methodology may be very helpful as a decision support tool for both plant director and maintenance personnel [57]. Artificial intelligence techniques may help in focusing on the most significant information for improving the diagnostic process and maintenance practices [58]. Moreover, prognostics methodologies would allow to forecast degradation evolution over time, based on current diagnostic state [59, 60], to optimize both production and maintenance, from both a technical and an economical point of view. In a wider sense, making prognostics, based on understanding machine deterioration and aging, may allow the estimation of machine remaining useful life.

3. Methodology for Measurement Validation

In this paper, two techniques are considered for measurement validation, both applied to the normalized measurement trend. Measurement normalization is performed by dividing each measured value by its expected value calculated in the same ambient and load conditions, in order to render measurements comparable to each other, though collected in different ambient and load conditions. Such expected value can be calculated by using a CP or functional relations. While the CP is rarely available to gas turbine users, functional relations can be obtained by only using measurements taken along the gas path and thus the user can always apply this methodology.

The functional relations can be obtained through identification techniques, as for instance by using a linear regression procedure [53] or neural networks [61, 62]. The tuning of the models requires the identification of a baseline condition across gas turbine life (e.g., the condition after overhaul maintenance). In this condition, some measurement sets taken at different loads and ambient conditions have to be available. Starting from these sets of measurements it is possible to establish relations in the form $Q_m = F(Q_{wp})$. The relations obtained relate thermodynamic measurements Q_m (such as pressure and temperature at the compressor outlet) to ambient and load conditions Q_{wp} (ambient pressure and temperature, relative humidity, rotational speeds, and power output) over the entire gas turbine operating range. The identification procedure has to be performed once for each Q_m measured quantity and has to be updated when the chosen baseline condition can be considered no longer representative for the engine. The obtained relations can be then used to normalize each measurement.

3.1. Measurement Acceptability Bands. Once measurements are normalized and the trend over time of each measured quantity is determined, measurement validation can be performed through measurement acceptability bands [43, 53]: a measurement lying within these bands will be considered acceptable, otherwise not. Band amplitude can be calculated by considering the following contributions.

(a) Measurement accuracy, that can be taken into account by considering band amplitude equal to measurement uncertainty.

(b) An engine fault, which may affect the value of the measurements along the gas path.

(c) Measurements noise along the data acquisition system. This error is usually included within measurement accuracy.

(d) Errors due to the CP accuracy. In fact, the CP is usually tuned on the engine type and not on the specific unit. This leads to an estimation error on the expected value of measurements and so to an error in the normalization process.

For the calculations performed in this paper, the uncertainty "d" was not considered since the CP was tuned on the specific unit and, therefore, this source of error was considered negligible. Two bands were instead considered: the first takes into account measurements uncertainty (contribution "a"), while the second considers the variation of the measurement due to an engine fault (contribution "b"). The total band amplitude is the sum of these two contributions.

Therefore, it is possible to identify three cases to establish whether a measurement is acceptable.

(i) Value beyond measurements accuracy.

(ii) Value out of measurements accuracy, but within the maximum estimated variation of the measurement due to a fault. A measurement can lie in this region either because of sensor fault or because of an incipient engine fault. In this situation, to obtain significant information, it is necessary to consider the engine behavior before and after the anomalous value and data have to be processed by the diagnostic tool.

(iii) Value out of the maximum variation due to a fault. The measurement should not be processed by the diagnostic tool since it is identified as an outlier.

As anticipated, the bands have to be referred to as reference value. The trend value of the normalized measurements seems a reasonable choice, since, in this manner, measurement variations due to aging can be taken into consideration.

3.2. Statistical-Based Method for Outlier Identification. An alternative way for measurement validation is the analysis of the statistical distribution of the normalized measurement trend in order to detect outliers. A wide number of techniques for outlier detection exist in the literature. In gas turbine applications, practical and easy-to-use techniques such as parametric test methodologies seem to offer a compromise solution with respect to simplicity and robustness [44].

A parametric test methodology is based on the definition of a test criterion. For any given element x_i of an N-dimensional sample, a test criterion can be written as

$$\frac{|x_i - x_m|}{S} \geq k, \quad i = 1, \ldots, N, \tag{1}$$

where x_m is the mean of the sample and S is the standard deviation. In the most general case, the coefficient k is a function of the sample size N and of the level of significance α. The latter parameter has to be chosen *a priori* and represents the probability of rejecting a good point. For practical purposes, three levels of significance are usually considered, namely $\alpha = 1\%$, 2%, and 5%. This means, for example, that if $\alpha = 5\%$ is considered, the odds against rejecting a good point are 20 to 1 (or less).

If (1) applies, x_i is to be considered an outlier. Different methods, based on (1), and characterized by different assumptions for the coefficient k, are available in the literature (e.g., the Thompson method, and the Grubbs method, the Chauvenet criterion).

These "traditional" methods proved only partially effective and not very robust for gas turbine data [44], in particular when data trends are not constant over time, such as in the case of measurements taken during a long period of time on a gas turbine. In order to overcome some of the limitations of these methods, a new method was developed [44]. The method is based on the application of a test criterion in the form of (1), but both the left-hand side term and the coefficient k are modified to meet the requirements that have been highlighted. The specificity of the method is that the coefficients introduced account for decreasing or increasing data trends, although they are also correctly defined when constant over time data trends are considered.

In other words, the new coefficients allow the test criterion to take the behavior-in-time of the considered quantity into account. In the new formulation, the test criterion is defined as

$$\frac{|x_i - x_m|}{k_B S} \geq t_\alpha k_A, \quad i = 1, \ldots, N. \tag{2}$$

The coefficient k_A is defined as

$$k_A = 1 + \frac{\lim_{N \to \infty} t_\alpha^2 + 1}{4N}. \tag{3}$$

This coefficient is defined in such a manner as to tend to one when N tends to infinity. The coefficient t_α represents the value of the quantile corresponding to a certain level of significance related to the Gaussian distribution. The coefficient k_B, defined as

$$k_B = (1+N)^{1/N}\left(1 + \left|S_{\text{ov}} - S_i\right|\right), \tag{4}$$

allows the scatter of the considered sample to be taken into account. In fact, a relationship between the overall standard deviation S_{ov} and the standard deviation calculated by considering the data range $[1, i]$, namely S_i, where i is the current data, which is under investigation, is introduced in the test criterion. In this manner, the left-hand side term of (2), instead of being constant as in the Thompson, Grubbs, and Chauvenet methods, is updated at each step and, thus, the behavior-in-time of the quantity is taken into account. However, this requires the availability of all data for the time period under examination, in order to estimate data overall standard deviation S_{ov}, and, thus, k_B dependence with time can be evaluated only in the case of data off-line processing. Otherwise, if data are processed on line, k_B only depends on sample size N, since, in this case, S_{ov} equals S_i.

4. Methodology for Health State Determination of Gas Turbines

4.1. Single Operating Point Analysis. Gas turbine operating state determination consists of the assessment of the modification, due to deterioration and fault, of performance and geometric data characterizing the machine components. One of the main effects of deterioration and fault is the modification of compressor and turbine performance maps. Since detailed information about actual modification of component maps is usually unavailable, many authors simulate the effects of deterioration and fault by scaling the map itself, that is, by multiplying the maps in *new and clean* condition by scaling factors F point by point [9, 12, 56]. Different scaling factors can be used; compressor and turbine maps are usually scaled by multiplying efficiency and corrected mass flow rate, at constant pressure ratio (or equivalent parameter, such as the ratio between isentropic enthalpy variation and turbine inlet temperature) and at constant corrected rotational speed [9, 12, 56]. The modification of compressor and turbine performance maps with respect to *new and clean* condition due to actual deteriorations and faults can be assessed by calculating the map scaling factors F in order to reproduce the measurements taken on the gas turbine [9, 12, 56].

Therefore, the scale factors of efficiency and corrected mass flow of compressor and turbines can be assumed as characteristic parameters of compressor and turbine health state. In addition to these parameters, two other important parameters for representing the gas turbine health state are the combustor efficiency and pressure drop. In particular, the combustor efficiency accounts for thermal losses along the gas path, which are considered as a fixed constant percentage of the thermal power introduced by the fuel, while the combustor pressure drop is usually defined as a constant percentage of the total pressure at the combustor inlet. This type of parameters is sensitive to the gas turbine health state, while it is not dependent on the gas turbine operating point. Parameters with this characteristic are usually called *health indices* (HIs).

The HIs can be calculated by solving in inverse mode the CP in order to reproduce the measurements taken on the gas turbine. In fact, the values of the measurable variables computed by the CP $Q_{m,c}$ are a function of the values assumed both by the health indices X and by the variables that unequivocally determine the operating point at which the gas turbine is working Q_{wp}:

$$Q_{m,c} = f\left(X, Q_{\text{wp}}\right). \tag{5}$$

By inverting (5), it is possible to calculate X starting from the measured variables:

$$X = f\left(Q_m, Q_{\text{wp}}\right). \tag{6}$$

The solution of (6), usually called "inverse" solution, has been performed by the authors through a minimization technique which determines the values of HIs that minimize the sum of the square differences, between measured and computed values of the measurable variables [12]. This problem is solved by using a nonlinear algorithm, which minimizes the objective function:

$$f_{\text{ob}}\left(X_1, \ldots, X_{N_X}\right) = \sum_{i=1}^{N_m} w_i \left(\frac{Q_{m,c} - Q_m}{Q_m}\right)_i^2, \tag{7}$$

where X_i ($i = 1, \ldots, N_x$) are the unknown values of HIs, $(Q_{m,c})_i$ and $(Q_m)_i$ are the computed estimates and the measured values of the measurable quantities, respectively, and w_i are the weights which can be assigned to each term of the objective function.

The minimization algorithm which was used is included in the IMSL math library [63] and was successfully utilized to solve the gas turbine mathematical model [3, 11, 12, 24, 29, 39, 41, 57, 64]. Once the HIs are calculated, the gas turbine health state is determined by evaluating the variations of the HI values with respect to the expected values in the "new and clean condition." This allows the faulty component to be localized and malfunctions to be identified and quantified.

From (6), the number and type of gas turbine HIs that can be determined for each operating point (i.e., for each

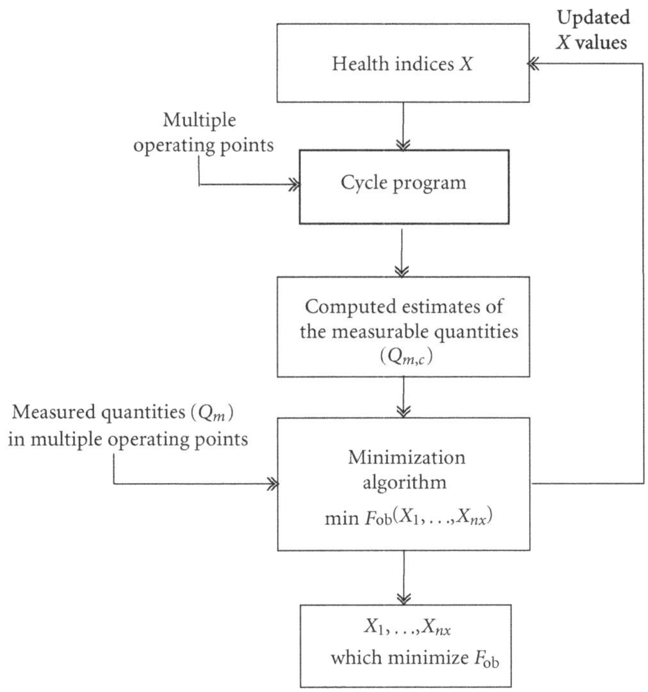

FIGURE 2: Multiple operating point analysis.

set of Q_{wp}) depend on the number and type of equations, which, in turn, depend on the number and type of the available measured quantities. In particular, the number of HIs is usually equal to the number of the Q_m measured variables. Thus, since the number of the Q_m available measured quantities is usually lower than the number of HIs to be estimated, some of them have to be kept constant during the calculations. Therefore, variations due to aging or deterioration which, in the actual machine, occur on the HIs, which were considered as fixed HIs, will cause an estimation error on the HIs to be determined [40, 41].

4.2. Multiple Operating Point Analysis. A direct consequence of (6) is that more equations can be obtained by using more than one Q_{wp} set. Thus, the number of gas turbine HIs that can be determined also depends on the number of gas turbine operating points (i.e., number of Q_{wp} sets) [24, 56]. Therefore, by using multiple operating points it is possible to evaluate a number of HIs higher than the number of the available Q_m measured quantities.

Therefore, as made in the case of single operating point analysis, the solution of the system of equations obtained by using more than one Q_{wp} set in (6) was performed through a minimization technique, which determines the values of HIs that minimize the sum of the square differences, between measured and computed values of the measurable variables in all the operating points considered. The objective function to be minimized becomes

$$f_{ob}(X_1, \ldots, X_{N_X}) = \frac{1}{N_{OP}} \sum_{j=1}^{N_{OP}} \left[\sum_{i=1}^{N_m} w_i \left(\frac{Q_{m,c} - Q_m}{Q_m} \right)_i^2 \right]_j, \quad (8)$$

where X_i $(i = 1, \ldots, N_x)$ are the unknown HIs, $(Q_{m,c})_i$ and $(Q_m)_i$ are the computed estimates and the measured values of the measurable variables respectively, and w_i are the weights assigned to each term of the objective function.

The main steps of the multiple operating point technique are sketched in Figure 2. The adopted minimization algorithm is the same as for that the single point analysis [63].

The different operating points have to be taken within a small time interval (for instance during the same day) so that the gas turbine health state can be considered the same for all the operating points and, thus, the solution X of (6) is the same for all the operating points. The most suitable choice of the operating points to be used was analyzed in [24].

5. Application of the Methodology to Gas Turbines Running in a Natural Gas Compression Plant

The comprehensive methodology for measurement validation and health state determination of gas turbines was applied to two compressor drive gas turbines working in a natural gas compression plant. In this plant, the pressure of the gas coming from the Adriatic Sea reservoirs is raised to the value required by the Italian Gas Supply Company by using four compression systems. Two of them are driven by 5.2 MW regenerative cycle two-shaft industrial gas turbines with variable power turbine nozzles (VN) and two by 1.2 MW simple cycle two shaft gas turbines. The methodology was applied to the two 5.2 MW gas turbines whose lay out is shown in Figure 3.

In any case, the methodology is completely general. In fact, it was also successfully applied to a single shaft gas

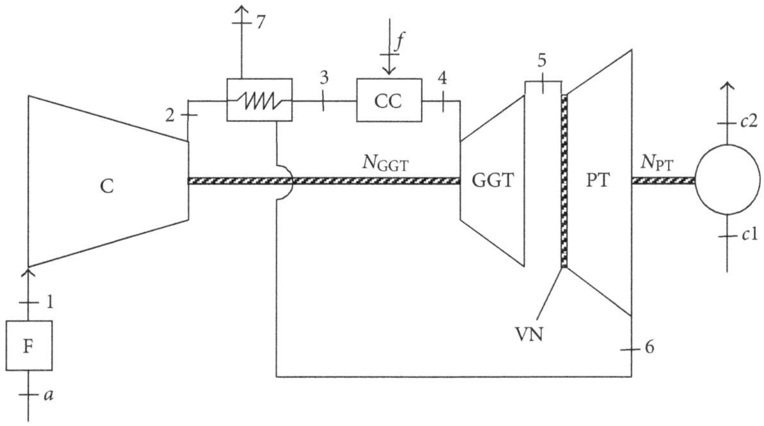

FIGURE 3: Layout of the 5.2 MW two-shaft regenerative cycle gas turbine.

TABLE 1: Available measurements.

Q_{wp} measurements	Q_m measurements
T_1: comp. inlet temp.	p_2: comp. outlet pressure
p_a: ambient pressure	T_2: comp. outlet temp.
RH: relative humidity	T_6: power turb. outlet temp.
Δp_{a-1}: filter pressure drop	
VN : var. nozzle position	
N_{GGT}: gas gen. turb. rot. speed	Other measurements
N_{PT}: power turbine rot. speed	T_3: recuperator outlet temp. (air side)
P_{PT}: power output (calculated from measured T_{c1}, T_{c2} and M_{c1})	T_7: recuperator outlet temp. (gas side)

turbine working in a cogenerative combined power plant [65].

Table 1 reports the measurements available on each 5.2 MW gas turbine system. As can be seen from Table 1, some very important measurements for a reliable diagnosis analysis are not available. They are the fuel mass flow rate (at present, only the total fuel mass flow feeding both the 5.2 MW gas turbines is measured), the pressure and temperature between the gas generator and the power turbine (p_5, T_5), the air side and exhaust side regenerator pressure drops ($\Delta p_{2\text{-}3}$, $\Delta p_{6\text{-}7}$), and the air inlet mass flow rate (M_1).

The methodology was applied to a poorly instrumented plant, since it represents a selective test for verifying the capabilities of the proposed diagnostic system. Moreover, poor instrumented plants are highly widespread and, thus, the application to such cases seems particularly interesting.

Before applying the methodology for gas turbine health state determination, a generalized CP developed by the authors was tuned to reproduce the gas turbine under consideration. This was made by using as reference values the performance curves provided by the gas turbine manufacturer to the user.

These curves provide the compressor outlet pressure (p_2) and temperature (T_2), power turbine outlet temperature (T_6), fuel and inlet air mass flow rates (M_f and M_1) at various gas turbine working points. After the tuning, the program estimates manufacturer data with a maximum error usually lower than 1% [3].

The following step of the procedure requires the identification of the optimal combination measurements/HIs. As shown in Table 1, there are only three available measurements to perform the operating state determination: p_2, T_2, and T_6. In fact, the other measurements are used to define the working point. Therefore, only three HIs can be unequivocally determined. The optimal HIs set (identified according to [43, 52]), which can be determined by using the set of available measurements, is composed of the following HIs: compressor efficiency (η_C), corrected mass flow (μ_C), and gas generator turbine efficiency (η_{GGT}).

Phase 1: Acquisition and Storage of Field Measurements. The available data were gathered once a day manually by an operator. This caused the presence of a number of wild points originating from sources such as reading errors, data taken in unstable and/or not representative conditions, and so forth, which were difficult to detect. This may be a common situation in practice and thus it could represent a test bed for measurement validation techniques. The results reported refer to data acquired in the period from 2nd November 1999 to 16th January 2000 on one of the two gas turbines under investigation.

Phase 2: Measurement Validation. Measurement validation was performed by means of the two previously described techniques, both applied to the normalized measurement trend. Measurement normalization was performed by

Table 2: Band amplitude: measurements accuracy and total band amplitude.

Measured Quantities	Measurements accuracy [% of reference value]	Confidence band [% of trend value]
T_2	±0.85	[−1; + 5.5]
p_2	±1.00	[−4; + 3]
T_6	±0.75	[−1; + 5.5]

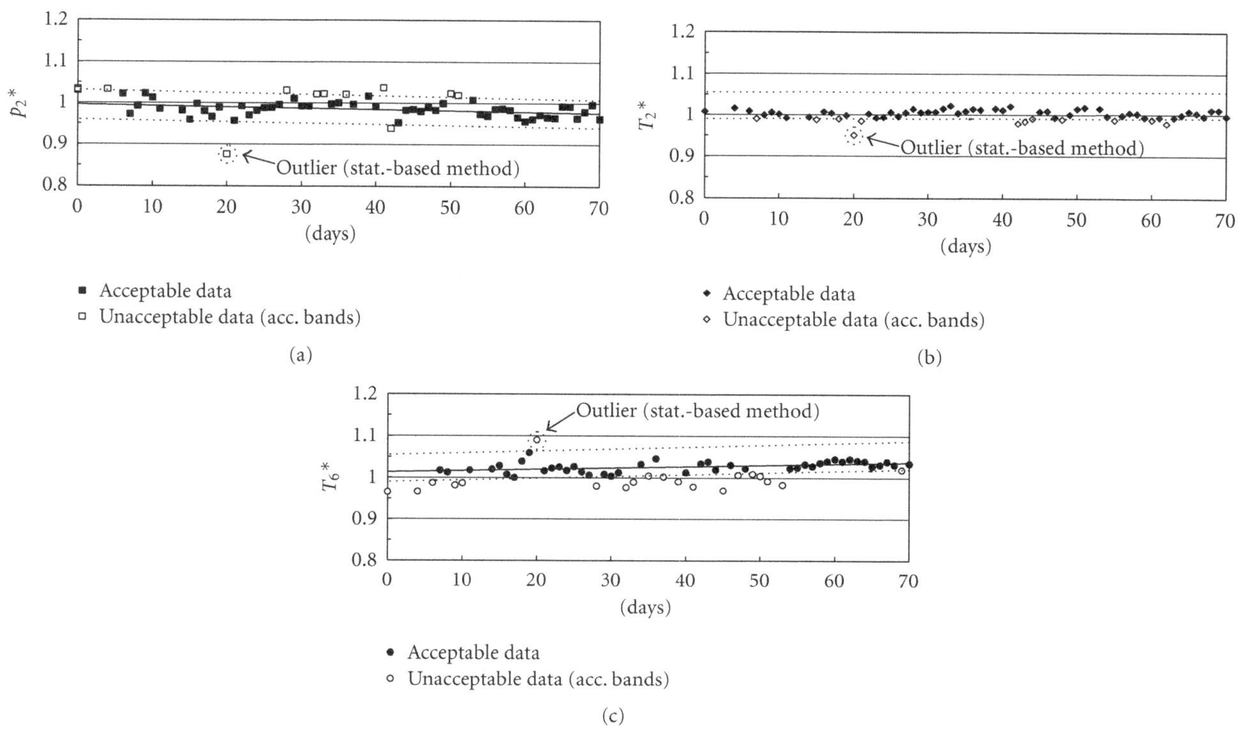

Figure 4: Normalized values of compressor outlet pressure (p_2*), compressor outlet temperature (T_2*) and power turbine outlet temperature (T_6*) versus time.

dividing each measured value by its expected value calculated in the same ambient and load conditions, by using the CP tuned on the machine under consideration. Measurement acceptability bands were applied to the normalized measurement trend and take into account both measurement uncertainty and maximum variations of measurements due to faults, according to Table 2 [43, 53].

Measurement uncertainty bands (second column in Table 2) were set according to the sensor accuracy reported in [66] and by considering the case in which the measurements were taken in the field with standard machine instrumentation during normal operation and not when conducting an acceptance test. In order to establish the amplitude of the bands deriving from measurement variation due to faults, the CP tuned on the considered machine was used to simulate some of the most common faults that can occur on a gas turbine (compressor fouling, compressor mechanical damage, gas generator turbine mechanical damage, gas generator turbine erosion, and power turbine erosion) [43]. Faults are to be considered as sudden faults, since measurement variations due to aging or deterioration are considered through the normalization process. Figure 4 reports the normalized values of the compressor outlet pressure (p_2) and temperature (T_2) and of the power turbine outlet temperature (T_6) versus time. The solid line indicates measurement trend over time, while the dashed lines are the acceptability bands. The figure highlights:

(i) the decreasing trend of the outlet compressor pressure and the slightly increasing trend of the compressor outlet temperature. These symptoms may be attributed to compressor fouling The increasing trend of the power turbine outlet temperature can be also observed.

(ii) the measurement scattering due to the uncertainties in field measurement readings.

(iii) the unacceptable measurements according to the use of acceptability bands (white symbols). Moreover, the application of the statistical-based method for outlier identification reveals that only one measurement set (i.e. the one at day no. 20) can be considered unacceptable also by using this second method. Thus, the use of the statistical-based method reveals less restrictive than the use of acceptability bands.

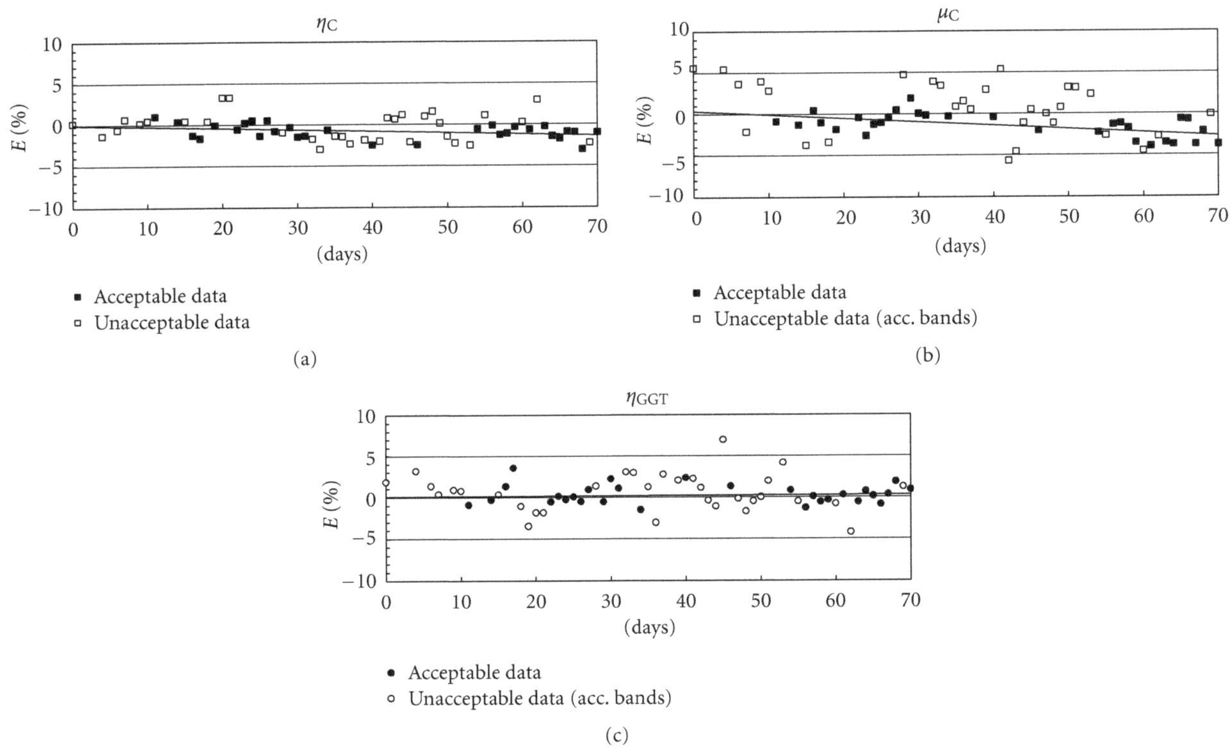

FIGURE 5: Computed and reference value percentage variations for compressor efficiency (η_C) and corrected mass flow (μ_C) and gas generator turbine efficiency (η_{GGT}).

Phase 3: Analysis of the Normalized Measurement Trend (Trend Analysis). The analysis is aimed at establishing relations among performance drops and normalized measurement trends of compressor outlet pressure (p_2) and temperature (T_2) and the power turbine outlet temperature (T_6) reported in Figure 4.

Form Figure 4, a reduction of about 2% in 70 days on the trend of normalized compressor outlet pressure can be noticed, while compressor outlet temperature remains almost constant. These trends highlight a normal compressor fouling. The analysis of $T_6{}^*$ alone, whose trend is slightly increasing, does not provide additional diagnostic information about the two turbines.

Phase 4: Gas Turbine Health State Determination. The percentage variation E between computed and reference values of compressor efficiency (η_C) and corrected mass flow (μ_C) HIs and of gas generator turbine efficiency HI (η_{GGT}) versus time is reported in Figure 5. The black and white symbols indicate the HIs evaluated by using the acceptable and unacceptable measurement sets, respectively.

Figure 5 highlights the decreasing trends of both compressor efficiency and mass flow function HIs, which can be attributed to compressor fouling, as already highlighted by the Trend Analysis in the previous section. Over a period of two months, the trend values of η_C, and μ_C HIs were reduced by 1.0% and 2.5%, respectively, showing that fouling is not severe [67, 68]. The trend of gas generator turbine efficiency HI is instead almost constant, indicating that this component is not suffering from significant changes in its health state. Figure 5 also highlights the remarkable reduction of the scattering of HI trends, obtainable by using the acceptable measurement sets only.

Phase 5–Gas Turbine Health State Determination on Multiple Operating Points. A multipoint analysis was also applied to the gas turbine unit considered by using measurements taken during special operating conditions, immediately before and after a maintenance stop. In these cases, measurements at different gas turbine loads were performed, so that it was possible to perform the multipoint analysis.

Two different calculations were performed by using different sets of variable HIs. In Figure 6(a), the normalized values of η_C, μ_C, η_{GGT}, and μ_{GGT}, calculated by using the multipoint analysis with three HIs (η_C, μ_C, η_{GGT}) as problem variables, are reported. In Figure 6(b), the normalized values of η_C, μ_C, η_{GGT}, μ_{GGT}, η_{PT}, and μ_{PT}, calculated by using the multipoint analysis with five HIs (η_C, μ_C, η_{GGT}, μ_{GGT}, and η_{PT}) as problem variables, are shown. Black and white symbols refer to the measurements taken before and after the maintenance stop, respectively. It can be highlighted how the multipoint analysis allows the determination of a number of HIs higher than available measured quantities, since it compensates for the lack of measured quantities with the measurements taken at different operating points.

The results obtained by using five HIs as problem variables (Figure 6(b)) seem more convincing than the ones obtained in the case of three variable HIs (Figure 6(a)). In

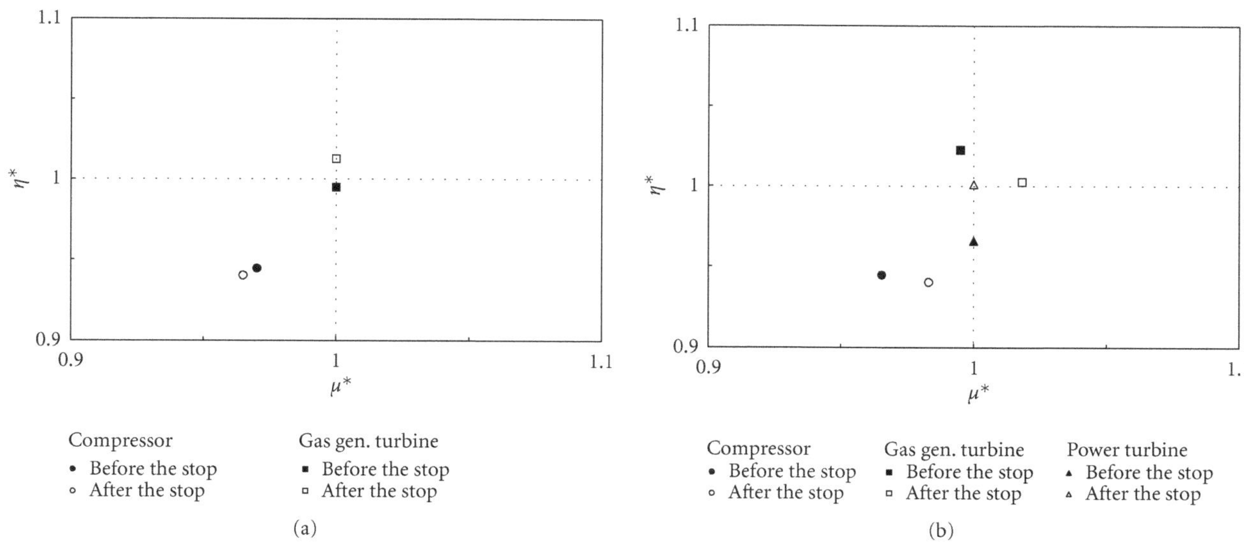

FIGURE 6: Normalized values of HIs in case of (a) three problem variables (η_C, μ_C, and η_{GGT}), and (b) five problem variables (η_C, μ_C, η_{GGT}, μ_{GGT}, η_{PT}, and μ_{PT}) (multiple operating point analysis).

this last case, in fact, it seems that there are no improvements due to maintenance, while, in the case of five variable HIs, an increase in the compressor and gas generator turbine corrected mass flows HIs can be noticed.

6. Software Tool for Gas Turbine On-Condition Monitoring and Diagnostics

For the considered natural gas compression plant, a software tool was developed and implemented [57], to allow the prompt visualization of the required information by means of a user-friendly interface, so allowing gas turbine health state analysis and supporting the decision for maintenance actions.

The developed software, whose structure is sketched in Figure 7, allows

(i) the measurement normalization,

(ii) the calculation of HIs for the evaluation of gas turbine health state,

(iii) the comparison of the measured delivery of compressed natural gas compared with the calculated maximum possible delivery. This latter value is estimated by assuming the gas turbine in new and clean conditions.

Moreover, the software system stores all data in a database and calculates the trend line in order to perform the trend analysis for all the parameters.

The kernel of the software system consists of the subroutines that allow the manual insertion of data, data direct acquisition from the control system, the visualization of the results, and the modification of calculation parameters. The user interface is provided by means of floating toolbars (buttons and/or text).

Results can be displayed in form of graphs and tables, which allows a user-friendly interpretation of the results. The graphical form shows plots of calculated values and the corresponding trend lines versus time; acceptability bands are visualized around the trend line to evidence unacceptable data.

The program also determines the net power output of the gas turbine starting from measurements acquired on the centrifugal compressor, the overall gas turbine efficiency, and the turbine inlet temperature. The ratio between the actual and the nominal value for both the net power output and the turbine inlet temperature can be used as indices of the gas turbine overall health state.

Moreover, starting from the indices of the gas turbine health state, the software calculates the value of the production losses due to the actual health state. In fact, if M_{gasM} is the measurement of the natural gas flow rate processed by the centrifugal compressor and M_{gasC} the flow rate that the centrifugal compressor could process when the driving gas turbine is in new and clean conditions, a lost production index LPI can be calculated as

$$\mathrm{LPI} = \frac{M_{gasC} - M_{gasM}}{M_{gasC}}. \qquad (9)$$

In Figure 8 a sample trend of LPI is reported. The trend is increasing due to gas turbine ageing: in fact, the more the gas turbine deterioration increases (and, thus, its performances decrease), the more the gas flow rate that the centrifugal compressor can process (M_{gasM}) reduces.

In conclusion, the software can actually represent a helpful support tool for the plant operation management. The persons which can take advantage of these analyses can be

(i) the plant manager; In fact LPI evaluation allows useful information on the actual economical benefit

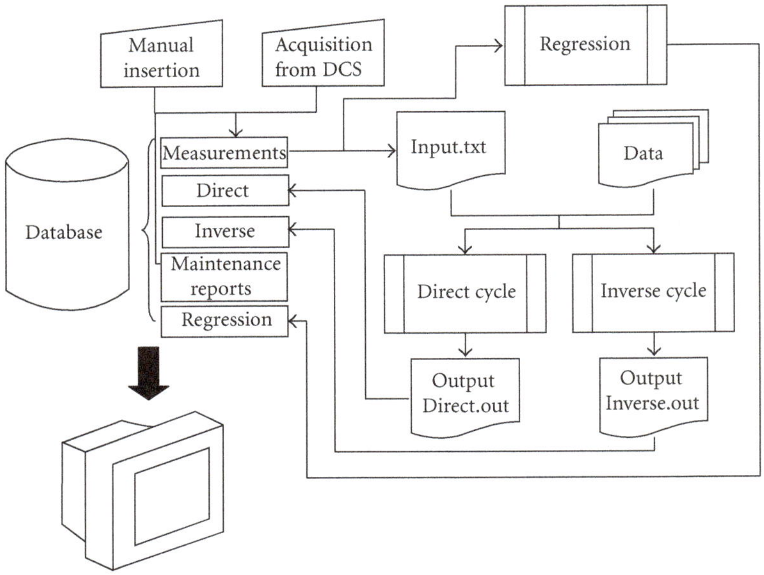

FIGURE 7: Software system architecture.

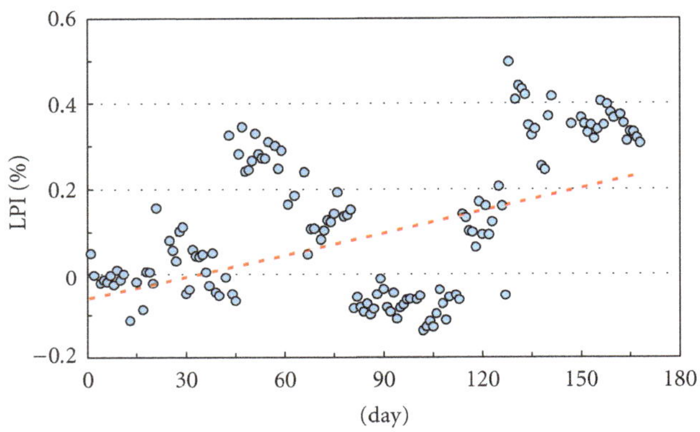

FIGURE 8: Lost production index trend.

of keeping the plant in operation rather than stop it to perform maintenance actions,

(ii) the maintenance manager, which, through the analysis of the normalized measurement and of the HIs trends, can obtain information on overall performance deterioration, on the components that are responsible for the deterioration, on the type and on the quantification of the deterioration;

(iii) the maintenance chief engineer, which, still through the analysis of the normalized measurement and of HIs trends, can obtain information on the performance recovery after maintenance actions.

7. Conclusions

In this paper, a comprehensive methodology for both measurement validation and health state determination of gas turbines was presented, discussed, and applied to a 5 MW gas turbine working in a natural gas compression plant.

The methodology demonstrated to be effective in supporting plant operation and maintenance management and some interesting results were presented. In particular.

(i) The application of the methodologies for measurement validation allowed the identification of outliers, the elimination of which allowed a remarkable reduction of the scattering of the trend-over-time of the measurements. The measurement sets composed of outliers were not used for estimating machine health indices.

(ii) Gas turbine health state determination, performed by applying the developed gas path analysis technique over a working period of two months, highlighted that compressor fouling (though not severe) was occurring, while gas generator turbine was not suffering from significant changes in health state.

(iii) The application of the multipoint analysis to measurements taken immediately before and after a maintenance stop allowed a more detailed analysis of the health state of the main gas turbine components.

Finally, the main features of a software, which was implemented in the considered compression plant to automate the diagnostic process and to support plant operation and management, were presented. As a sample application, the loss of production, due to gas turbine deterioration, was reported.

Nomenclature

E: Variation
F: Map scaling factors
F_{ob}: Objective function
k, k_A, k_B: Test criterion coefficients

LPI: Loss of production index
M: Mass flow rate
N: Sample size, rotational speed
N_m: Dimension of Q_m vector
N_{op}: Number of operating points
N_x: Dimension of X vector
P: Total pressure
P: Power
Q_m: Vector of measured variables
$Q_{m,c}$: Vector of computed estimates of the measured variables
Q_{wp}: Vector of measured variables necessary to define the working point
RH: Relative humidity
S: Standard deviation of the sample
T: Total temperature
t_α: T-student distribution quantile
VN: Variable nozzle angular position
w: Weight
x: element of the sample
X: $[X_1, \ldots, X_{Nx}]^T$ vector of health indices
α: Level of significance
Δ: Variation
η: efficiency
μ: $= M\sqrt{T}/p$ mass flow function
$*$: Normalized value.

Subscripts

a: Ambient
C: Compressor
f: Fuel
m: Mean value
GGT: Gas generator turbine
GasC: Compressed gas (calculated)
GasM: Compressed gas (measured)
ov: Overall
PT: Power turbine.

Acronyms

C: Compressor
CC: Combustion chamber
CP: Cycle Program
F: Filter
GGT: Gas generator turbine
GPA: Gas path analysis
HI: Health index
PT: Power turbine.

References

[1] R. F. Hoeft, "Heavy duty gas turbine operating and maintenance considerations," in *Proceedings of the 39th GE Turbine State-of-the-Art Technology Seminar*, GER-3620D, 1996.

[2] T. P. Schmitt and C. G. Petroff, "Gas turbine performance monitoring and testing," in *Proceedings of the 39th GE Turbine State-of-the-Art Technology Seminar*, GER-3958, 1996.

[3] R. Bettocchi, M. Pinelli, P. R. Spina, M. Venturini, and S. Sebastianelli, "A system for health state determination of natural gas compression gas turbines," in *Proceedings of the Turbine Technical Conference and Exposition*, ASME Paper 2001-GT-223, 2001.

[4] D. Therkorn, "Remote monitoring & diagnostic for combined-cycle power plants," in *Proceedings of the Turbine Technical Conference and Exposition*, ASME Paper GT2005-68710, pp. 697–703, June 2005.

[5] H. R. Depold and J. Siegel, "Using diagnostics and prognostics to minimize the cost of ownership of gas turbines," in *Proceedings of the Turbine Technical Conference and Exposition*, ASME Paper GT2006-91183, pp. 845–851, May 2006.

[6] E. Hindle, R. Van Stone, C. Brogan, J. Vandike, K. Dale, and N. Gibson, "A prognostic and diagnostic approach to engine health management," in *Proceedings of the Turbine Technical Conference and Exposition*, ASME Paper GT2006-90614, pp. 673–680, May 2006.

[7] L. C. Jaw, "Recent advancements in aircraft Engine Health Management (EHM) technologies and recommendations for the next step," in *Proceedings of the Turbine Technical Conference and Exposition*, ASME Paper GT2005-68625, pp. 683–695, June 2005.

[8] L. A. Urban, "Gas path analysis applied to turbine engine condition monitoring," in *Proceedings of the AIAA/SAE 8th Joint Propulsion Conference*, New Orleans, La, USA, 1972.

[9] A. Stamatis, K. Mathioudakis, and K. D. Papailiou, "Adaptive simulation of gas turbine performance," *ASME Journal of Engineering for Gas Turbines and Power*, vol. 112, pp. 168–175, 1990.

[10] E. Benvenuti, R. Bettocchi, G. Cantore, G. Negri di Montenegro, and P. R. Spina, "Gas turbine cycle modeling oriented to component performance evaluation from limited design or test data," in *Proceedings of the 7th ASME COGEN-TURBO*, vol. 8, pp. 327–337, IGTI, Bournemouth, UK, 1993.

[11] E. Benvenuti, R. Bettocchi, G. Cantore, G. Negri di Montenegro, and P. R. Spina, "Experimental validation of a gas turbine cycle model based on a simultaneous solution method," in *Proceedings of the 8th ASME COGEN-TURBO*, vol. 9, pp. 245–255, IGTI, Portland, Ore, USA, 1994.

[12] R. Bettocchi and P. R. Spina, "Diagnosis of gas turbine operating conditions by means of the inverse cycle calculation," in *Proceedings of the Turbine Technical Conference and Exposition*, ASME Paper 99-GT-185, 1999.

[13] R. Bettocchi, M. Pinelli, P. R. Spina, and M. Venturini, "Artificial intelligence for the diagnostics of gas turbines—part 1: neural network approach," *Journal of Engineering for Gas Turbines and Power*, vol. 129, no. 3, pp. 711–719, 2007.

[14] R. Bettocchi, M. Pinelli, P. R. Spina, and M. Venturini, "Artificial intelligence for the diagnostics of gas turbines—part II: neuro-fuzzy approach," *Journal of Engineering for Gas Turbines and Power*, vol. 129, no. 3, pp. 720–729, 2007.

[15] R. Bettocchi, M. Pinelli, P. R. Spina, M. Venturini, and G. A. Zanetta, "Assessment of the robustness of gas turbine diagnostics tools based on neural networks," in *Proceedings of the Turbine Technical Conference and Exposition*, ASME Paper GT2006-90118, pp. 603–613, May 2006.

[16] M. Bianchi, A. Peretto, and P. R. Spina, "Modular dynamic model of multi-shaft gas turbine and validation test," in *Proceedings of the Winter Annual Meeting of ASME*, vol. 38, pp. 73–81, AES, Anaheim, Calif, USA, 1998.

[17] M. Venturini, "Development and experimental validation of a compressor dynamic model," *Journal of Turbomachinery*, vol. 127, no. 3, pp. 599–608, 2005.

[18] M. Morini, M. Pinelli, and M. Venturini, "Development of a one-dimensional modular dynamic model for the simulation of surge in compression systems," *Journal of Turbomachinery*, vol. 129, no. 3, pp. 437–447, 2007.

[19] M. Morini, M. Pinelli, and M. Venturini, "Application of a one-dimensional modular dynamic model for compressor surge avoidance," in *Proceedings of the Turbine Technical Conference and Exposition*, ASME Paper GT2007-27041, pp. 1425–1434, May 2007.

[20] M. Morini, G. Cataldi, M. Pinelli, and M. Venturini, "A model for the simulation of large-size single-shaft gas turbine start-up based on operating data fitting," in *Proceedings of the Turbine Technical Conference and Exposition*, ASME Paper GT2007-27373, pp. 1849–1856, May 2007.

[21] M. Morini, M. Pinelli, and M. Venturini, "Analysis of biogas compression system dynamics," *Applied Energy*, vol. 86, no. 11, pp. 2466–2475, 2009.

[22] M. Venturini, "Simulation of compressor transient behavior through recurrent neural network models," *Journal of Turbomachinery*, vol. 128, no. 3, pp. 444–454, 2006.

[23] M. Venturini, "Optimization of a real-time simulator based on recurrent neural networks for compressor transient behavior prediction," *Journal of Turbomachinery*, vol. 129, no. 3, pp. 468–478, 2007.

[24] M. Pinelli, P. R. Spina, and M. Venturini, "Optimized operating point selection for gas turbine health state analysis by using a multi-point technique," in *Proceedings of the Turbine Technical Conference and Exposition*, ASME Paper GT2003-38191, pp. 43–51, June 2003.

[25] Y. G. Li, "Gas turbine diagnosis using a fault isolation enhanced GPA," in *Proceedings of the Turbine Technical Conference and Exposition*, pp. 361–369, June 2004.

[26] A. I. Zwebek and P. Pilidis, "Application of GPA to combined cycle gas turbine plants," in *Proceedings of the Turbine Technical Conference and Exposition*, ASME Paper GT2004-53026, pp. 225–232, June 2004.

[27] O. Córdoba, "Gas path analysis study for overhaul engines," in *Proceedings of the Turbine Technical Conference and Exposition*, ASME Paper GT2005-68137, pp. 497–505, June 2005.

[28] M. Morini, M. Pinelli, P. R. Spina, and M. Venturini, "Influence of blade deterioration on compressor and turbine performance," *Journal of Engineering for Gas Turbines and Power*, vol. 132, no. 3, Article ID 032401, 2010.

[29] P. R. Spina, "Gas turbine performance prediction by using generalized performance curves of compressor and turbine stages," in *Proceedings of the Turbine Technical Conference and Exposition*, ASME Paper GT-2002-30275, pp. 1073–1082, June 2002.

[30] M. Bagnoli, M. Bianchi, F. Melino, and P. R. Spina, "Development and validation of a computational code for wet compression simulation of gas turbines," *Journal of Engineering for Gas Turbines and Power*, vol. 130, no. 1, Article ID 012004, 2008.

[31] M. Bagnoli, M. Bianchi, F. Melino et al., "Application of a computational code to simulate interstage injection effects on GE frame 7EA gas turbine," *Journal of Engineering for Gas Turbines and Power*, vol. 130, no. 1, Article ID 012001, 2008.

[32] M. Morini, M. Pinelli, P. R. Spina, and M. Venturini, "Computational fluid dynamics simulation of fouling on axial compressor stages," *Journal of Engineering for Gas Turbines and Power*, vol. 132, no. 7, Article ID 072401, 2010.

[33] M. Morini, M. Pinelli, P. R. Spina, and M. Venturini, "Numerical analysis of the effects of non-uniform surface roughness on compressor stage performance," *Journal of Engineering for Gas Turbines and Power*, vol. 133, no. 7, Article ID 072402, 2011.

[34] M. Pinelli, M. Morini, P. R. Spina, M. Venturini, and C. Ferrari, "Analysis of the effects of simulated fouling on an axial compressor stage through CFD modeling," in *Proceedings of the 9th European Conference on Turbomachinery*, M. Sen, G. Bois, M. Manna, and T. Arts, Eds., vol. 1, pp. 261–270, Istanbul, Turkey, 2011.

[35] F. Melino, M. Morini, A. Peretto, M. Pinelli, and P. R. Spina, "Compressor fouling modeling: relationship between computational roughness and gas turbine operation time," in *Proceedings of the Turbine Technical Conference and Exposition*, ASME Paper GT2011-46089, 2011.

[36] R. Bettocchi, M. Pinelli, and P. R. Spina, "A multistage compressor test facility: uncertainty analysis and preliminary test results," *Journal of Engineering for Gas Turbines and Power*, vol. 127, no. 1, pp. 170–177, 2005.

[37] M. Morini, M. Pinelli, and M. Venturini, "Acoustic and vibrational analyses on a multi-stage compressor for unstable behaviour precursor identification," in *Proceedings of the Turbine Technical Conference and Exposition*, ASME Paper GT2007-27040, pp. 1415–1423, May 2007.

[38] R. Bettocchi, M. Morini, M. Pinelli, P. R. Spina, M. Venturini, and G. Torsello, "Setup of an experimental facility for the investigation of wet compression on a multistage compressor," *Journal of Engineering for Gas Turbines and Power*, vol. 133, no. 10, Article ID 102001, 2011.

[39] R. Bettocchi, P. R. Spina, and E. Benvenuti, "Set-up of an adaptive method for the diagnosis of gas turbine operating state by using test-Bench measurements," in *Proceedings of the Turbine Technical Conference and Exposition*, ASME Paper 2000-GT-0309, 2000.

[40] A. Stamatis, K. Mathioudakis, and K. Papailiou, "Optimal measurement and health index selection for gas turbine performance status and fault diagnosis," *Journal of Engineering for Gas Turbines and Power*, vol. 114, no. 2, pp. 209–216, 1992.

[41] M. Pinelli and P. R. Spina, "Gas turbine field performance determination: sources of uncertainties," *Journal of Engineering for Gas Turbines and Power*, vol. 124, no. 1, pp. 155–160, 2002.

[42] Y. G. Li, P. Pilidis, and M. A. Newby, "An adaptation approach for gas turbine design-point performance simulation," in *Proceedings of the Turbine Technical Conference and Exposition*, pp. 95–105, usa, June 2005.

[43] M. Pinelli and M. Venturini, "Improvement of the accuracy in gas turbine health determination," in *Proceedings of the Turbine Technical Conference and Exposition*, ASME Paper 2001-GT-476, 2001.

[44] M. Pinelli, M. Venturini, and M. Burgio, "Statistical methodologies for reliability assessment of gas turbine measurements," in *Proceedings of the Turbine Technical Conference and Exposition*, ASME Paper GT2003-38407, pp. 787–793, June 2003.

[45] H. DePold, A. Volponi, J. Siegel, and J. Hull, "Validation of diagnostic data with statistical analysis and embedded knowledge," in *Proceedings of the Turbine Technical Conference and Exposition*, ASME Paper 2003-GT-38764, pp. 573–579, June 2003.

[46] P. C. Chen and H. Andersen, "The implementation of the data validation process in a gas turbine performance monitoring system," in *Proceedings of the Turbine Technical Conference and Exposition*, ASME Paper GT2005-68429, pp. 609–616, June 2005.

[47] P. Hartner, J. Petek, P. Pechtl, and P. Hamilton, "Model-based data reconciliation to improve accuracy and reliability of performance evaluation of thermal power plants," in

Proceedings of the Turbine Technical Conference and Exposition, ASME Paper GT2005-68937, pp. 195–200, June 2005.

[48] S. C. Gulen and R. W. Smith, "A simple mathematical approach to data reconciliation in a single-shaft combined-cycle system," in *Proceedings of the Turbine Technical Conference and Exposition*, ASME Paper GT2006-90145, pp. 479–491, May 2006.

[49] K. Mathioudakis and P. Kamboukos, "Assessment of the effectiveness of gas path diagnostic schemes," in *Proceedings of the Turbine Technical Conference and Exposition*, ASME Paper GT2004-53862, pp. 723–731, June 2004.

[50] S. W. Butler, K. R. Pattipati, A. Volponi, J. Hull, R. Rajamani, and J. Siegel, "An assessment methodology for data-driven and model-based techniques for engine health monitoring," in *Proceedings of the Turbine Technical Conference and Exposition*, ASME Paper GT2006-91096, pp. 823–831, May 2006.

[51] C. Romessis, P. Kamboukos, and K. Mathioudakis, "The use of probabilistic reasoning to improve least squares based gas path diagnostics," in *Proceedings of the Turbine Technical Conference and Exposition*, ASME Paper GT2006-90619, pp. 681–689, May 2006.

[52] R. Bettocchi, M. Pinelli, P. R. Spina, and M. Venturini, "Statistical analyses to improve gas turbine diagnostics reliability," in *Proceedings of the IGTC*, Paper IGTC2003-Tokyo TS-004, 2003.

[53] M. Pinelli and M. Venturini, "Operating state historical data analysis to support gas turbine malfunction detection," in *Proceedings of the Turbine Technical Conference and Exposition*, ASME IMECE2001/AES-23665, pp. 457–462, November 2001.

[54] P. R. Spina, "Reliability in the determination of gas turbine operating state," in *Proceedings of the 39th IEEE Conference on Decision and Control*, Paper CDC00-INV5805, Sydney, Australia, 2000.

[55] R. Bettocchi, M. Pinelli, M. Venturini, P. R. Spina, S. Bellagamba, and G. Tirone, "Procedura di calibrazione del programma per la diagnosi funzionale dei turbogas della centrale a ciclo combinato di la spezia," in *Proceedings of the Atti del 57th Congresso Nazionale ATI*, Pisa, Italy, 2002.

[56] A. Gulati, M. Zedda, and R. Singh, "Gas turbine engine and sensor multiple operating point analysis using optimization techniques," in *Proceedings of the 36th AIAA/ASME/ SAE/ASEE Joint Propulsion Conference and Exhibit*, AIAA 2000-3716, 2000.

[57] R. Bettocchi, M. Pinelli, M. Venturini, P. R. Spina, S. Sebastianelli, and F. Bezzi, "A software tool for gas turbine on-condition monitoring and diagnostics," in *Proceedings of OMC The Mediterranean: New Oil and Gas Routes for an Ancient Sea*, Ravenna, Italy, 2003.

[58] P. R. Spina, G. Torella, and M. Venturini, "The use of expert systems for gas turbine diagnostics and maintenance," in *Proceedings of the Turbine Technical Conference and Exposition*, ASME Paper GT-2002-30033, pp. 127–134, June 2002.

[59] N. Puggina and M. Venturini, "Development of a statistical methodology for gas turbine prognostics," in *Proceedings of the Turbine Technical Conference and Exposition*, ASME Paper GT2011-45708, 2011.

[60] A. Cavarzere and M. Venturini, "Application of forecasting methodologies to predict gas turbine behavior over time," in *Proceedings of the Turbine Technical Conference and Exposition*, ASME Paper GT2011-45708, 2011.

[61] R. Bettocchi, P. R. Spina, M. Pinelli, M. Venturini, and M. Burgio, "Set up of a robust neural network for gas turbine simulation," in *Proceedings of the Turbine Technical Conference and Exposition*, ASME Paper GT2004-53421, pp. 543–551, June 2004.

[62] P. R. Spina and M. Venturini, "Gas turbine modeling by using neural networks trained on field operating data," in *Proceedings of the ECOS*, vol. 1, pp. 213–222, Padova, Italy, 2007.

[63] Visual Numerics, Inc., "IMSL MATH/LIBRARY: FORTRAN Subroutines for Mathematical Applications," Houston, Texas, USA, 1994.

[64] R. Bettocchi, P. R. Spina, and S. Alliney, "Resolution method for gas turbine mathematical models," in *Proceedings of the 8th ASME COGEN-TURBO*, vol. 9, pp. 361–369, Portland, Ore, USA, 1994.

[65] M. Pinelli and M. Venturini, "Application of methodologies to evaluate the health state of gas turbines in a cogenerative combined cycle power plant," in *Proceedings of the Turbine Technical Conference and Exposition*, ASME Paper GT-2002-30248, pp. 187–195, June 2002.

[66] ISO 2314, "Gas Turbine—Acceptance Tests," International Standard, 1989.

[67] I. S. Diakunchak, "Performance deterioration in industrial gas turbines," *Journal of Engineering for Gas Turbines and Power*, vol. 114, no. 2, pp. 161–168, 1992.

[68] H. I. H. Saravanamuttoo and B. D. Maclsaac, "Thermodynamic models for pipeline gas turbine diagnostics," *Journal of Engineering for Power*, vol. 105, no. 4, pp. 875–884, 1983.

6

Numerical Simulation of Surge in Axial Compressor

Wenhai Du[1,2] and Olivier Léonard[1]

[1] *Turbomachinery Group, University of Liège, Chemin des Chevreuils 1, 4000 Liège, Belgium*
[2] *Institute of Engineering Thermophysics, Chinese Academy of Sciences, Beijing 100190, China*

Correspondence should be addressed to Wenhai Du, wenhai.du@gmail.com

Academic Editor: Sung Ro

The object of this paper is to provide a reliable tool to simulate the stationary and transient operation performances in multistage axial compression systems, especially poststall behavior. An adapted version of the 1D Euler equations with additional source terms is solved by a time marching and control volume method. The equations are discretized at midspan both inside the blade rows and the nonbladed regions, along the real flow path geometry. The source terms express the blade-flow interactions and are estimated by calculating the velocity triangles for each blade row. Loss coefficient and deviation models are supplied by empirical correlations and are compared to experimental data in all flow regions. Transient simulations are carried and compared to the experimental results for several values of parameter *B*. The flow mechanism inside the compressor during a surge cycle is also shown.

1. Introduction

The compression system is an important component of a gas turbine engine. During the operation of axial multistage compressors, instabilities such as surge and rotating stall were observed in many experimental tests. However, such tests are expensive and not practical for preliminary design purpose. Numerical simulations of poststall performances of compression system are needed and useful in the design phase. In this paper, a reliable numerical tool is presented to simulate the instability behavior and to analyze the flow mechanism inside the compressor during poststall process.

This tool is only based on the compressor geometry and empirical correlations to simulate the poststall performance. The mesh cells are not only distributed between blade rows, but also inside of each blade row, allowing for a detailed representation of the flow field. An adapted version of Euler equations, mass, meridional momentum, circumferential momentum, and energy balances being applied for each control volume at mean radius. These equations are solved by a time-marching, finite-volume method [1]. The source terms expressing the blade-flow interactions are estimated by calculating the velocity triangles for each blade row with the help of empirical correlations.

The test compressor rig of Day is chosen to validate this model because of the detailed experiment results [2]. The steady characteristic curve of the compressor is predicted and is compared to the experiments. Dynamic simulations of the compression system are also carried out to predict the poststall performance and to validate the present tool. The flow mechanism inside of the compression system during one surge cycle is studied and the frequency of the surge is analyzed.

2. Background

Rotating stall and surge are two kinds of poststall phenomena in compression systems [3]. Rotating stall is a severe non-axisymmetric flow which includes one or several stall cells, propagating in the circumferential direction at a fraction of the rotor speed. Rotating stall can lead to serious mass flow and pressure ratio reductions. However, the annulus average mass flow and pressure ratio remain almost constant once the stall cells are fully developed. Surge is an unsteady

1D phenomenon with mass flow oscillations in the axial direction, even the reversed flow inside the compressor at severe conditions. Both rotating stall and surge can lead destructive to consequences to the compressor; however the appearance of the nonrecovery rotating stall in the aero-engine system, make it important to study the poststall behavior during the transient operations.

If only the global behavior is concerned, a lumped parameter technique can be implemented, which considers each component in the compression system as one control volume on which mass, momentum and energy balances are applied. The interactions of the compressor and the transient flow are included through source terms which are derived from steady characteristics. This method can provide fast and reliable results, but it lacks the ability to simulate the flow details inside the compressor. Greitzer developed such a lumped-volume method to study surge and rotating stall phenomena, based on the compressor steady characteristics [4, 5]. Davis applied a lumped volume method for each stage of the compressor and carried out parametric studies [6]. Morini developed a nonlinear modular dynamic model to study the deep surge of compression systems through an imaginary steady-state performance map [7].

If the mechanisms of stall inception and rotating stall development are the main focus, 2D and 3D models are necessary. Takata and Nagano developed a 2D model to simulate rotating stall in axial compressors, with the assumption that blade rows are replaced by semiactuator disks [8]. Moore and Greitzer proposed a 2D, unsteady, lumped parameter model to study the dynamic response of compression systems and the coupling between rotating stall and surge [9]. Lindau and Owen developed a quasi-3D actuator-disk model to simulate rotating stall and surge, with source terms representing blade losses and deviations [10]. Escuret used 3D Euler equations with source terms estimated from empirical equations; however, it could not simulate reversed flows [11]. Tauveron proposed new empirical correlations to estimate the steady compressor performance in all regions and carried out the dynamic simulations [12].

Most of these models rely on steady compressor characteristics for the entire flow regions, which can only be obtained from experimental data or derived from generic performance curves. Therefore, for preliminary design purpose, it is worthy to use empirical correlations to predict the physical trends of compressor steady characteristics and use them further for transient simulations. Such a 1D model can be regarded as a good compromise between the accuracy and computational load for dynamic simulations in multistage compressors. It can provide more accuracy and more details of the flow field than the lumped method and requires much less computational resources than 3D calculations.

3. Model Description

It is assumed that the flow path meanline is a streamline and that the flow is axisymmetric. The fluid is treated as an ideal gas. The quasi-1D Euler equations are applied in a curvilinear (m, n, θ) coordinate system. The mass flow, meridional momentum, circumferential momentum and energy equations can be described in vector, nondimensional, conservative form, as follows:

$$\frac{\partial U}{\partial t} + \frac{\partial G}{\partial m} = Q,$$

$$U = \begin{bmatrix} \rho A \\ \rho V_m A \\ \rho r V_t A \\ \rho E^0 A \end{bmatrix}, \qquad G = \begin{bmatrix} \rho V_m A \\ (P + \rho V_m^2) A \\ \rho r V_m V_t A \\ \rho H^0 V_m A \end{bmatrix}. \tag{1}$$

The cross-section area is calculated through the simple form $A = 2\pi r$.

The source terms represented by Q in the Euler equations are used to represent the influence of the blade rows and the flowpath on the fluid. They can be splitted into four contributions: the isentropic action of the blades Q_b, the influence of friction Q_f, and the influence of the flowpath geometry Q_g:

$$Q = Q_b + Q_f + Q_g. \tag{2}$$

3.1. Friction Source Term. In the Euler equations, the entropy production is due to a separate distributed friction force F_f in the momentum equations. This friction force is aligned with the relative velocity but has the opposite direction:

$$\vec{F_f} = -\left|\vec{F_f}\right| \frac{\vec{W}}{\left|\vec{W}\right|}. \tag{3}$$

The entropy production can be written as

$$\rho T \frac{ds}{dt} = -\vec{W} \cdot \vec{F_f}, \tag{4}$$

where $\vec{W}$ is the relative velocity vector. The magnitude of the friction force can be rewritten as

$$F_f = \rho T \frac{W_m}{W} \frac{\partial s}{\partial x}. \tag{5}$$

Only the tangential component of the friction force can change the energy balance, and the friction source term can be described as following:

$$Q_f = \begin{bmatrix} 0 \\ F_{f,m} A \\ r F_{f,t} A \\ \Omega r F_{f,t} A \end{bmatrix}. \tag{6}$$

3.2. Blade Source Term. This term is implemented to represent the isentropic influence of blade rows. An angular momentum balance over a blade row provides the tangential component of the corresponding (inviscid) blade force $F_{b,t}$:

$$r\left(F_{f,t} + F_{b,t}\right) = \frac{q_m}{\text{Vol}} \Delta(r V_t). \tag{7}$$

It is intended that this force is applied in an isentropic way, the viscous contribution being applied separately through the friction source term. Therefore this blade force is perpendicular to the relative velocity:

$$\vec{F_b} \cdot \vec{W} = 0. \tag{8}$$

Once again, only the tangential component of the blade force appears in the energy balance:

$$Q_b = \begin{bmatrix} 0 \\ F_{b,m}A \\ rF_{b,t}A \\ \Omega r F_{b,t}A \end{bmatrix}. \tag{9}$$

3.3. Geometrical Source Term. This term represents the effect of the variation of the cross-section area and the meanline radius. The axial component can be derived as

$$\frac{\partial \rho V_m A}{\partial t} + \frac{\partial (P + \rho V_m^2)A}{\partial x} = \frac{\rho A V_t^2}{r}\frac{\partial r}{\partial x} + P\frac{\partial A}{\partial x}. \tag{10}$$

This geometrical force does not contribute to the energy balance and also does not create entropy.

In the rotating stall and reversed flow regions, the axial velocity may become extreme small, especially around the zero mass flow point. It will generate unphysical magnitudes of both blade and friction forces when they are decomposed in the axial and tangential directions. In this case the blade and the friction forces are combined together and derived from the steady Euler momentum equations applied between the inlet and the outlet of the blade row. These forces and the corresponding shaft work are distributed linearly across the control volumes inside the blade row. The geometrical source term is kept unchanged.

4. Empirical Correlations

The accuracy of the simulations is highly dependent on the source terms in the equations, which are derived from empirical correlations, particularly loss coefficient and deviation angle models. Different empirical correlations have been chosen for all flow conditions, such as normal forward flow, stalled flow, and reversed flow (Table 1).

The separation of boundary layers in compressors has not been fully understood. There are several basic criteria to detect it, such as the static-total pressure coefficient, incidence angle, or the diffusion rate. All these methods are based on experimental observations and their accuracy is highly dependent on the configuration of the compression system. In order to eliminate the uncertainty in numerical simulations, the stall inception and the stall recovery point are obtained from experimental results so that only the loss and deviation models can affect the steady performance. According to the experimental results of Day [2], the stalled point is imposed when the mass flow coefficient is less than 0.45 and the recovery point from stalled flow is set when the mass flow coefficient is more 0.50. The reversed flow is defined when the average axial velocity in the compressor duct becomes negative.

5. The Equation Solver

The computational domain is the real compressor flow path and discretized by a one-dimensional structured mesh. The mesh cells are distributed not only in the nonbladed regions but also inside of blade rows, to make sure that the flow deflection is smooth.

For each control volume, the integral form of the 1D axisymmetric Euler equations is written with the conservative variables located at the cell centers. A quadratic reconstruction scheme with a hybrid limiter is applied to extrapolate the variables at the interfaces. Numerical fluxes are based on Roe's flux difference. Both explicit and implicit schemes are implemented. The explicit scheme is based on a fourth-order Runge-Kutta algorithm. The implicit scheme is based on an Euler single-step forward in time differentiation. This implicit time-marching method is used for stationary simulations, allowing large local CFL numbers to speed up the calculation.

6. Steady Simulations

This model has been applied to a low-speed multistage axial compressor studied by Day. The test rig consists of four repeating stages with an inlet guide vane and a constant cross-sectional annulus area. The hub/casing radius ratio is 0.75, and the rotation speed is 3000 rpm, giving a maximum Reynolds number of 1.7×10^5, based on the chord [2].

The steady compressor characteristic is simulated using the current model and is compared to the experimental results. The empirical correlations are used to provide loss coefficients and deviation angles for each blade row in all flow regions, such as normal forward flow, stalled flow, and reversed flow. During the numerical simulations, a throttle is gradually turned down in order to obtain the different operating points. When the throttle is set across the instability point, the compressor operates in the stalled region. Meanwhile the reversed flow is obtained by imposing lower static pressure for inlet boundary, stagnation pressure, and temperature for outlet boundary. The comparison of the numerical results and the experimental data are presented in Figure 1.

It can be seen from Figure 1 that the current model underestimates the compressor performance in all flow regions; however the trend is correctly captured. It is found that a discontinuity exists at zero mass flow for both the experimental and numerical results, which is due to the radial circulation [18]. It should be emphasized that all empirical correlations introduced into the model come from the open literature. As far as these correlations can be replaced by a manufacturer's own correlations, this simulation tool would certainly be able to provide better predictions. At last, it should be pointed out that the purpose of empirical correlations is to provide steady compressor performance that is required for the transient simulations. Once chosen, they do not affect the transient simulation results for a variety of compressor/plenum configuration.

TABLE 1: Selected empirical correlations.

	Loss coefficient	Deviation angle
Forward, unstalled flow	Reference loss coefficient: Koch and Smith model Off-design loss coefficient: Creveling model [14]	Reference deviation angle: Lieblein method [13] Off-design deviation angle: Lieblein model [13]
Stalled flow	Moses and Thomason modified loss model [15, 16]	Tauveron deviation angle model [12]
Reversed flow	Longley modified loss model [12, 17]	Exit flow parallel to the stagger direction

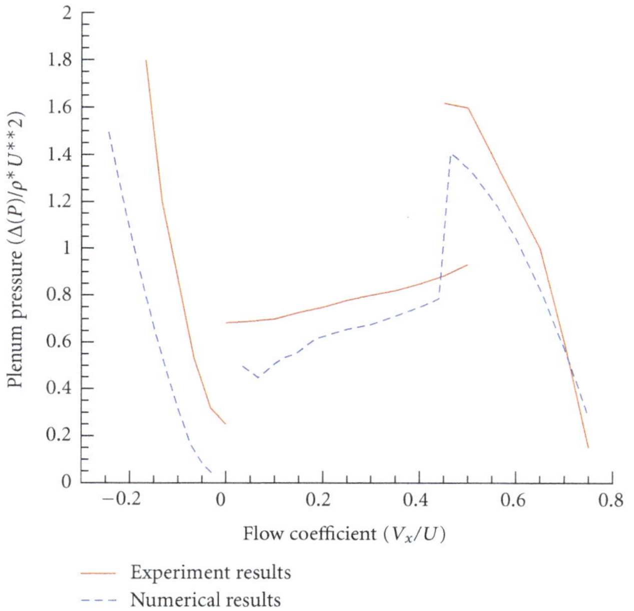

FIGURE 1: Experiment and numerical characteristic curves [2].

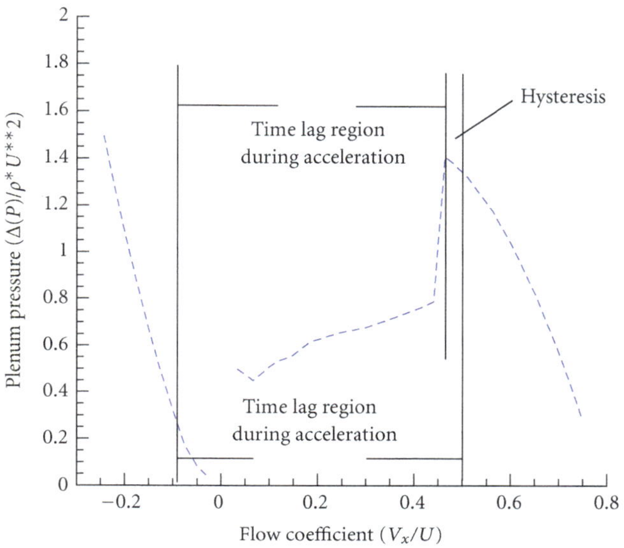

FIGURE 2: Application of time lag for the dynamic force.

FIGURE 3: Typical mesh of the compression system.

7. Transient Simulations

The time needed for stall cells to be fully developed can be several rotor revolutions and the mass flow undergoes significant changes during this time. The blade forces extracted from empirical correlations are steady and are not adequate for transient simulations. Therefore a first-order time lag equation has been included in the model in order to provide the dynamic blade force [3]. It should be emphasized that only the axial force is modified. The shaft work, and the tangential force are assumed constant

$$\tau \frac{dF}{dt} + F = F_{ss}. \quad (11)$$

The time constant τ is approximated to be two rotor revolutions and will be kept constant in all transient simulations. The time lag equation is used only during rotating stall. The experimental observations suggest that the transition between rotating stall and reversed flow does not really occur at zero mass flow [18]. Thus the time lag equation is applied until the flow coefficient reaches the value of -0.09; the steady blade force is used during the rest of reversed flow. The flow range for the application of the time lag equations is shown in Figure 2.

A throttle is imposed at the outlet according to [19]

$$P = P_{\text{ambient}} + \frac{1}{2} K_{\text{out}} \rho V_x^2. \quad (12)$$

The throttle is further turned down with larger coefficient K_{out}.

A characteristic method is implemented for the boundary conditions in order to avoid numerical spurious reflections. For positive flow operation the stagnation pressure and temperature are imposed as inlet boundary conditions, and ambient pressure is applied at outlet boundary. In reversed flow operation, it seems unreasonable to expect surrounding air to flow back into plenum through throttle and ambient pressure is always fixed at outlet boundary. Meanwhile, it is likely that the flow can be discharged from the compressor through the inlet boundary. Therefore the ambient pressure is imposed at the inlet when the inlet average velocity becomes negative, otherwise the stagnation pressure and temperature are imposed.

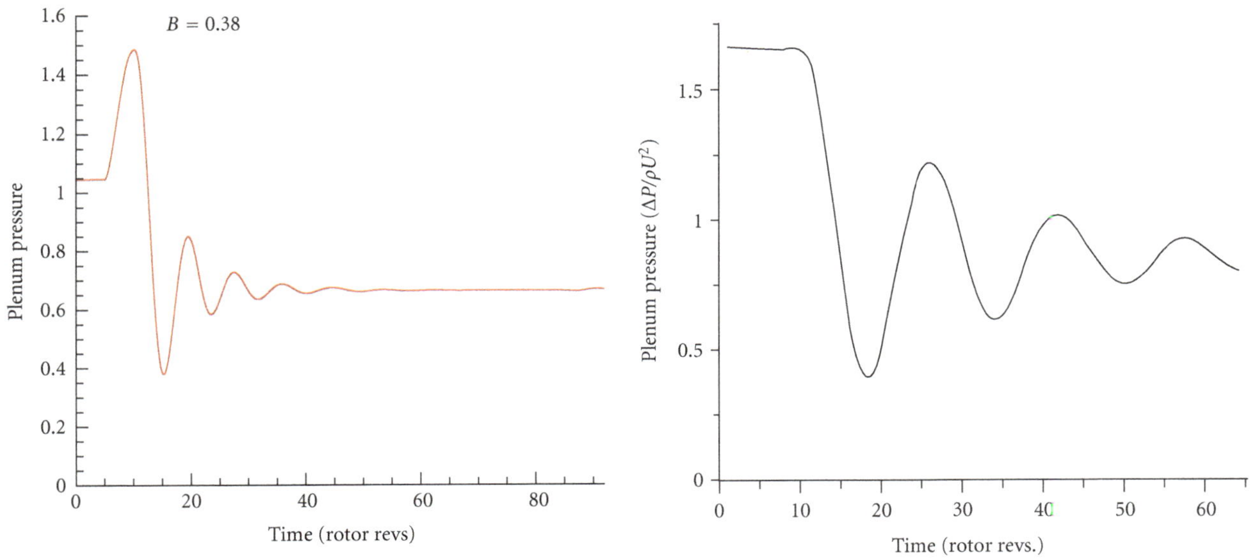

FIGURE 4: Numerical and experimental plenum pressure coefficients during rotating stall [2].

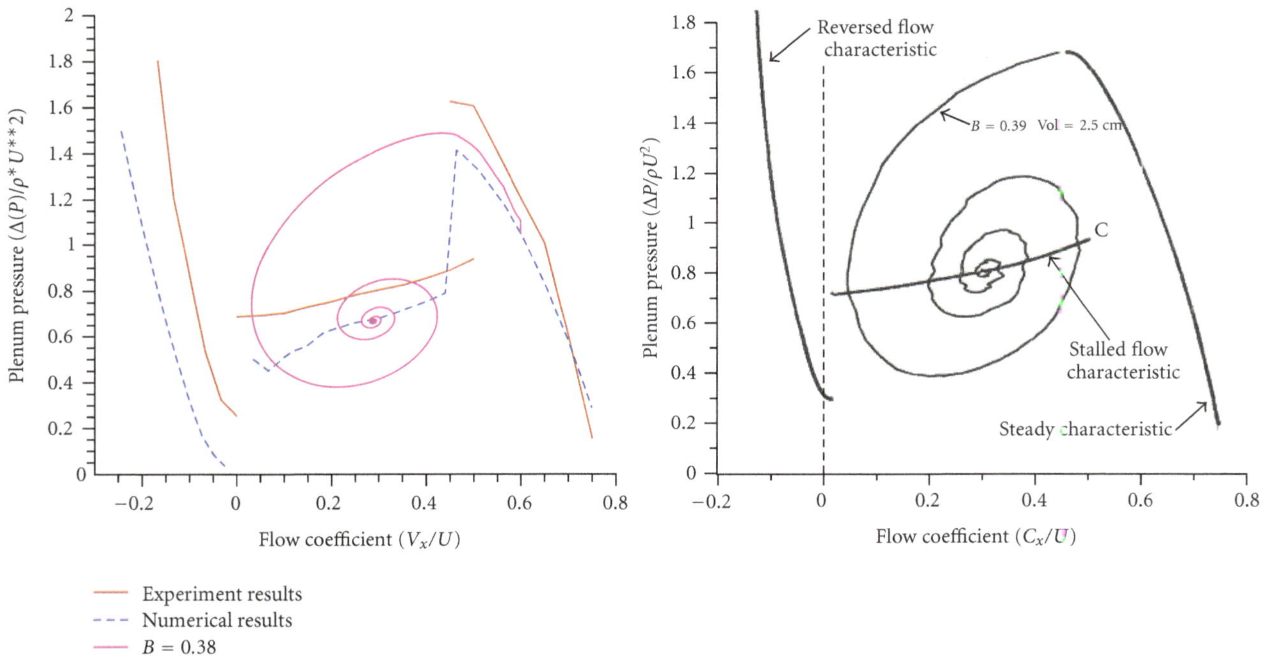

FIGURE 5: Numerical and experimental trajectory of compressor operating point during rotating stall [2].

A fourth-order Runge-Kunta algorithm is used for time integration in order to obtain time accurate transient performance. The plenum and throttle are added to form the compression system. A typical mesh is shown in Figure 3.

$B = (U/2a)\cdot\sqrt{V_p/(A_cL_c)}$ is used as a reference parameter to determine whether rotating stall or surge may occur in a particular compression system. Different compressor/plenum geometry configurations have been tested in the numerical simulations.

8. Rotating Stall

Although rotating stall is a 2D phenomenon which includes a strong nonaxisymmetric distribution of mass flow and

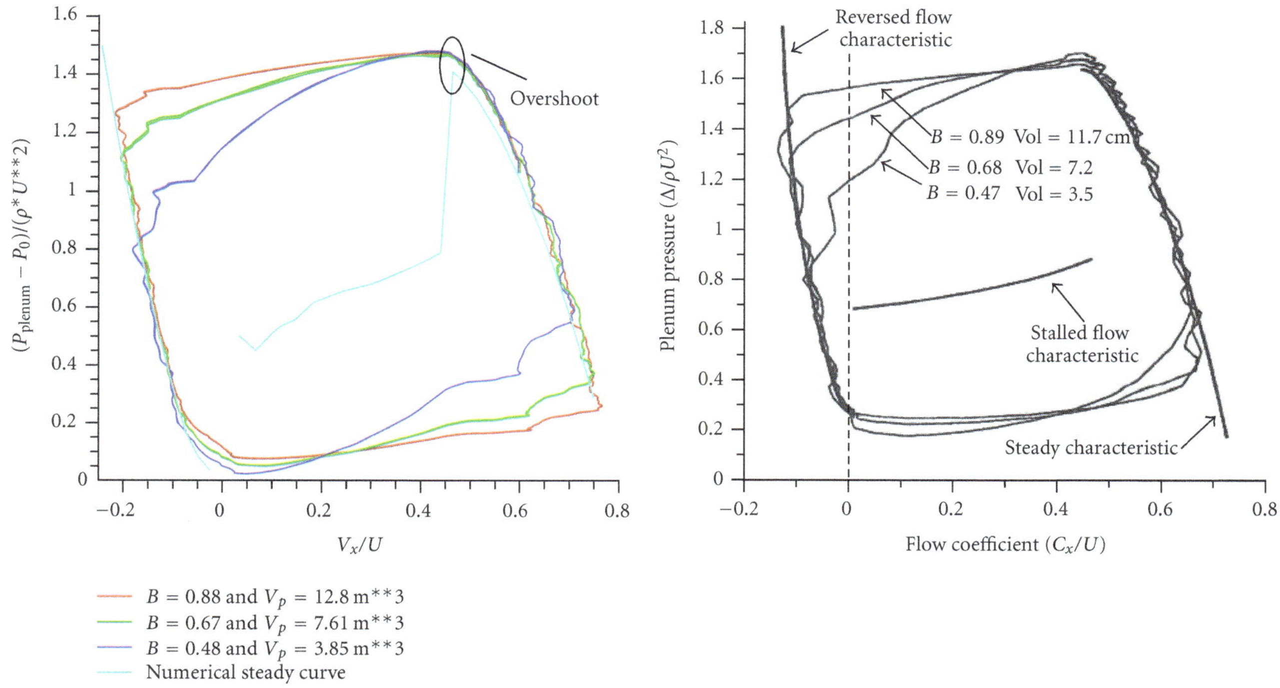

Figure 6: Numerical and experimental performance characteristics during surge [2].

pressure, the present model is able to predict the overall performance when the rotating stall is maturely developed. Since empirical correlations are applied to calculate annular average performance through blade rows, average compressor performances can be predicted by the current model.

The plenum is designed to be 2.41 cubic meters and the compressor rotating speed is fixed at 3000 RPM, and the corresponding parameter B is equal to 0.38 [2]. In the numerical simulation, the compressor operates in the normal flow region and it moves into rotating stall as the throttle is gradually turned down. In the experimental test, the compressor operates in a stable manner near the peak of pressure and the throttle is slowly turned down to cause the instability. Figure 4 shows the plenum pressure coefficient versus time (rotor revolutions) for both experimental results and numerical results. It can be seen that the plenum pressure oscillates several cycles before it reaches a stable condition. In the experimental test, the plenum pressure almost keeps constant before the collapse because the compressor starts around the maximum pressure ratio, while the plenum pressure increases in the numerical test since the compressor operates far from the peak point. This is confirmed by Figure 5, which shows the trajectory of the compressor operating point during rotating stall. As the compression system becomes unstable, the operating point settles down on the fully stalled characteristics after a large mass flow excursion. The operating point does not cross the reversed flow neither the recovery point, thus deep surge or classic surge are excluded. This is basically due to a small plenum volume which can be emptied or filled in a short time.

9. Surge

The plenum volume is, respectively, increased to 3.85, 7.61, 12.80 cubic meters in order to obtain parameter B to be 0.48, 0.67 and 0.88 at the same rotating speed 3000 RPM. Figure 6 shows the numerical and experimental surge cycles including the steady state curve for the three different plenum volumes. The surge trajectory follows the steady reversed flow curve for a short period. The excursion along the reversed steady curve decreases as the plenum volume is reduced. The smaller the volume is, the faster the flow can be discharged through the throttle and the compressor, and the surge trajectory intersects the reversed curve at lower pressure ratio. For the same reason, the surge trajectory recovers to higher pressure ratios with smaller plenum volumes. Another interesting point is the overshoot of pressure ratio between the surge trajectory and the steady curve around the peak point. It is due to the dynamic deceleration of the flow during the surge, which means that the higher plenum pressure is used to decrease the mass flow in the compressor duct.

In order to observe the flow mechanism in the compression system during the surge cycle, the flow parameters are displayed for the compression system $B = 0.67$. Figure 7 shows the numerical and experimental static pressure differences between the compressor exit and the plenum inlet. Figure 8 shows the mass flow coefficient at different locations. The static pressure difference between compressor exit and plenum inlet can be seen to accelerate or decelerate the compressor flow during the surge cycle. The deceleration 1 takes place when the compressor stalls (curve AB in Figure 8) and indicates a fast drop of the mass flow. The deceleration

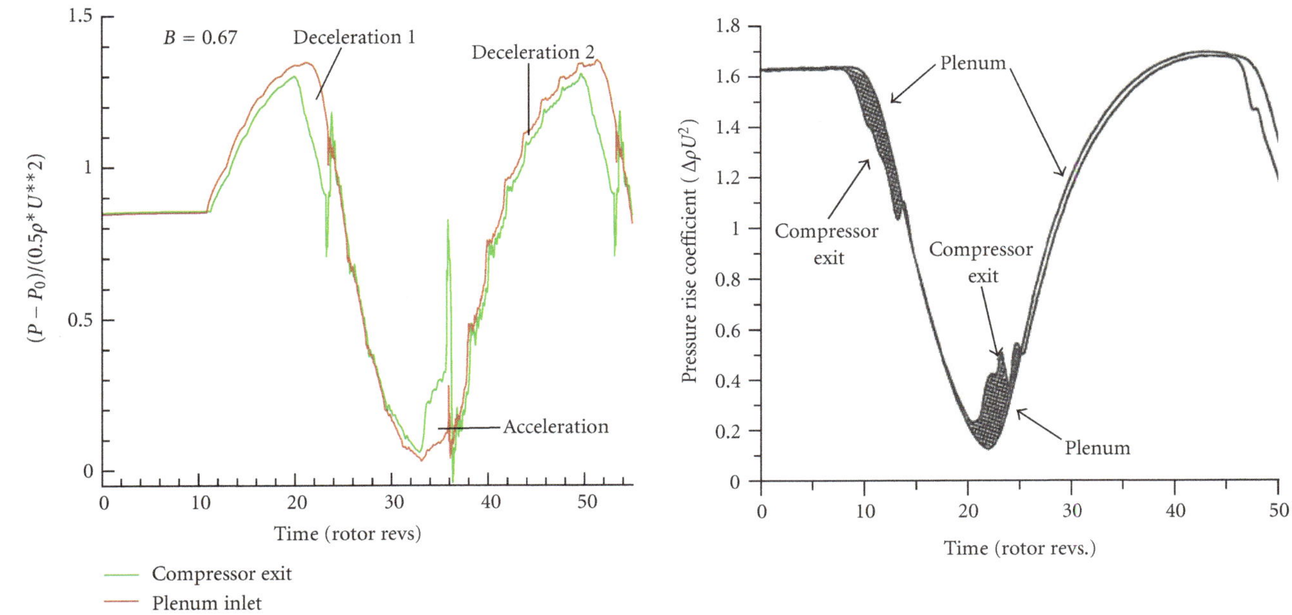

FIGURE 7: Numerical and experimental pressure differences between compressor exit and plenum inlet [2].

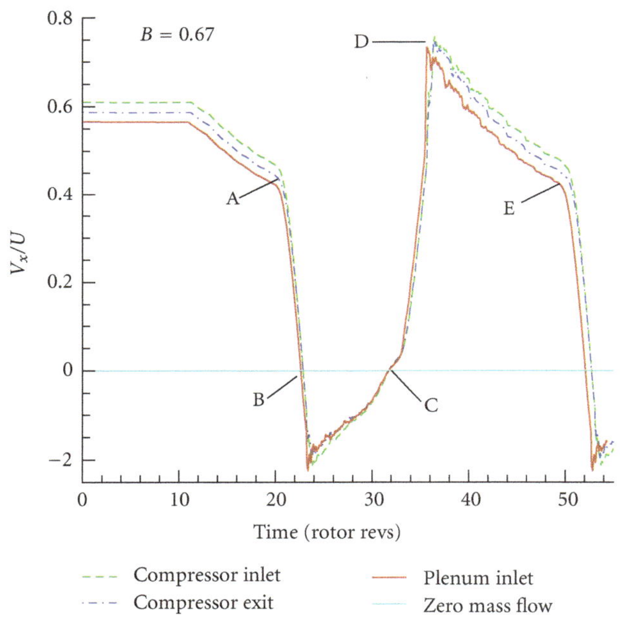

FIGURE 8: Numerical mass flow coefficients at different locations.

2 takes place when the plenum is repressurized (curve DE in Figure 8), which means that the mass flow decreases gradually and the compressor operating point rises up along the unstalled part of steady-state characteristics. The static pressure at the inlet of the plenum is somewhat higher than at the exit of the compressor, which explains the overshoot in Figure 6. The acceleration in Figure 7 (curve CD in Figure 8) corresponds to the recovery of the compressor from reversed flow to normal flow. The larger the static pressure difference in Figure 7, the faster the flow acceleration and deceleration in Figure 8. The curve BC in Figure 8 shows that the compressor operates in the reversed mode, which can be seen as a steady process for the compressor. Thus the static pressures at compressor exit and at plenum inlet are overlapped during that portion of time.

The static pressure at the compressor exit exhibits some spikes (Figure 7). Compared to experimental results, these spikes are stronger in simulations, due to the empirical correlations which fail to provide the correct pressure ratio for these regions, but the physical trend is similar. This major spike is caused when the compressor recovers from reversed to forward flow (around point C in Figure 8). It is speculated that the mass flow with high pressure and high temperature stored in compressor inlet moves downstream during the transient process from reversed to forward flow.

This can be demonstrated by Figure 9, which shows the total temperature and static pressure coefficients at three different locations during a surge cycle. The total temperature spike at compressor inlet is generated at rotor revolution 24, by the high temperature fluid stored in the plenum and flowing back during the reversed operation. The work input by the compressor in the flow direction increases the total temperature at the compressor inlet compared to the compressor outlet. At the same time, the pressure at the compressor inlet is significantly increased but still much smaller than at the compressor outlet due to the high resistance of the compressor during the reversed flow. This explains the total temperature spike in the inlet duct (point 1). During the reversed operation, the ambient static pressure is imposed at the inlet as shown in the Figure 9. As the mass flow is discharged through the throttle and through the compressor inlet during the reversed process, the temperature and the pressure gradually decrease in the

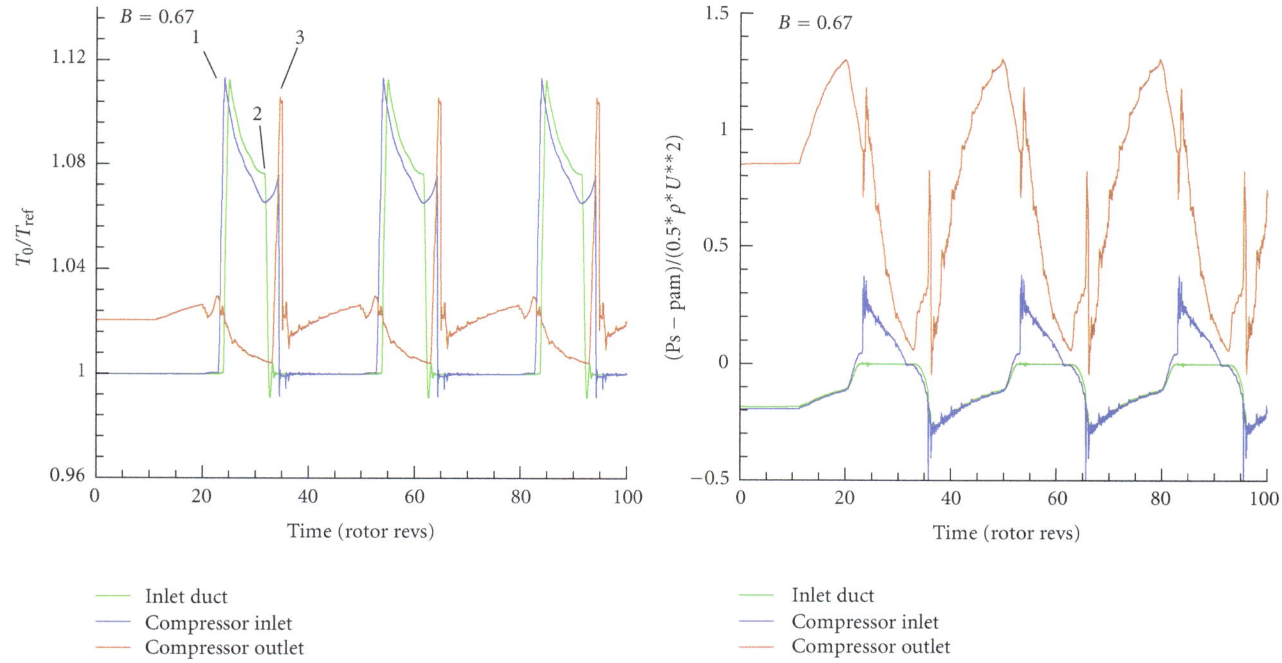

FIGURE 9: Numerical total temperature and total pressure at different locations.

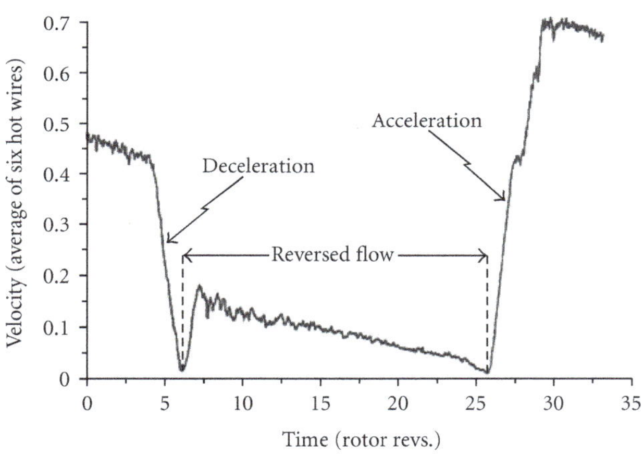

FIGURE 10: Average velocity from six hot wires downstream of compressor exit showing details of flow reversal during deep surge cycle in Day's experiment tests [2].

plenum, and so the temperature and the pressure drop at the compressor inlet (from point 1 to point 2).

When the compressor switches from reversed to forward flow, the high temperature fluid stored in the inlet duct is ingested by the compressor, which can lead to even a higher temperature at the compressor exit reflecting the high temperature flow plus the shaft work (from point 2 to point 3). When the plenum is almost emptied, the compressor gets momentum to repressurize the plenum and the static pressure difference between compressor exit and plenum inlet acts as a force that pushes the compressor to unstalled operating point (Figure 7).

A typical deep surge cycle can thus be divided into four portions: (1) deceleration; (2) reversed flow; (3) acceleration; (4) normal flow region. Figure 10 shows the average velocity at compressor exit during deep surge cycle in Day's experiments. The proportion of time required for each part is roughly 4%, 28%, 3%, and 65%. In the numerical simulations (Figure 8), the time proportion for each part is about 7.8%, 31.1%, 11.3%, and 50% for $B = 0.67$. Although it is not clear that what was the parameter B in the experiments, the simulation results are quite similar to the experimental results. Obviously, the reversed flow phase is significantly shorter than the positive flow phase since the plenum is discharged through the throttle and the compressor in the reversed flow.

In the experiments the compressor total duct length was either 2.0 or 3.5 m. In the above numerical results, the compressor effective length was imposed to be 2.0 m. The compressor effective length has been extended to 3.0 m, as well as the plenum volume to keep $B = 0.67$. Figure 11 displays the surge cycle and the mass flow coefficient versus time. The resulting surge cycles have almost the same shape, the parameter B playing a predominant role. However, the frequency becomes smaller as the length increases. This can be explained by Figure 12, which shows the experimental frequency, numerical frequency, and Helmholtz frequency versus the plenum volume. It was concluded by Day that the compressor is more akin to a relaxation oscillation than a Helmholtz resonator-type phenomenon in deep surge [2]. It also can be found that the frequency becomes smaller with larger plenum volume at the same compressor geometry.

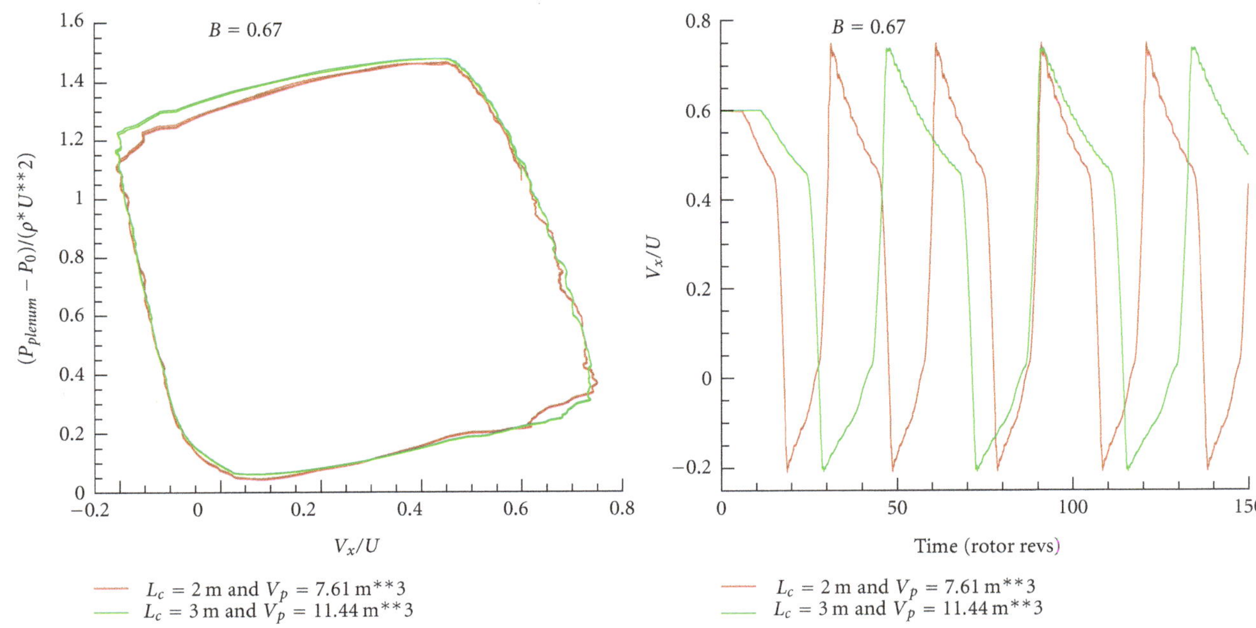

Figure 11: Numerical surge performances with two different compressor lengths at $B = 0.67$.

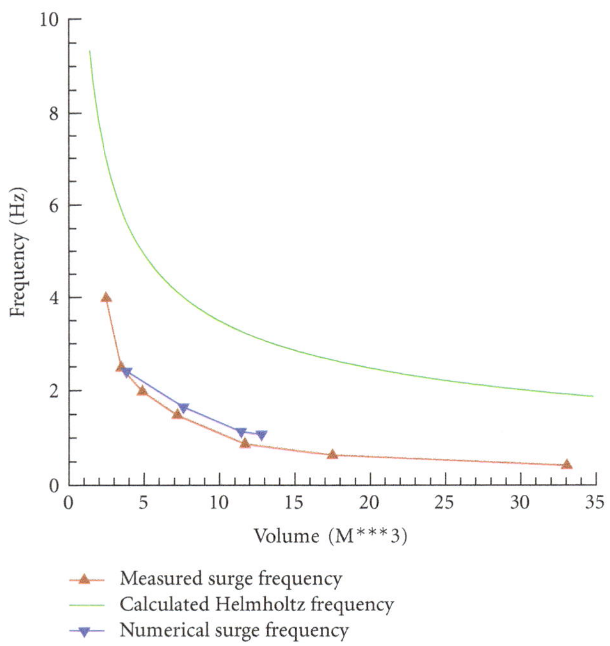

Figure 12: Experimental and numerical surge frequencies versus plenum volumes [2].

10. Conclusions

A reliable prediction tool is developed and implemented to simulate the transient performance of poststall behavior in compression systems. The model is based on an adapted version of the 1D Euler equations with additional source terms, which are estimated by empirical correlations. These empirical correlations are only used to provide the steady state performance and can be replaced by the user's own ones if the accurate steady state performance is needed.

This model can be used for preliminary design purpose as it is only based on the compressor geometry: the blade angles at mid span and the hub and shroud geometry. The model does neither rely on existing experimental performance curves nor on generic ones.

The comparison of numerical steady curve and experimental steady curve shows that the current correlations can capture the main trends of the compressor characteristics with an acceptable accuracy.

For rotating stall, the numerical results can be compared to the experiments for $B = 0.39$, and indicate that the current quasi-one-dimensional model has the ability to predict the overall performance of the multistage compression system when the rotating stall is fully developed.

For surge, the shape of the surge cycles and the frequency of the surge oscillations in the numerical simulations are tested against experimental results for $B = 0.47$, $B = 0.68$, and $B = 0.89$. The comparisons indicate that the model correctly captures the surge performance. Meanwhile, it is found that the static pressure difference between compressor exit and plenum inlet accelerates or decelerates the flow during the surge cycle, and the high temperature fluid stored in the inlet duct during reversed flow moves downstream when the compressor switches from reversed to forward flow, which can lead to the quite high temperature flow at the compressor exit. Finally, the influences of the compressor effective length on the surge shape and frequency are studied.

Nomenclature

a: Speed of sound
A: Flow section area
B: Nondimensional number
E: Specific internal energy
F: Force term
H: Specific enthalpy
K_{out}: Throttle coefficient
L: Effective length of equivalent duct
P: Pressure
Q: Source term
q_m: Mass flow
r: Radius
s: Entropy
SW: Shaft work
T: Temperature
U: Rotor velocity at mean radius
V_p: Plenum volume
V: Absolute velocity
W: Relative velocity
Vol: Blade row volume
ρ: Density
τ: Constant time scale
Ω: Rotating speed.

Subscripts

0: Ambient condition
1: Blade leading edge
2: Blade trailing edge
m: Meridional direction
t: Tangential direction
p: Plenum
c: Compressor.

Superscript

0: Total condition.

References

[1] O. Léonard and O. Adam, "A quasi-one-dimensional CFD model for multistage turbomachines," *Journal of Thermal Science*, vol. 17, no. 1, pp. 7–20, 2008.

[2] I. J. Day, "Axial compressor performance during surge," *AIAA Journal of Propulsion and Power*, vol. 10, no. 3, pp. 329–336, 1994.

[3] E. M. Greitzer, "Review—axial compressor stall phenomena," *Journal of Fluids Engineering, Transactions of the ASME*, vol. 102, no. 2, pp. 134–151, 1980.

[4] E. M. Greitzer, "Surge and rotating stall in axial flow compressors. Part I: theoretical compression system model," *Journal of Engineering for Power*, vol. 98, pp. 190–217, 1976.

[5] E. M. Greitzer, "Surge and rotating stall in axial flow compressors. Part II: experimental results and comparison with theory," *Journal of Engineering For Power*, vol. 98, no. 2, Article ID article id, pp. 199–211, 1976.

[6] M. W. Davis, *A stage-by-stage post-stall compression system modeling technique: methodology, validation, and application*, Ph.D. thesis, Virginia Polytechnic Institute and State University, 1987.

[7] M. Morini, M. Pinelli, and M. Venturini, "Development of a one-dimensional modular dynamic model for the simulation of surge in compression systems," *Journal of Turbomachinery*, vol. 129, no. 3, pp. 437–447, 2007.

[8] H. Takata and S. Nagano, "Nonlinear analysis of rotating stall," *Journal of Engineering for Power*, vol. 94, no. 4, pp. 279–293, 1972.

[9] F. K. Moore and E. M. Greitzer, "A theory of post-stall transients in axial compression system: Part I: development of equations," ASME paper 85-GT-171, 1985.

[10] J. W. Lindau and A. K. O. Owen, "Nonlinear quasi-three-dimensional modeling of rotating stall and surge," AIAA paper 97-2772, 1997.

[11] V. Garnier and J. F. Escuret, "Numerical simulations of surge and rotating stall in multi-stage axial-flow compressors," AIAA paper 94-3202, 1994.

[12] N. Tauveron, M. Saez, P. Ferrand, and F. Leboeuf, "Axial turbomachine modelling with a 1D axisymmetric approach. Application to gas cooled nuclear reactor," *Nuclear Engineering and Design*, vol. 237, no. 15-17, pp. 1679–1692, 2007.

[13] S. Lieblein, "Experimental flow in two-dimensional cascades," NASA SP-36, 1965.

[14] H. F. Creveling and R. H. Carmody, "Axial-flow compressor computer program for calculating off-design performance," NASA CR-72472, 1968.

[15] H. L. Moses and S. B. Thomason, "An approximation for fully stalled cascades," *Journal of Propulsion and Power*, vol. 2, no. 2, pp. 188–189, 1986.

[16] I. J. Day, E. M. Greitzer, and N. A. Cumpsty, "Prediction of compressor performance in rotating stall," *ASME Journal of Engineering for Power*, vol. 100, no. 1, pp. 1–14, 1978.

[17] J. P. Longley, "Calculating stall and surge transients," ASME paper GT-2007-27378, 2007.

[18] R. N. Gamache and E. M. Greitzer, "Reverse flow in multistage axial compressors," AIAA paper 86-1747, 1986.

[19] R. V. Chima, "A three-dimensional unsteady CFD model of compressor stability," NASA TM 2006-214117, 2006.

7

Empirical Design Considerations for Industrial Centrifugal Compressors

Cheng Xu and Ryoichi S. Amano

Department of Mechanical Engineering, University of Wisconsin-Milwaukee, Milwaukee, WI 53212, USA

Correspondence should be addressed to Ryoichi S. Amano, amano@uwm.edu

Academic Editor: Ashwani K. Gupta

Computational Fluid Dynamics (CFD) has been extensively used in centrifugal compressor design. CFD provides further optimisation opportunities for the compressor design rather than designing the centrifugal compressor. The experience-based design process still plays an important role for new compressor developments. The wide variety of design subjects represents a very complex design world for centrifugal compressor designers. Therefore, some basic information for centrifugal design is still very important. The impeller is the key part of the centrifugal stage. Designing a highly efficiency impeller with a wide operation range can ensure overall stage design success. This paper provides some empirical information for designing industrial centrifugal compressors with a focus on the impeller. A ported shroud compressor basic design guideline is also discussed for improving the compressor range.

1. Introduction

New compressor designs always must meet the customers' needs with the shortest time to market, low cost, and improved performance. To push the design to state of the art aerodynamic performance, the structure design also needs to meet a suitable performance life of the compressors. Mechanical integrity is one of the important parts of the centrifugal compressor design. Mechanical constraints are usually negative factors for aerodynamic design, for example, mechanical constraints require thick blade for reliability but hurt impeller efficiency. The purposes of the mechanical analyses are to provide all compressor components within a reasonable time duration to sustain the aerodynamic and centrifugal force, and eigen frequencies do not match critical excitation frequencies [1]. The safety factors of the mechanical design had been reduced dramatically compared with "old fashioned" design. Due to the nature of the Finite Element Analysis (FEA) tools and material property improvements, the safety factor of a modern industrial compressor design normally is set to 7 to 12%. The mechanical requirements need structure designers to have better practice to allow more freedom to aerodynamic designers and to keep all the components at the lowest weight and the lowest cost.

Design of a long lifetime single component of compressors is not a goal for designers. Emphasis on improving efficiency has been a primary issue, but this also is not as important as in the past. The development cost and development time is also a key factor that needs to be considered for a modern compressor design. Industrial compressor design expects a state-of-the-art performance compressor without making a second build for less cost and short development time. For achieving this goal, compressor design engineers need to have multidiscipline knowledge of centrifugal compressor design. Detailed design considerations can reduce the time to perform the advance design studies and laboratory investigations. The wide variety of design subjects represents a very complex design world for compressor designers. One purpose of this paper is to provide information in an aerodynamic point of view to understand the overall design before starting the detailed design process of a centrifugal compressor. The paper also summarizes important aspects of the centrifugal compressor design for industrial compressor designers and scientists.

The compressor market and business model has changed in the last few decades. Industrial compressor design now requires designing for success in the marketplace, not just for scientific experiments. In the past, compressor designers developed a new compressor in the development group and passed the design to manufacturing. The manufacturing group would evaluate how to make it at the lowest cost, and some designs were rejected because they could not meet the market requirements. The new development model requires the compressor designer to design for market, manufacturing, and end users. New business concepts have been proposed [2, 3] in which the design also considers an integrated system of manufacturers and end users. The new compressor developments become a complex system task. Minimizing manufacturing cost of the compressor design is not enough. The compressor design must consider all aspects of the manufacturing and end users. If surplus is defined as the total profit of manufacturing, end users, and aftermarket, the compressor's new development will focus on the design for maximum surplus. Therefore, in the compressor design stage, many choices of design options need to be considered before the final design, and discussions must consider the surplus value. It is essential that design engineers begin to perform a compressor design with full understanding of all aspects of the design considerations [4–8].

To reduce manufacturing cost, many high volume compressor manufacturers, for example turbocharger, often use the flow cut for different applications. The flow cut uses the same defined blade geometry for multiple flows and a similar or lower pressure ratio. This is different from scaling, in that the impeller blade and any diffuser vane geometry maintain the same definition. A brief introduction of the flow cut is also discussed in this paper.

With the development of computational science and computer hardware, design engineers rely on quality models to establish the physical relationships among diverse thermodynamic, geometric, and fluid dynamic parameters that govern turbomachinery performance [9–26]. Although CFD has helped to design many successful industrial compressors and has become an important tool in industrial compressor design, multidomain optimisation is still very time consuming. Most CFD optimisations still focus on the component [22]. When the compressor inlet flow is reduced, the compressor experiences an unsteady flow phenomena surge and rotating stall. These instabilities can cause noise nuisance and critical operating conditions with strong dynamical loading on the blades. Therefore, they cannot be tolerated during compressor operation. With the reduction of the compressor inlet flow, a rotating stall occurs in the impeller, or diffuser or scroll. If the compressor inlet flow continues to reduce, the rotating stall eventually will drive the compressor into a surge [27–32]. A rotating stall is an unsteady and three-dimensional flow phenomenon. CFD simulation is still a big challenge. The flow range of a centrifugal compressor can be extended by allowing gas to bleed from a ring of holes or a circular groove port around the compressor casing at a point slightly downstream of the compressor inlet. This type of the compressor called ported shroud compressor. Ported shroud forces a simulation of impeller stall to occur continuously, allowing some air to escape at port location inhibits the onset of surge and widens the operating range. The flow inside port is unsteady and complicated when compressor stalls. CFD guides ported shroud design is still very time consuming and less reliable. Some design practices for a ported shroud impeller casing [33] are discussed for improving the compressor operational range.

2. Industrial Centrifugal Compressor

Centrifugal compressors are widely used in automotive, marine turbocharging, oil and gas, aerospace, and distributed power applications because of their compact design and high stage pressure ratio. With different types of applications, the structural characteristics of the compressors have two basic types, that is, horizontal split and vertical split, as shown in Figure 1. Horizontal split type compressors are applied for low-to-medium pressure service, as shown in Figure 1(a). This type of casing is split along the rotor shaft and bolted at the split line. The bearing and seal sections allow easy disassembly and assembly via the inspection cover, without having to remove the upper casing. A vertical split compressor is easy to access the gears, bearings, seals, and be repaired on site. However, due to the large crossing area in the splitting surface, it is difficult to prevent gas and lubricant oil leakage. A vertical split compressor is applied for medium- and high-pressure service, as shown in Figure 1(b). This type of compressor consists of an inner casing and an outer casing. The inner casing forms a single unit with the head, bearing, and seal and is fixed to the outer casing by shearing. The nozzle can be attached to the top, bottom, or side in accordance with client specifications. Both bearings and seals can be inspected without removing the inner casing. However, the manufacturing cost and installation cost may be higher than for a vertical split compressor. It is also not easy to access the gear, bearings, and seal. For combining the advantages of both split types, a hybrid split has become popular. For some applications, the gearbox can be a vertical split, and the compressor stage can use a horizontal split, as shown in Figure 1(c).

3. Impeller Design Methodologies

The impeller is a key component to influence overall performance of a centrifugal compressor [5]. The efficiency of centrifugal compressors has increased dramatically, especially low-pressure ratio centrifugal compressors. A major challenge for a centrifugal compressor design is to keep a high efficiency level at a state of the art and to increase the compressor operating range [27–33]. Increasing the compressor operating range without sacrificing compressor peak efficiency is difficult to achieve. Aerodynamic engineers not only need to understand the surge physics but also need to apply design experience to the design. Another important objective for impeller design is to reduce the manufacturing cost. Manufacturing cost could be reduced when designs for manufacturability are effectively considered. The impeller should meet requirements to be easily withdrawn from a casting mold without destruction and disassembly of the mold for a casting impeller. This requires the lean angle of the blade to change linearly with the impeller radius and axial direction, as shown in Figure 2. All these considerations

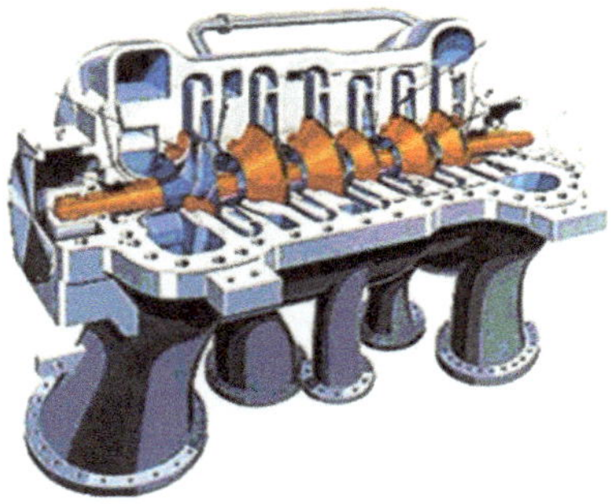

(a) Horizontal split compressor

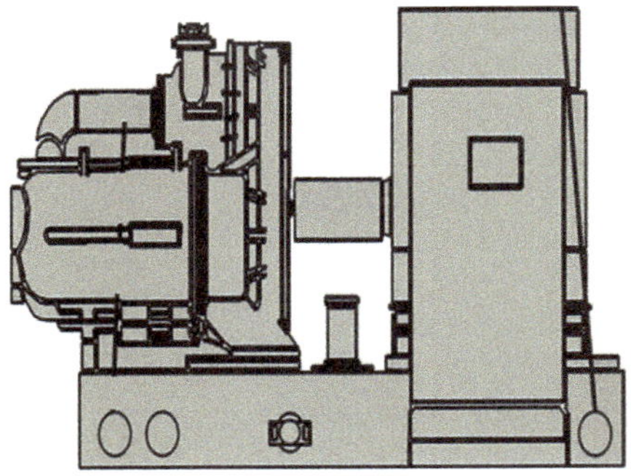

(b) Vertical split compressor

(c) Hybrid split compressor

Figure 1: Compressor split types.

Figure 2: Impeller for manufacturing.

for design will help the final design to meet the compressor design target with fewer design iterations.

Different designers may have different methodologies for impeller developments. What kind of area distributions, curvatures, velocity, or pressure profile will lead to a good design is strongly dependent on the designers' practice and experience. Two totally different design philosophies could produce similar performance. For example, two impellers designed by Garret and Pratt Whitney [7], as shown in Figure 3, had different shapes with similar performance at the design point. The impeller designed by the author also presented different features, which also provide a good performance. It is shown that, if design follows basic design guideline, a wide range of solutions to the design can be used.

The authors' design, as shown in Figure 3, is an example of a recently developed single-stage centrifugal compressor. At the design point, the total to static stage pressure ratio was about 3.7, and the flow coefficient was about 0.12. The running clearance at the impeller tip was 4.5% of the impeller exit blade width. Six builds were assembled and tested based on the ASME PTC-10 test procedure [34]. The compressor performance obtained from an average of six build tests is shown in Figure 4. The differences of test results for different builds for adiabatic efficiency and head coefficient were within ±0.5% and ±0.75%, respectively. The test uncertainties for total pressure (in psi), static pressure (in psi), and temperature (in Fahrenheit) were ±0.25%, ±0.2% and ±0.5%, respectively, based on uncertainty analysis [14]. Test results showed that the compressor performance was encouraging at both the design and the off-design point. The design met the low-cost target and allows large manufacturing tolerances. The insensitivity of the impeller surface finish and large tip clearance makes it easy to assemble.

4. Impeller Geometry

The initial design of a centrifugal compressor always begins with requirements from customers or marketing analysis. Designers select basic configurations and provide basic performance to customers or marketing by using their experience data. Aerodynamic designers also need to provide an estimation for the compressor's basic geometry to engineers in other disciplines. For example, rotor-dynamic engineers and bearing designers rely on the impeller geometry information to perform their work. Although the basic geometry design is not intended to yield optimization of the impeller, it can accelerate the overall design process and reduce the development cost.

Before aerodynamic designers determine the basic impeller geometry, the rotational speed of the impeller needs to be selected. If there are no special requirements for rotational speed, we normally optimize rotational speed based on the Balje's charts [35, 36] by using optimal specific speed. Although Balje's charts are not very accurate tools, they are sufficient enough to provide the initial estimate for impeller geometries.

During the initial design, the important information needed for bearing designers and rotor-dynamic engineers is impeller weight. Aerodynamic designers can estimate impeller sizes based on the required gas flow, pressure ratio, and impeller rotational speed. Our design practices showed that the weight of the impeller is the function of the impeller diameter. Figure 5 summarizes the relationship between the impeller diameter and the weight for sixteen ASTM A564 stainless-steel unshrouded impellers. The impeller weight mainly is determined by the impeller disk; the blades only contribute a very small portion of the weight. Therefore, we plotted impeller weight and diameter relation in one figure

Garret design Pratt Whitney design Author's design

FIGURE 3: Different impeller designs.

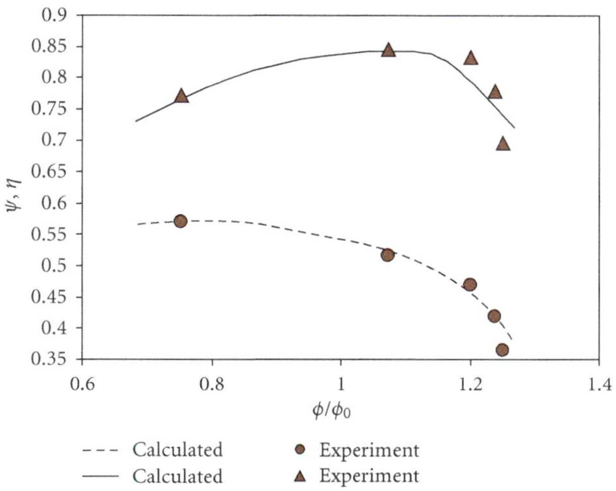

FIGURE 4: Single stage compressor efficiency and head coefficient versus flow coefficient (upper curve is the adiabatic efficiency, and lower curve is head coefficiency).

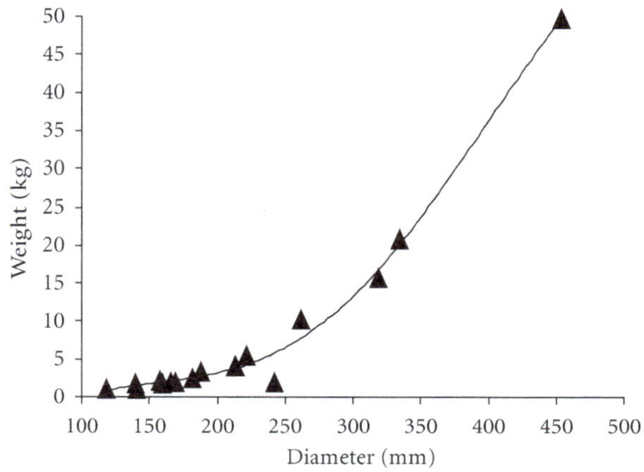

FIGURE 5: Impeller weight with diameter.

for all designed unshrouded impellers with different blade counts and with or without splitters.

In an initial stage of compressor design, selections of the impeller inlet, the outlet velocity vectors, and the choice of blade numbers are the key initial design decisions. Velocity vectors may be obtained through a mean-line program. The experience data show that both the inlet blade numbers and the exit blade numbers are a function of stage pressure ratio. Relationships between the numbers of blades and the stage pressure ratio are shown in Figures 6 and 7 for with and without splitter impellers, respectively. In general, high-stage pressure causes a blade-loading increase, and the impeller needs more blades to distribute loading. Variations of the numbers of blades at a similar pressure ratio were due to the size of the impellers. For a smaller-sized impeller, manufacturing capabilities may limit the number of blades. Impeller sizes plotted in Figures 6 and 7 are in the range from 2 inches to 45 inches. The machine performance requirements and manufacturing feasibilities are factors to determine whether to use splitters or not.

The inlet blade height is determined by the design inlet flow rate and the impeller hub radius. The inlet hub radius is determined by the attachment of the impeller. For overhung impellers, the inlet hub radius normally is selected in a range between 10% and 20% of the impeller tip radius. For the shaft and bolt through the impeller, selections of the inlet hub radius are based on stress requirements.

The blade thickness at the inlet and the discharge was determined mainly by tensile and bending root stresses at the leading edge and the blade exit. FEA calculations and stress tests showed that blade root stresses are mainly caused by the centrifugal force. The blade high was a key factor to impact the blade root stresses. The mean-line thickness at the inlet and the exit was determined by the blade heights at the inlet and the exit, as shown in Figures 8 and 9. Experience showed that the blade thickness changed linearly with the blade height.

The three-dimensional features of the impeller blade are dependent on the engineers' experience and on stress limitations. The modern impeller is normally a three-dimensional design. The wrap angle, lean angle, and back sweep angle use a larger value than those in the past. The large wrap angle can reduce the camber of the blade but increase the frictions of the fluid. A large lean angle permits blade design at all blade sections with the desired shape. Leaning the blades creates a back-sweep and retains purely radial fibers, which are beneficial for bending moments. Experience showed that impellers with a back sweep generally have high efficiency.

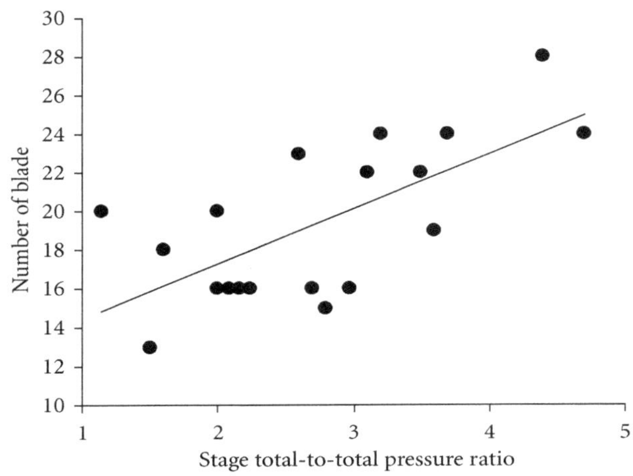

FIGURE 6: The number of blades at the inlet versus stage pressure ratio.

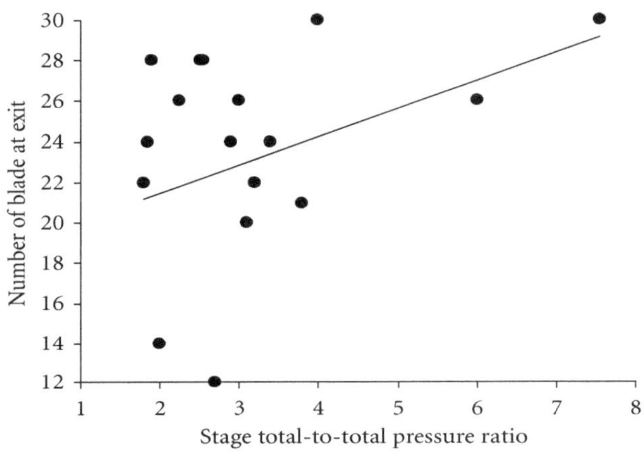

FIGURE 7: The number of blades at the exit versus stage pressure ratio for split impellers.

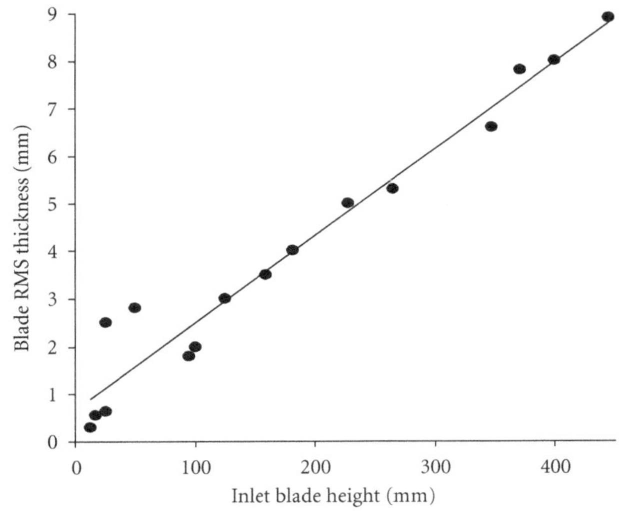

FIGURE 8: Relationship between impeller inlet RMS thickness and inlet width.

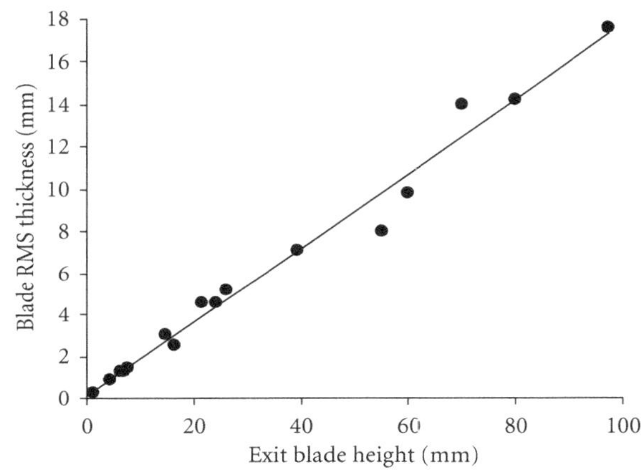

FIGURE 9: Relationship of impeller exit RMS thickness and tip width.

5. Reynolds Number and Surface Finish

The Reynolds number or the Ross by number has significant impacts on the impeller maximum surplus value. Fundamental fluid dynamics theory [37, 38] shows that the flow inside a pipe for a different Reynolds number represents different flow patterns. This is also true for flow inside impeller blades. Experience showed that if there is a flat velocity profile at the inlet between two impeller blades or diffuser vanes, the flow development along the flow channel presents different profiles with different Reynolds numbers. For low Reynolds number flows, the exit velocity profile is almost parabolic and only with a small portion of a flat profile. For high Reynolds number flows, the exit velocity profiles have large flat profiles.

The peak meridional velocities for high Reynolds number flows are normally located at the hub pressure sides of the blade due to potential flow effects. Low Reynolds number flow regions are located near the suction side of the blades. The viscous jet and wake interaction causes flow separations. The Reynolds number also strongly influences secondary flow patterns. Increasing the Reynolds number increases the strength of the clockwise secondary passage flow circulation. Reducing pressure on the suction velocity gradient increases the flow circulation of the counter clockwise secondary flow. The optimum design should try to offset each other to minimize the secondary flow losses.

The machining and casting of centrifugal compressor impellers and other components result in an inherent surface roughness. The sizes and forms of roughness depend on the manufacturing process. The levels of the surface finish represent the manufacturing cost. It is very important to balance manufacturing cost and performance. Surface finish requirements for different designs have different requirements. A detailed discussion on surface finish and Reynolds number can be found in reference [39, 40]. Loss due to the

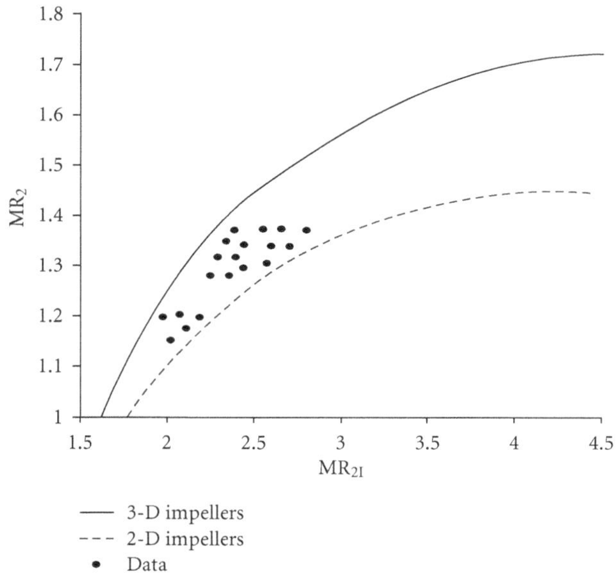

FIGURE 10: Relationship of MR_2 versus MR_{2I}.

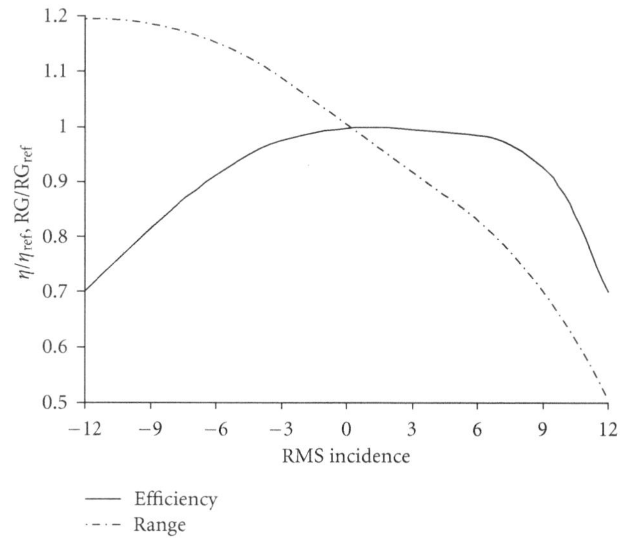

FIGURE 12: Incidence versus efficiency and range.

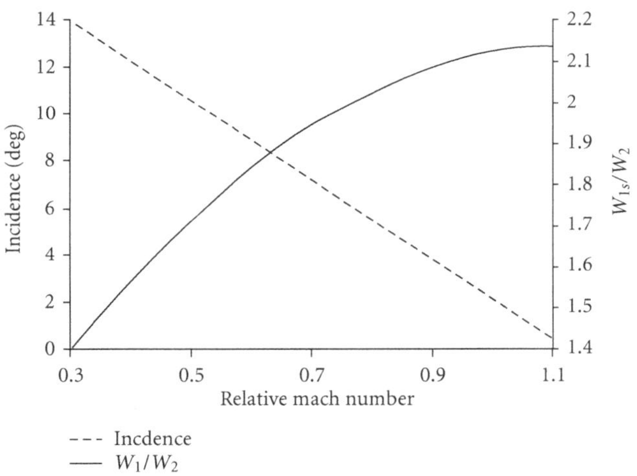

FIGURE 11: Stall incidence and relative velocity ratio versus Mach number.

surface finish can be represented as wall friction. Wall friction is the function of Reynolds number and can be written as

$$\frac{1}{\sqrt{f}} = 1.74 - 2\log_{10}\left(\frac{k}{B_2} + \frac{18.7}{\mathrm{Re}\sqrt{f}}\right). \tag{1}$$

This equation can be solved by using a simple computer program or spreadsheet.

6. Impeller Aerodynamic Design

One important guideline for impeller aerodynamic design is to set a reasonable diffusion ratio. The diffusion of the impeller can be represented by velocity ratio, diffusion factor, and relative Mach number ratio. The ratio of the relative Mach number was used in this discussion because it can avoid the one-dimensional assumption at the inlet. MR2 is defined as the ratio of the relative Mach number at the impeller inlet to the average Mach number at the impeller exit. Figure 10 shows the upper and lower boundaries for maximum deceleration likely to be achieved for two-dimensional and three-dimensional impellers [8]. The experience data fell inside the theory boundaries, and the Mach number ratio MR2 fell between 1.15 and 1.4, giving a good overall performance. The upper boundary of MR2 of 1.4 for an industrial compressor and 1.7 for jet engine impellers are reasonable expectations. Our experience also indicated that a large diffusion might cause a large loss. The ratio of the Mach number can be selected within a large range. An important factor to impact the selection of diffusion level is the inlet Mach number. Figure 11 is a relationship between the relative velocity ratio, the incidence, and the inlet relative Mach number for a typical industrial impeller. It can be seen that diffusion is not an absolute parameter, which influences the stall of the compressor. It is worthwhile to point out that this test impeller was stalled first at the inducer. The inducer shroud velocity represented the rotational speed.

Traditionally, the impeller inlet incidence is set to zero at the design condition. Modern impeller designs need not only to consider maximum efficiency at the design point, but also to consider the manufacturing cost and the off-design performance for the whole operating range [8]. Inlet blade angles are not necessary the same as the inlet relative flow angles. Experience data in Figure 11 shows that changes of the inlet flow incidence impact both the efficiency and the operating range of the impeller. Figure 12 shows that little negative incidences could raise the impeller operating range. However, when the negative incidence increased to a certain level, the operating range did not enlarge and efficiency dropped significantly. The impeller design should avoid this situation.

Estimations of the impeller exit width are critical for both the primary performance estimation and the basic dimension setup. The major impacts of the impeller blade exit width are the flow capacity and the pressure ratio of the stage. It is difficult to calculate the impeller exit width accurately in a simple way. The Rodgers diffusion factor equation [41] provided a good estimated value for the impeller exit width (B_2). If the mean meridional blade length can be estimated as

$$L = \frac{2\pi(r_2 - r_1)}{4}, \tag{2}$$

then the impeller exit width can be estimated as

$$B_2 = \frac{10(D_2 - D_{s1})}{1 + (W2/W_1)}\left[\mathrm{DF} - 1 + \frac{W2}{W_1} - \frac{\pi D_2 C_{\tau 2}}{2LzW_1}\right] - \frac{(D_{s1} - D_{h1})}{2}. \tag{3}$$

Secondary flows inside the impeller are caused by an imbalance of the static pressure and the kinetic energy. One of the typical secondary flows, the horseshoe vortex, has been well documented. It is shown that the strength of the secondary flows is governed by the vortex starting conditions. The further development of the vortex is determined by the conservation of angular momentum. The impeller meridional blade profiles influence the secondary flow loss level and the laminar viscous dissipation function can estimate the secondary flow loss due to the blade profiles [8]. We have

$$\Delta H_l = N\int \mu\left[2\left(\frac{\partial u}{\partial x}\right)^2 + 2\left(\frac{\partial v}{\partial y}\right)^2 + \left(\frac{\partial u}{\partial y} + \frac{\partial v}{\partial x}\right)^2\right]dV. \tag{4}$$

Tip clearance cannot be avoided for unshrouded impellers. Bearing clearances and manufacturing tolerances of the impeller and the intake ring control the minimum impeller tip clearance. The minimum tip clearance is normally defined at the maximum rotational speed with hot weather conditions for most motor-driven compressors. For compressors installed in the same shaft with a gas turbine, the minimum compressor tip clearance was estimated when the compressor was operating at maximum rotational speed with hot weather and with the machine overall net axial thrust load towards the compressor. The tip clearance increases quadratically with the impeller rotational speed if other operating conditions do not change. The tip clearance impacts the overall compressor performance because it increases the magnitude of the secondary flow inside the impeller blades and produces strong tip vortices. The tip clearance flow transports a low momentum fluid from the suction side to the pressure side of blades. The circumferential center of the secondary flow is dependent on the size of the tip clearance. Secondary vortices are located near the shroud side for small clearance, whereas secondary vortices may spread to the center and even the hub of the flow channel for a large tip clearance impeller. The clearance distribution affects the wake formation and the location at the impeller exit.

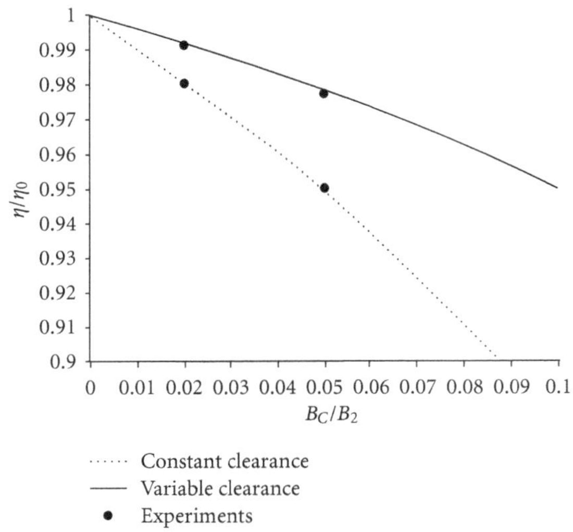

FIGURE 13: Compressor efficiency change versus clearance.

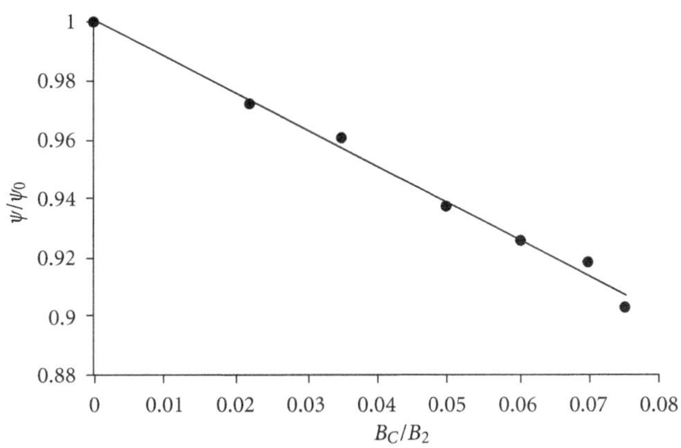

FIGURE 14: Head coefficient change with tip clearance.

A large clearance at the leading edge results in a low-energy center close to the suction side of the blade. Reducing the clearance at the leading edge, the wake moves towards the pressure side of the blade. The tip clearance setting depends on the compressor's maximum surplus value. There are several methods to reduce the tip clearance losses. Figure 13 shows that variable clearance could significantly improve the stage efficiency of compressors. The tip clearance changes the compressor stage head and capacity. Test data indicated that the head coefficient almost changed linearly with the tip clearance, as shown in Figure 14, and the flow coefficient followed a secondary order curve with clearance, as shown in Figure 15.

With the manufacturing technology improvement and design system improvement [42, 43], modern impellers always are designed in a three-dimensional shape. It is very important to understand the impact of the three-dimensional feature on performance and structure. Lean is one of the critical three-dimensional features. In general, both negative and positive lean improves the peak efficiency. One of the design examples for lean effects is shown in Figure 16.

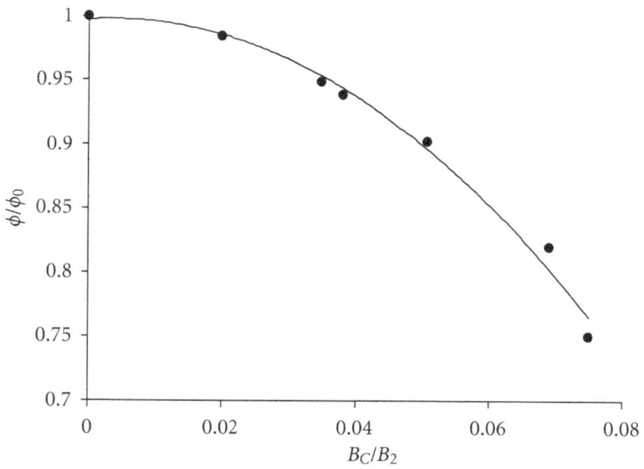

FIGURE 15: Flow coefficient change with tip clearance.

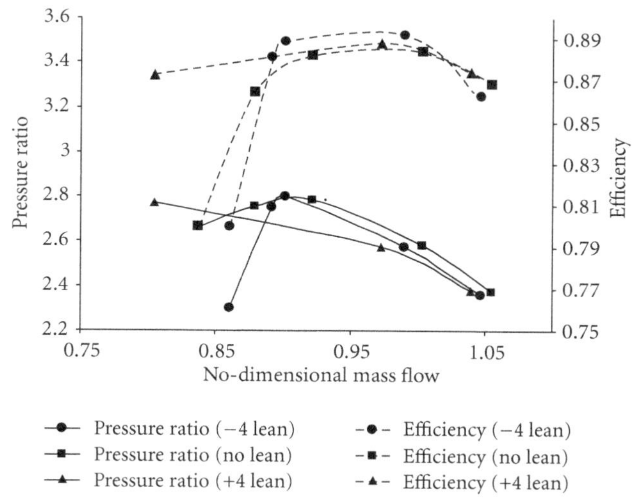

FIGURE 16: Compressor performance for a lean blade.

Negative lean has the best peak efficiency, while positive lean has a wide operating range. It can be seen from Von Mises stress contour plots in Figures 17 and 18 that the Von Mises stress is the highest for a negative four-degree lean on a suction surface, and the stress is the lowest for a four-degree lean. However, the pressure side stress contours show the lowest leading edge stress area for a negative four-degree lean design compared with other cases. The negative lean has maximum stress for both the leading edge and the trailing edge while the positive lean has the lowest maximum stress in all locations. The highest stress areas for the positive lean are less than the negative lean and no lean cases. The bore stress and back face stresses are similar because the blade thickness distribution is similar. It can be seen that a small positive lean can reduce the peak stress.

7. Impeller Flow Cut

For large volume low-cost compressors, for example, automobile turbochargers, the impeller flow cut is always used for meeting different flow and pressure ratio requirements with minimum manufacturing cost. A flow cut is defined as a change to the impeller blade height or a change to the shroud contour while following the same hub contour and blade angle definition as the original impeller design. This allows a single manufacturing method to be used for the base impeller, plus a machining operation to adjust the blade height and the flow capacity of the impeller. Achieving the objectives for high efficiency and end users' cost of a compressor installation are always in conflict with each other. The flow cut of a compressor stage is one of the important activities of centrifugal compressor manufacturers to achieve a certain performance level with minimum cost. A theoretical equation provides the basic information, but the empirical performance effects on the compressor stage are very important to the compressor manufacturers. Figure 19 shows the original impeller (Figure 19(a)), the shroud contour cut (Figure 19(b)), and a diameter cut (Figure 19(c)). Modern impeller design normally has a bigger back sweep angle, as shown in Figures 2 and 3. The diameter cut, as shown in Figure 19(c), affects the impeller performance at a significant level. Therefore, the diameter cut is less popular than the shroud contour cut.

Here we present some test experience for both the shroud and the diameter contour cut. The shroud contour cut (or extend) is defined as the percentage of the local blade height. Most of time, manufacturers extend or trim the shroud contour the same percentage of the impeller inlet and exit blade heights. For large industrial compressors (choke flow larger than 1000 ICFM (0.472 m^3/s)), the flow changes normally near linear to the contour change. The major impacts of the flow cuts on the compressor performance are the compressor stage efficiency and the surge flow or the operating range (RG). Figure 20 is the head coefficient at surge point changes with the impeller contours, where the surge point defined here is the operation point at which the compressor stage adiabatic efficiency is 55%. It is shown that the head coefficient at the surge increased as the contour increased. This may be because when the contour increases, the cold tip clearance remains the same during the tests for all the contours. For a larger contour, the tip clearance percentage relative to the impeller exit blade height is smaller, and the surge margin improves. This may be for the same reason. When the contour increases a little, the stage efficiency increases until the design penalty away from the design point is larger than the clearance loss reduction, the compressor efficiency starts to drop, as shown in Figure 21. Test results also suggest at an increased contour, a broader range of high efficiency exists. Testing also indicates that the point of peak efficiency is not proportional to the change in the area of the impeller inducer and the exit geometry. It is interesting to see that the flow corresponding to the peak efficiency increases as the impeller contour increased, as shown in Figure 22. It is important to make the flow cut to ensure the peak efficiency at the operating flow.

Although the diameter cut is less popular than the shroud contour cut, understanding fundamental information and

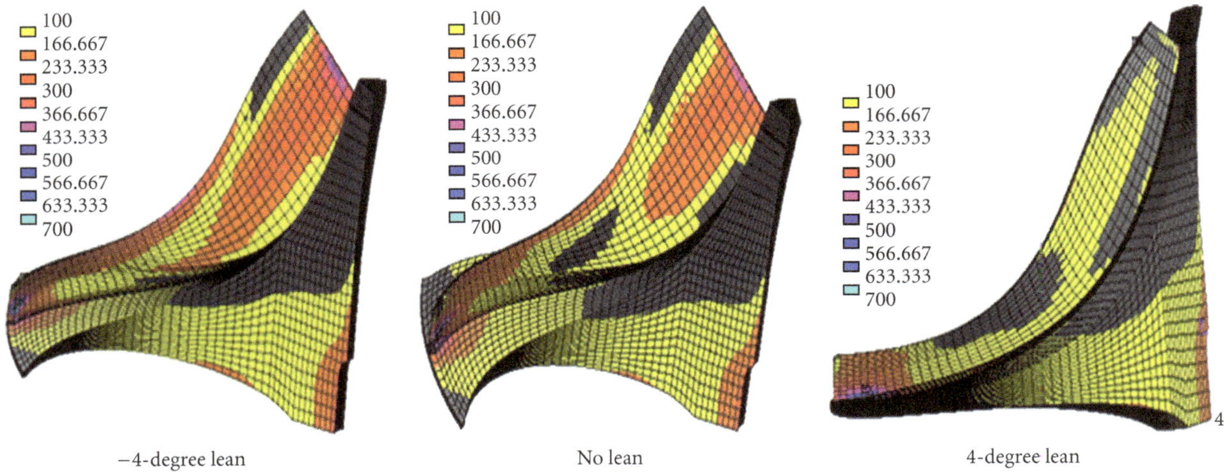

Figure 17: Von Mises stress at the suction side of the wheel.

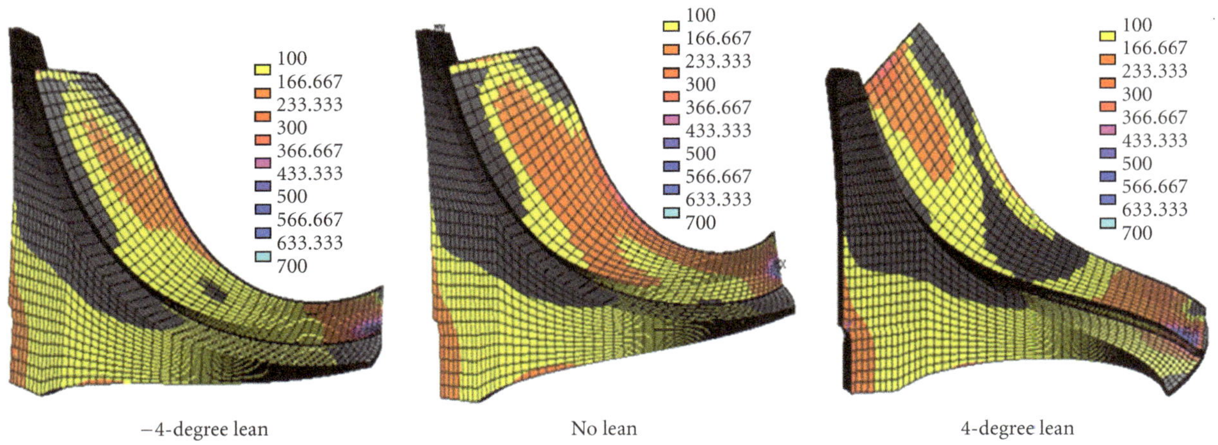

Figure 18: Von Mises stress at the pressure side of the wheel.

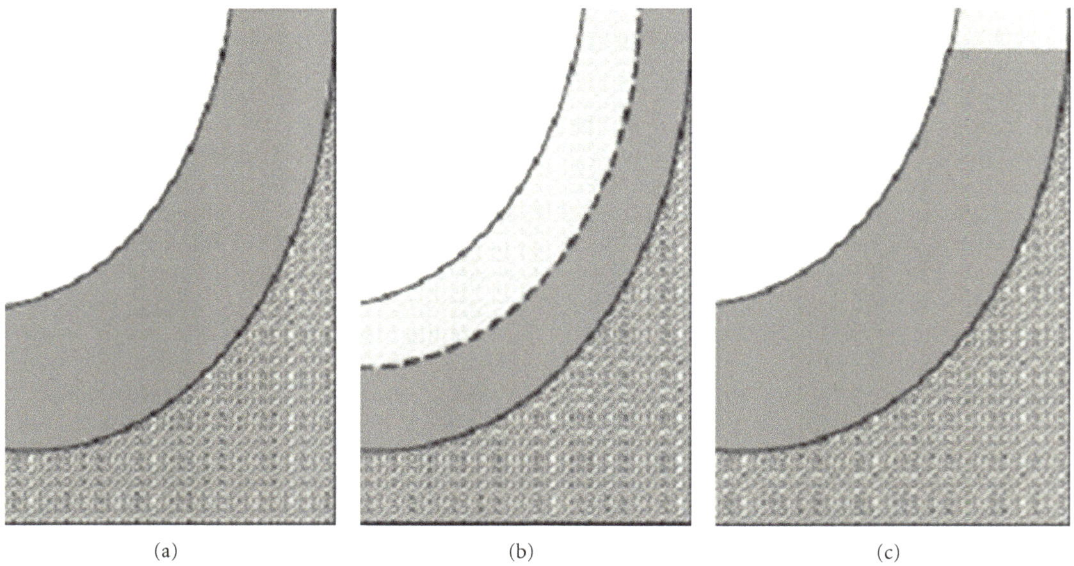

Figure 19: Flow cut of the impeller.

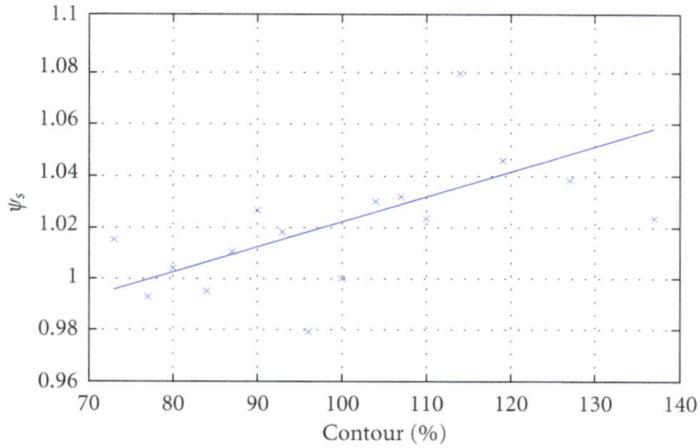

FIGURE 20: Head coefficient at the surge as a function of the contour.

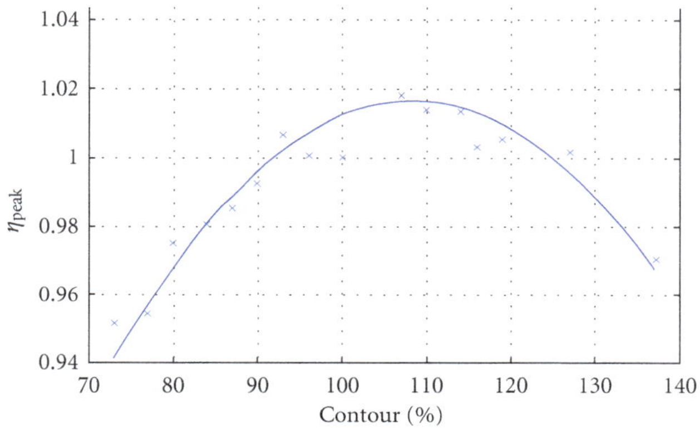

FIGURE 21: Peak efficiency change with flow cuts.

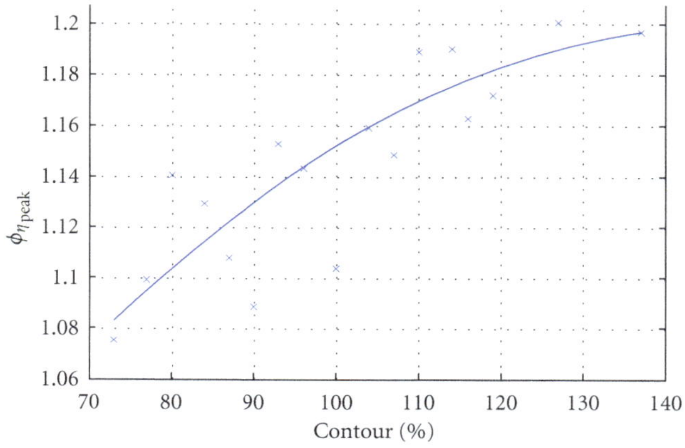

FIGURE 22: Flow coefficient at the peak efficiency versus the flow contours.

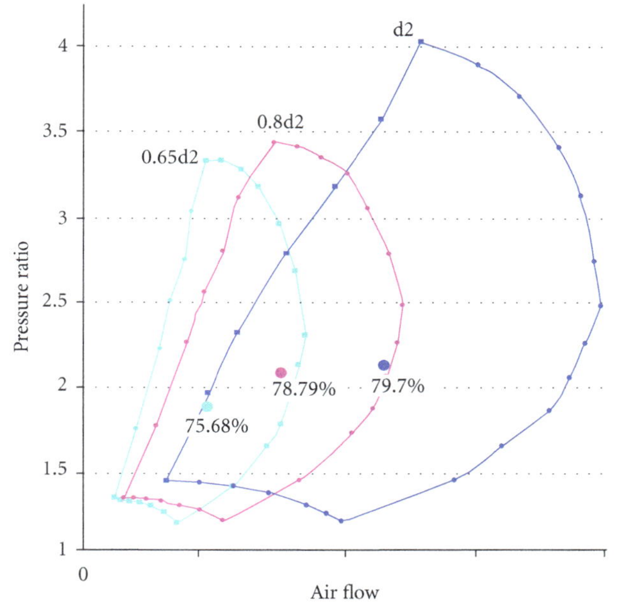

FIGURE 23: Performance characteristics of an impeller diameter cut.

basic compressor characteristics for the diameter cut is also very helpful. Figure 23 is the medium back-curvature impeller compressor (about 25 degree back-curvature) characteristics of a diameter cut. It can be seen that the diameter cut not only affect the peak efficiency but also affect the peak efficiency location. The impeller peak efficiency of a small diameter impeller is located in the lower pressure ratio operating point. It also can be seen that the compressor operating range is reduced after the diameter cut. It is interesting to notice that the surge boundary slope increases with the diameter decrease.

8. Ported Shroud Impeller Casing

A compressor surge is a system-unstable phenomenon that is influenced by all components of the compressor. The physics of surge and stall are still not fully understood. We still cannot find any tool that can capture all features of the surge and stall, as shown in Figure 24. Many theoretical studies [27, 30, 31, 44] have focused on a better understanding of the surge and stall, but none can be used as a design tool yet. More theoretical work and experimental studies need to be done in order to incorporate stall in the design system. Designs for a wide operating range were mainly dependent on engineers' and manufacturers' experience and their understanding of the stall and surge. Rotating stall and surge are violent limit cycle-type oscillations in compressors, which result when perturbations (in flow velocity, pressure, etc.) become unstable. Originally treated separately, these two phenomena are now recognized to be coupled oscillation modes of the compression system—surge that is the zeroth order or planar oscillation mode, while a rotating stall is the limit cycle resulting from higher-order, rotating-wave disturbances, as shown in Figure 24. The compressor normally starts with a stall and then eventually become a surge, as shown in Figure 25. During the surge, the compressor can experience a reversing flow. System resistance releases after the flow reversal, and the flow starts flowing into the system. This surge cycle continues, as shown in Figure 25. The pressure variation rate with the mass flow rate variation is much larger

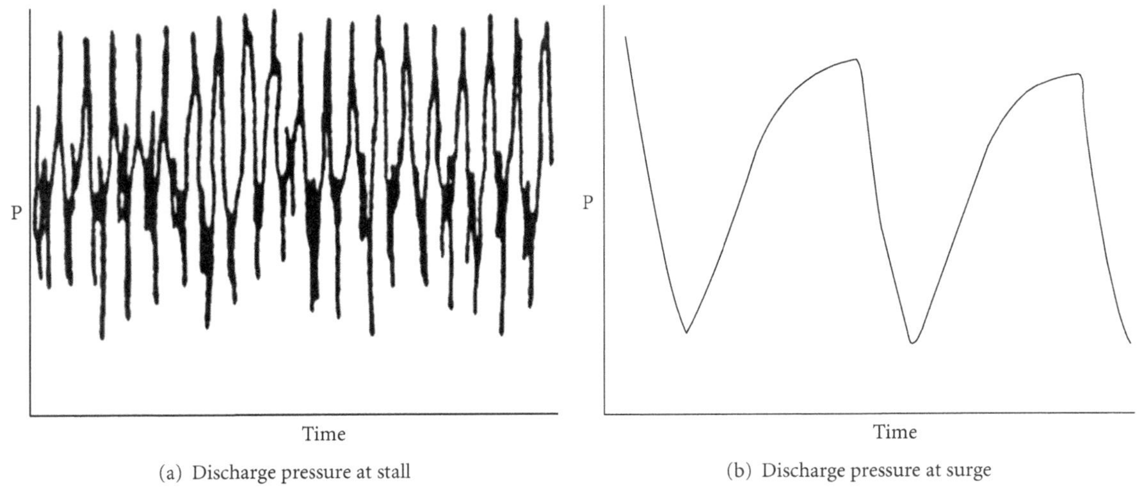

FIGURE 24: Stall and surge pressure variation with time.

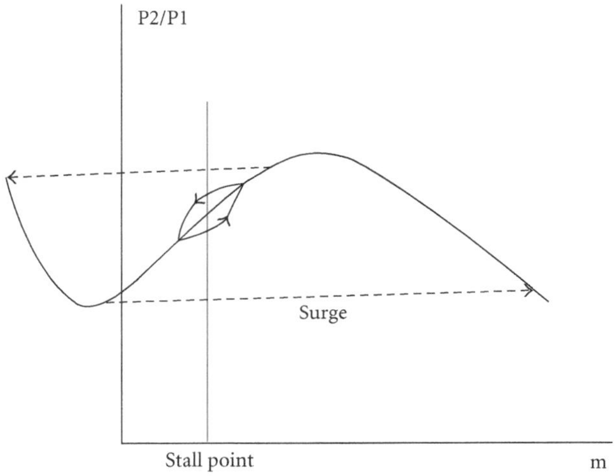

FIGURE 25: Surge of the centrifugal compressor.

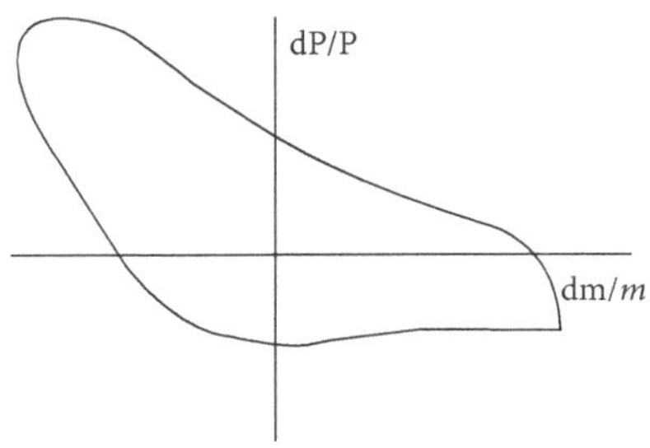

FIGURE 26: Variation of mass flow rate versus pressure fluctuations.

than the near choke, as shown in Figure 26. To extend the surge margin, it is important to optimize the impeller, the vaned diffuser, the vaneless diffuser, and the volute design. The surge control can extend the surge margin further. The simplest the way is to have a bypass surge relief valve. When the centrifugal compressor pressure rises beyond a certain level, the valve bypasses the flow from the discharge to the inlet to prevent surge. A wise valve choice can help engineers pay for the rest of an advanced surge-control system without exceeding the budget. The valve needs to have a fast stroking speed when open, high capacity, low noise, and a very stable throttling control. For example, the oil and gas industry often chooses high-performance rotary valves.

Casing treatments are other popular methods to extend the centrifugal compressor range. One of the key casing treatments is a ported shroud casing, as shown in Figure 27. The ported shroud design is entirely passive, having no moving parts, control valves, and so forth, The ported slot can be vertical to the inlet flow or angled, as shown in Figures 27(a) and 27(b). There are vanes or ribs to support the stationary shroud in the leading edge inducer region. The vanes or ribs are not only a support structure but also an aerodynamic device to the flow inside the port, bleeding out with minimum losses. The ported shroud passage provides the bleed path when the compressor is near the surge, as shown in Figure 28(a) and also a secondary air inlet to the impeller when the compressor needs more air, as shown in Figure 28(b). The vanes or ribs in the passage are tangentially slanted in the direction of the impeller rotation, to preferentially augment the airflow into the impeller at high speed, while in part-speed discouraging the airflow out of the impeller through the secondary inlet. The advantage of the ported shroud can increase in the part-load surge margin and increase the choke flow at full load, as shown in Figure 29. It can be seen that for a ported shroud compressor, at a lower mass flow the pressure ratio increases. This is because the separation and reverse flow were reduced near the inducer compared with an unported case. When the compressor is near choke, the impeller can bypass the impeller throat and draw an extra flow from the port, as shown in Figure 28(b). Figure 30 shows that the flow passed the port at different impeller mass flow conditions when the port width and location were optimized to have the best surge margin for a gas width equal to bg.

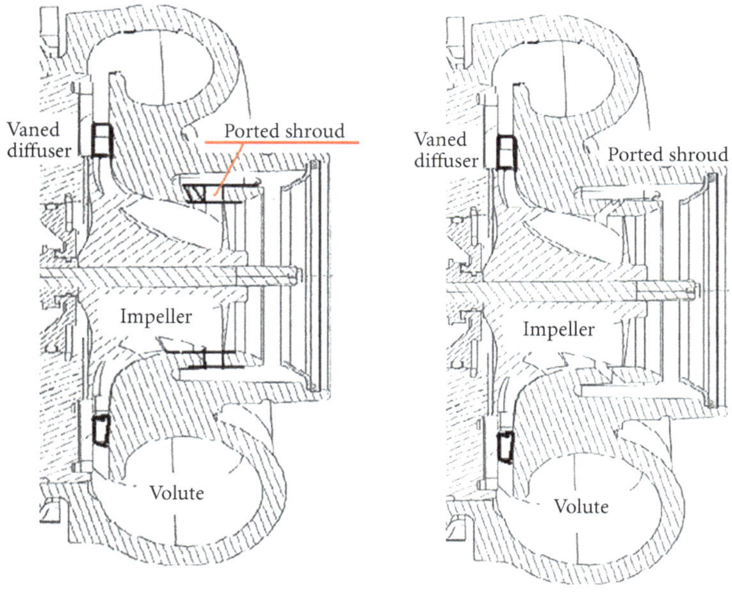

(a) Straight ported shroud (b) Angle ported shroud

FIGURE 27: Ported shroud compressor.

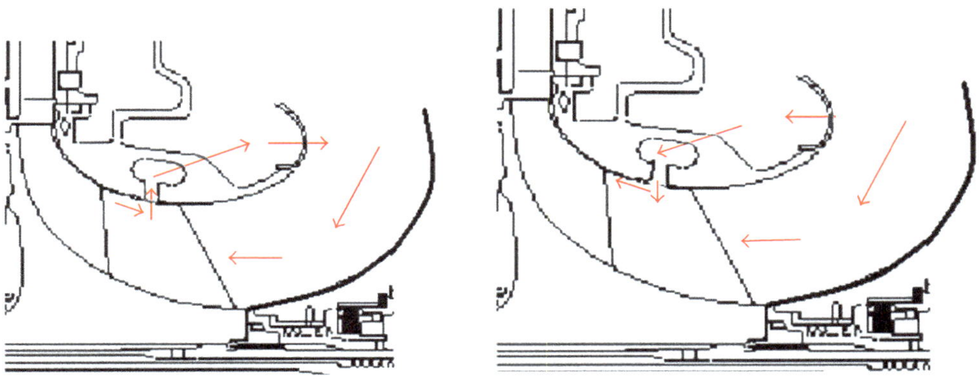

(a) Near surge (b) Near choke

FIGURE 28: Ported shroud flows.

It can be seen that the port recirculation flow increases when the compressor mass flow reduces. The port can pass as much as 40% of the impeller design flow. It can be seen from Figure 30 that the port recirculation flow reduces when the port width reduces. Figure 31 is the port width affecting the compressor performance for a ported compressor with the port location at 16% of the shroud meridional length from the impeller inducer. It can be seen that when the port location is fixed, the port width increases from 0 to $0.6b_2$, and both the surge margin and choke margin increase. The com-pressor adiabatic efficiency remains almost unchanged until the port width is larger than $0.1b_2$. The efficiency drops almost negligibly up to a port width larger than $0.2b_2$. Figure 31 also shows that the angle port (70 degree with axial direction) has an advantage compared with a straight port. The port design goal is to improve the surge and choke margin without greatly affecting the compressor performance. In some applications, for example, the automobile and aviation industry, the surge margin, and choke margins are both defined by certain efficiency levels. If the efficiency is too low, the system cannot perform properly. For those applications, the ported shroud applications are very popular. The ported shroud not only can extend the operational margin but also can improve the efficiency near the choke and surge. For those applications, the compressor map for a ported shroud demonstrated more advantages than an unported compressor, as shown in Figure 32.

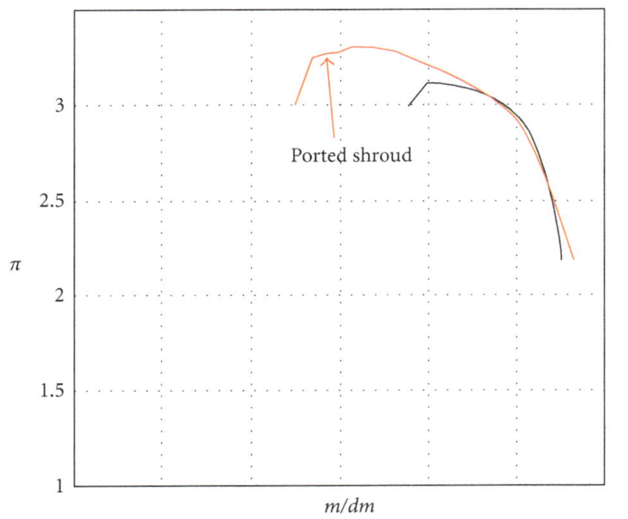

Figure 29: Compressor characteristic with and without ported shroud.

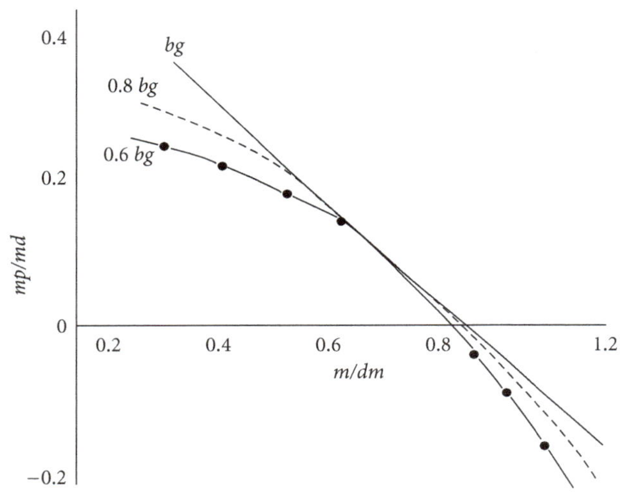

Figure 30: Flow inside bleed out from the port.

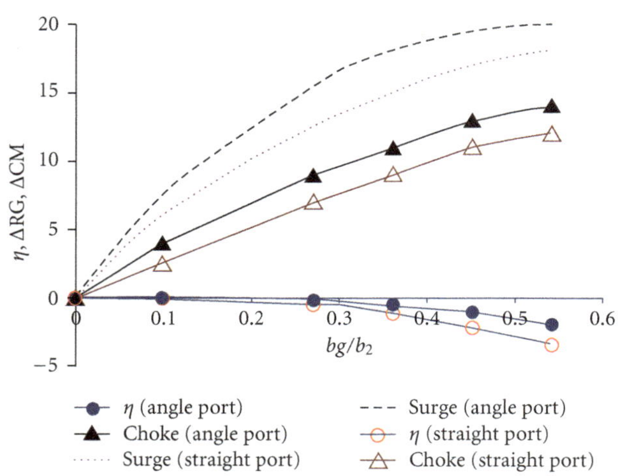

Figure 31: The compressor performance versus port width.

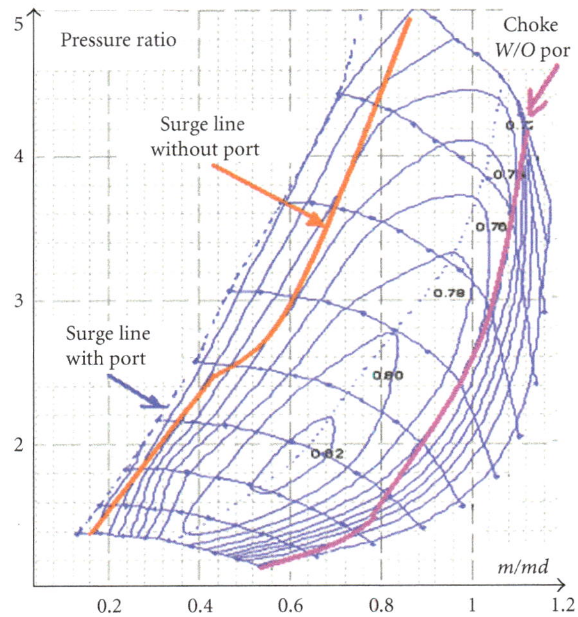

Figure 32: Ported shroud compressor map.

9. Conclusions

This paper provides several empirical consideration points in designing centrifugal compressors, focusing on the impeller design.

(1) The different structure split compressors are discussed.

(2) In an initial stage of compressor design, selections of the impeller inlet and the outlet velocity vectors and choice of blade numbers are the key initial design decisions.

(3) Leaning the blades creates a back sweep and retains purely radial fibers, which are beneficial for bending moments. Experience showed that impellers with a back sweep generally have high efficiency. But lean blade design needs to consider the impacts of the impeller structure.

(4) The velocity profile through the impeller blades largely depends on the flow Reynolds number. For low Reynolds number flows, the exit velocity profiles are almost parabolic and only with a small portion of the flat profile. For high Reynolds number flows, the exit velocity profiles have large flat profiles.

(5) The tip clearance affects the overall compressor performance because it increases the magnitude of the secondary flow inside the impeller blades and produces strong tip vortices.

(6) The tip clearance setting depends on the compressor's maximum surplus value. There are several methods to reduce the tip clearance losses. A variable tip clearance is one of the potential designs.

(7) Compressor impeller flow cuts are widely used by compressor manufacturers. Some performance impacts need to be considered for doing the flow cut.

(8) The ported shroud compressor has advantages for the compressor performance. Many new ported shroud structures have been proposed by many compressor manufactures [24]. But some fundamentals discussed in this paper can help designers to make the right decisions.

Nomenclature

B_2, b_2: Impeller tip width (m)
C_1: Inlet axial velocity (m/s)
C_P: Pressure coefficient = $p/(0.5\rho d_2 U_2{}^2)$
CM: Choke margin
DF: Diffusion factor
D_s: Specific diameter
dm: Flow fluctuations (kg/s)
dp: Pressure fluctuations (pa)
F: Function
f: Friction factor
H: Head (m)
ICFM: Cubic foot per minute at inlet
K: Sand grain roughness (μm)
L: Blade meridional length (m)
m: Mass flow rate (kg/s)
N: Rotational speed (RPS)
r: RMS radius (m)
Re: Reynolds number
ICFM: Compressor inlet flow in ft/min.
M: Mach number
MR: Mach number ratio
Mu: Head coefficient = $H/(D_2N)^2$
Q: Volume metric flow rate (m^3/s)
RG: Operating range = $(Q_c - Q_s)/Q_c$
RMS: Root mean square
u, v: Velocities (m/s)
V: Velocity vector
W: RMS-relative velocity (m/s)
x, y: Cartesian coordinates.

Greek Symbols

μ: Viscosity (pa·s)
φ: Flow coefficient = $Q/(ND^3)$
η: Adiabatic efficiency (total to static) = $(\pi^{(\gamma-1/\gamma)} - 1)/(T_2/T_{01} - 1)$
Ψ: Head coefficient = $h/u_2{}^2$
π: Pressure ratio
γ: Specific heat ratio.

Subscripts

1, 2: Impeller inlet and exit
C: Chock or clearance
l: Loss
0: Zero clearance
ref: Reference
S: Surge.

References

[1] M. P. Singh and J. J. Vargo, "Reliability evaluation of shrouded blading using the SAFE interference diagram," *Journal of Engineering for Gas Turbines and Power*, vol. 111, no. 4, pp. 601–609, 1989.

[2] D. P. Kenny, The history and future of the centrifugal compressor in aviation gas turbine, SAE/SP-804/602, 1984.

[3] P. D. Collopy, "Surplus value in propulsion system design optimization," AIAA-97-3159, 1997.

[4] D. Japikse, "Decisive factors in advanced centrifugal compressor design and development," in *Proceedings of the International Mechanical Engineering Congress & Exposition (IMechE '00)*, November 2000.

[5] C. Xu and R. S. Amano, "Development of a low flow coefficient single stage centrifugal compressor," *International Journal of Computational Methods in Engineering Science and Mechanics*, vol. 10, no. 4, pp. 282–289, 2009.

[6] C. Rodgers and L. Sapiro, "Design considerations for high-pressure ratio centrifugal compressor," 72-GT-71, 1972.

[7] R. H. Aungier, *Centrifugal Compressors—A Strategy for Aerodynamic Design and Analysis*, ASME Press, New York, NY, USA, 2002.

[8] C. Xu, "Design experience and considerations for centrifugal compressor development," *Proceedings of the Institution of Mechanical Engineers, Part G*, vol. 221, no. 2, pp. 273–287, 2007.

[9] C. Xu and R. S. Amano, "A hybrid numerical procedure for cascade flow analysis," *Numerical Heat Transfer, Part B*, vol. 37, no. 2, pp. 141–164, 2000.

[10] C. Xu and R. S. Amano, "Flux-splitting finite volume method for turbine flow and heat transfer analysis," *Computational Mechanics*, vol. 27, no. 2, pp. 119–127, 2001.

[11] C. Xu and R. S. Amano, "Computational analysis of pitch-width effects on the secondary flows of turbine blades," *Computational Mechanics*, vol. 34, no. 2, pp. 111–120, 2004.

[12] C. Xu and R. S. Amano, "Numerical prediction of swept blade aerodynamic effects," in *Proceedings of ASME Turbo & Expo*, pp. 1–9, Vienna, Austria, June 2004, GT2004-53008.

[13] C. Xu and R. S. Amano, "Computational analysis of swept compressor rotor blades," *International Journal of Computational Methods in Engineering Science and Mechanics*, vol. 9, no. 6, pp. 374–382, 2008.

[14] C. Xu and R. S. Amano, "The development of a centrifugal compressor impeller," *International Journal of Computational Methods in Engineering Science and Mechanics*, vol. 10, no. 4, pp. 290–301, 2009.

[15] C. Xu and R. S. Amano, "Computational analysis of scroll tongue shapes to compressor performance by using different turbulence models," *International Journal of Computational Methods in Engineering Science and Mechanics*, vol. 11, no. 2, pp. 85–99, 2010.

[16] C. Xu and R. S. Amano, "Study of the flow in centrifugal compressor," *The International Journal of Fluid Machinery and Systems*, vol. 3, no. 3, pp. 260–270, 2010.

[17] H. Krain, B. Hoffmann, and H. Pak, "Aerodynamics of a centrifugal compressor impeller with transonic inlet conditions," in *Proceedings of the International Gas Turbine and Aeroengine Congress and Exposition*, June 1995, 95-GT-79.

[18] H. Krain, "Flow in centrifugal compressor," von Karman Institute for Fluid Dynamics, Ls 17, 1984-07, 1984.

[19] D. Japikse, *Centrifugal Compressor Design and Performance*, Concepts Eti, Wilder, Vt, USA, 1996.

[20] C. Xu and R. S. Amano, "On the development of turbomachine blade aerodynamic design system," *International Journal of Computational Methods in Engineering Science and Mechanics*, vol. 10, no. 3, pp. 186–196, 2009.

[21] E. Benini and A. Toffolo, "Centrifugal compressor of a 100 kW microturbine: part 2—numerical study of impeller-diffuser interaction," in *Proceedings of the ASME Turbo Expo*, pp. 699–705, June 2003, GT-2003-38153.

[22] E. Benini and A. Toffolo, "Centrifugal compressor of a 100 kW microturbine: part 3—optimization of diffuser apparatus," in *Proceedings of the ASME Turbo Expo*, pp. 707–714, June 2003, GT-2003-38154.

[23] J. Hoffren, T. Siikonen, and S. Laine, "Conservative multiblock Navier-Stokes solver for arbitrarily deforming geometries," *Journal of Aircraft*, vol. 32, no. 6, pp. 1342–1350, 1995.

[24] T. Siikonen, "An application of Roe's flux-difference splitting for k-ε turbulence model," *International Journal for Numerical Methods in Fluids*, vol. 21, no. 11, pp. 1017–1039, 1995.

[25] M. M. Rai, "A relaxation approach to patched-grid calculations with the euler equations," *Journal of Computational Physics*, vol. 66, no. 1, pp. 99–131, 1986.

[26] S. Niazi, A. Stein, and L. N. Sankar, "Development and application of a CFD solver to the simulation of centrifugal compressors," AIAA paper 98-0934, 1998.

[27] W. Jansen, "Rotating stall in a radial vaneless diffuser," *ASME Journal of Basic Engineering*, vol. 86, pp. 750–758, 1964.

[28] Y. Senoo and Y. Kinoshita, "Influence of inlet flow conditions and geometries of centrifugal vaneless diffusers on critical flow angle for reverse flow," *ASME Journal of Fluids Engineering*, vol. 99, no. 1, pp. 98–103, 1977.

[29] H. S. Dou and S. Mizuki, "Analysis of the flow in vaneless diffusers with large width-to-radius ratios," *ASME Journal of Turbomachinery*, vol. 120, no. 1, pp. 193–201, 1998.

[30] S. Ljevar, H. C. de Lange, and A. A. van Steenhoven, "Two-dimensional rotating stall analysis in a wide vaneless diffuser," *International Journal of Rotating Machinery*, vol. 2006, Article ID 56420, 11 pages, 2006.

[31] K. Iwakiri, M. Furukawa, S. Ibaraki, and I. Tomita, "Unsteady and three-dimensional flow phenomena in a transonic centrifugal compressor impeller at rotating stall," in *Proceedings of the ASME Turbo Expo*, pp. 1611–1622, June 2009, GT2009-59516.

[32] E. M. Greitzer, "Review—axial compressor stall phenomena," *ASME Journal of Fluids Engineering*, vol. 102, no. 2, pp. 134–151, 1980.

[33] M. T. Barton, M. L. Mansour, J. S. Liu, and D. L. Palmer, "Numerical optimization of a vaned shroud design for increased operability margin in modern centrifugal compressors," *ASME Journal of Turbomachinery*, vol. 128, no. 4, pp. 627–631, 2006.

[34] ASME, Performance Test Code on Compressors and Exhausters, PTC 10-97, 1997.

[35] R. S. Amano and B. Sunden, *Thermal Engineering in Power Systems*, WIT Press, 2008.

[36] O. E. Balje, *Turbamachines*, John Wiley & Sons, New York, NY, USA, 1981.

[37] K. Brun and R. Kurz, "Analysis of secondary flows in centrifugal impellers," *International Journal of Rotating Machinery*, vol. 2005, no. 1, pp. 45–52, 2005.

[38] R. L. Street, G. Z. Watters, and J. K. Vennard, *Elementary Fluid Mechanics*, John Wiley & Sons, New York, NY, USA, 1996.

[39] T. Wright, "Comments on compressor efficiency scaling with Reynolds number and relative roughness," in *Proceedings of the American Society of Mechanical Engineers*, June 1989, 89-GT-31.

[40] P. R. N. Childs and M. B. Noronha, "The impact of machining techniques on centrifugal compressor impeller performance," in *Proceedings of the International Gas Turbine & Aeroengine Congress & Exposition*, June 1997, 97-GT-456.

[41] C. Rodgers, A diffusion factor correlation for centrifugal impeller stalling, 78-GT-61, 1978.

[42] C. Xu and R. S. Amano, "A turbo machinery blade design and optimization procedure," GT-2002-30541, 2002.

[43] D. P. Kenny, The history and future of the centrifugal compressor in aviation gas turbine, SAE/SP-804/602, 1984.

[44] R. van den Braembussche, "Rotating stall in centrifugal compressors," VKI Preprint 1987-16, 1987.

8

Application of a Fault Detection and Isolation System on a Rotary Machine

Silvia M. Zanoli and Giacomo Astolfi

D.I.I., Università Politecnica delle Marche, Ancona, Italy

Correspondence should be addressed to Silvia M. Zanoli; s.zanoli@univpm.it

Academic Editor: Zhaoye Qin

The paper illustrates the design and the implementation of a Fault Detection and Isolation (FDI) system to a rotary machine like a multishaft centrifugal compressor. A model-free approach, that is, the Principal Component Analysis (PCA), has been employed to solve the fault detection issue. For the fault isolation purpose structured residuals have been adopted while an adaptive threshold has been designed in order to detect and to isolate the faults. To prove the goodness of the proposed FDI system, historical data of a nitrogen centrifugal compressor employed in a refinery plant are considered. Tests results show that detection and isolation of single as well as multiple faults are successfully achieved.

1. Introduction

In oil and gas plants many processes such as refinery, natural gas extraction/compression, energy production, and gasification, involve the use of rotary machines like centrifugal pumps, turbines, and centrifugal compressors. Given the importance and the crucial role played by these machines, in the literature several approaches for their control and supervision have been proposed. For control purposes, the compressor regulation distant from the surge point has been discussed in [1–5] while the influence of compressors working conditions which can greatly affect the overall efficiency of the plant is considered in [6]. In recent years the attention has been focused on the prevention of possible recurrent malfunctions and potential faults which may cause equipment inactivity or even their complete break. In fact a prompt identification of perturbations on the working conditions by a diagnostic system may increase the availability of the machines and improve plant safety while achieving functioning costs reduction as well. For the fault detection on pumps machines several studies have been carried out: Dalton and Patton in [7] have proposed a model-based fault diagnosis system of a two-pump systems. Nold and Isermann in [8] have developed a fault detection system for centrifugal pumps and AC drives also derived from a model-based approach. Dealing with other rotary machines, various works can be found in the literature to diagnose possible faults on centrifugal compressor: in [9] the authors suggest using the time series analysis with neural network to realize a fault diagnosis system for monitoring a centrifugal compressor; in [10] the monitoring of the bearing temperature employed in a centrifugal compressor has been achieved. Another study concerning the detection of the fault occurring on the compressor bearings has been discussed in [11]. A comprehensive discussion on fault diagnosis system applied on rotary machines can be found in [12–14].

In the present work a multivariable data-driven and model-free approach, that is, Principal Component Analysis (PCA) [15], has been adopted. The focus is the design of a Fault Detection and Isolation (FDI) system to be applied on a multishaft centrifugal compressor. The main reason for choosing a model-free technique rather than a model-based one is that, as often verified for large scale chemical processes, it was difficult to develop a detailed physical model for the compression process. The presented FDI system deals with the functioning of the complete centrifugal compressor system, and this feature differentiates it by typical works on centrifugal compressors which focus on particular aspects like surge control or bearing temperature monitoring as in[1–5, 9–11].

In the literature, other authors suggested using a PCA technique for industrial plant applications: in [16] the authors use the PCA to detect possible faults on Tennessee Eastman (TE) Process while Bezergianni and Kalogianni in [17] propose the PCA approach on a hydrotreating process.

The contribution of the paper is the adoption of the adaptive thresholds approach, suitably implemented, to compute the gains associated with the inputs of the thresholds.

The paper is organized as follows. In Section 2 the background of the Principal Component Analysis is briefly summarized. In Section 3, the developed adaptive thresholds approach is described. In Section 4 the rotary machine considered for the actual implementation of the FDI system is briefly described. Results on single and multiple faults detection and isolation are reported in Section 5. Section 6 gives conclusions and future developments.

2. Theoretical Background on PCA

A brief overview of the Principal Components Analysis and its use in fault detection and isolation is given here. The main steps of the procedure useful for the understanding of the proposed FDI system are reported. Further detailed descriptions of the method are available in published articles (see among others [15, 17–19]).

The Principal Component Analysis can be considered a subspace decomposition technique by which the process measurement space is divided into two orthogonal subspaces, that is, the principal components (PC) subspace and residual subspace. The PC subspace contains the components that account for a maximal amount of total variance in the observed variables. In practice, given a data matrix X of N samples of m variables, an optimal linear transformation of X is operated by which the process measurement space is divided into the two orthogonal subspaces mentioned above. In this way possibly correlated variables of the process are mapped into a smaller number $l < m$ of uncorrelated variables called Principal Components. The main steps of the PCA procedures can be summarized as in the following [18].

Consider the transformation of the data matrix X of normal variables (zero mean and unit variance) given by

$$T_{[N\times m]} = X_{[N\times m]} P_{[N\times m]}. \tag{1}$$

Matrix T in (1) is named the *score matrix* while P is the *loading matrix*. The columns of matrix P are the eigenvectors associated with the eigenvalues of the following matrix A proportional to the data correlation matrix:

$$A_{[m\times m]} = X^{T}{}_{[m\times N]} X_{[N\times m]}. \tag{2}$$

Once the number of the l most significant components is determined, the loading matrix is partitioned as follows:

$$P = \left(\widehat{P} \quad \widetilde{P}\right), \quad \widehat{P} \in R^{m\times l} \tag{3}$$

and, considering the linear transformation (1), the data matrix X is partitioned into two parts: $\widehat{X}$, the *principal* part of the data explained by the first l eigenvectors and the *residual* part $\widetilde{X}$ explained by the remaining components:

$$\widehat{X}_{[N\times m]} = X_{[N\times m]} \widehat{P}_{[m\times l]} \widehat{P}^{T}{}_{[l\times m]} = X_{[N\times m]} C_{l[m\times m]}, \tag{4}$$

$$\widetilde{X}_{[N\times m]} = X_{[N\times m]} - \widehat{X}_{[N\times m]} = X_{[N\times m]} \left(I_{[m\times m]} - C_{l[m\times m]}\right). \tag{5}$$

Equation (4) is sometimes called *Back-Transformation process*; by this transformation it is possible to obtain the original variables with only significant variances; that is, if the number of Principal Components is correctly chosen, insignificant noise effects are removed.

When approaching a PCA problem, the number of the Principal Components retained in the model is an essential parameter that strongly determines its performance. When too few components are retained, the model will not capture all of the information in the data and a poor representation of the process will result. On the other hand, if too many components are chosen, the model will be overparameterized and it will include noise. Several approaches for selecting the optimal number of PCs have been developed. The approach followed in this paper is based on the ANalysis Of Variance, ANOVA test (see [20, 21]), that was proven to be a reliable and objective method.

Stored dataset is processed offline applying the ANOVA-based test so that principals components are determined; the loading matrix P is then computed and partitioned according to the selected PCs as in (3). Hence, back-transformation (4) is calculated online for each new sampled data vector $\mathbf{x}_{[1\times m]}$. The reconstructed vector $\mathbf{x}^*$ is thus obtained:

$$\widetilde{x}^* = x(k) \widetilde{P}\widetilde{P}^{T}. \tag{6}$$

Then the signal of the original variables $\mathbf{x}$ is compared with the reconstructed one $\mathbf{x}^*$; the difference between the two signals is called *residual*. Comparing the actual residual with its trends in standard operative conditions, detection of faults can be achieved. However, based on this residual quantity, isolation of the fault cannot be performed.

3. Fault Isolation

Developing a (good) diagnostic system, the interest in general is not just to accomplish the faults detectability but rather to specify the kind of the fault that has occurred, thus realizing the primary function of isolation. At this regard various methods have been developed for the generation of particular structured residuals. The authors have chosen to apply the so-called *structured residual approach* as described by Tharrault et al. in [18]. These structured residuals are based on the reconstruction principle; the founding idea of this approach is to reconstruct variables using the PCA model with the remaining variables of the model. Different subsets R of variables to be reconstructed can be considered. The reconstruction of variables consists in estimating, for each new sampled data, the reconstructed vector $\widehat{x}_R$ by eliminating the effect of possible faults occurred on the selected variables.

To achieve isolation the reconstructed variables $\widehat{x}_R$ are projected on the Residual subspace by a new projection matrix generating $\tilde{x}_R$. Given the propriety of the projection matrix which states that its product with the matrix that identifies the directions of reconstruction is null, it is possible to isolate potential faulty components. In fact from the above property it can be deduced that the components of $\tilde{x}_R$ are not sensitive to the components of x belonging to the subset R.

This property can be used to identify which components of x are disturbed by faults. Through the analysis of all the residual trends, computed considering different (structured) variables subsets, faults directions are identified.

Finally, considering the large dimensions of the problem, instead of checking all the components of computed residual vectors separately, it may be convenient to group them in to a single index called square prediction error (SPE) calculated as the square Euclidean norm of the computed residuals:

$$\mathrm{SPE}_R = \|\tilde{x}_R\|^2. \quad (7)$$

Fault isolation is based on the threshold violation of one or a few SPE variables.

3.1. Adaptive Thresholds. To perform the fault detection and isolation, it is necessary to compare the SPEs values to threshold values which can be fixed or adaptive. Values for static threshold are typically computed, analyzing the SPEs signal in normal operative conditions. This solution works well if the process stays in a steady state or if the operating point does not change. Furthermore, when operating with real processes data, it is highly possible that due to the presence of noisy data and/or due to the influence of external conditions, the generated residuals exceed the set fixed threshold even without faults. A possible solution is to enlarge threshold values. To set the tolerances, compromises have to be made between the threshold size for avoiding misdetection problems because of too large threshold values and the generations of false alarms because of normal fluctuations of the system.

In order to improve the isolation feature, the authors have chosen to employ an adaptive threshold approach. In fact, the process variability is such that the adoption of a fixed threshold would imply many spurious false alarms.

In the literature, different approaches to the design of adaptive thresholds have been introduced. Isermann in [15, 19] observed that deviations of the generated residuals frequently depend on the amplitude and frequencies of the input excitation and proposed an adaptive threshold scheme based on a first-order high-pass filter of the input signal with a possible additional proportional enlargement. A low-pass filter is then introduced to smooth the dynamic of the threshold. Conversely Clark in [22] proposes a scheme where the use of the high-pass filter is substituted by static gains.

Following Clark's approach, the adaptive threshold scheme, proposed by the authors, is constituted by a term proportional to the amplitude of the input signals and of a constant term for a tight tuning (see Figure 1). The proportional term is obtained, multiplying the input data signal by a gain matrix which is diagonal.

For the actual computation of the gains no clear suggestions are given in [22] so that a new method based on the frequency analysis of the input signals has been proposed, and implemented by the authors. The single gains elements have been computed by taking into account the input variables that actually contribute to the generation of each SPE. The significance of each input signal has been associated to its energy in terms of its power spectrum in the frequency domain. A variable thresholds approach has thus been adopted which resulted effective and more efficient than applying fixed ones. In particular, it has been chosen to assign low gain weights to signals "poor" in frequency while high weights have been assigned to signals with the highest energy contents.

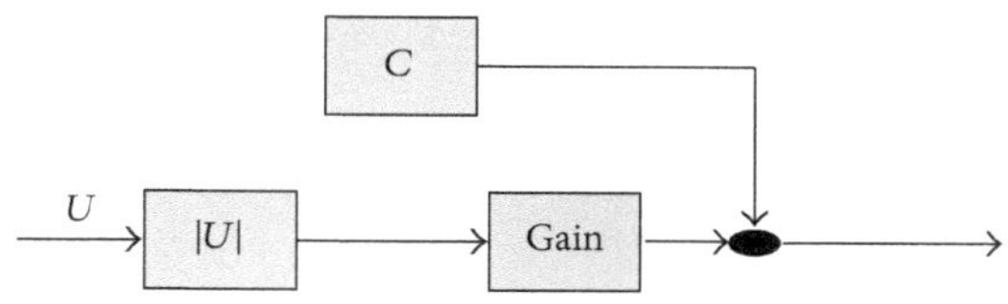

FIGURE 1: The scheme employed by the authors to realize the adaptive threshold.

4. The Multishaft Centrifugal Compressor

In the present paper, a multishaft centrifugal compressor is considered. Centrifugal machines are critical equipments; their essential characteristics have been the large pressure rises and flow rates involved. Given the importance of and the crucial role played by compressor machines, in recent years an increasing attention has been given to the prevention of possible frequent malfunctions and potential faults which may cause inactivity of compressor or even its complete break.

The machine, called BLNC (base load nitrogen compressor), is located in the air separation unit (ASU) of a refinery plant and is employed for nitrogen compression in the dilution of a particular gas, the Syngas, which is forwarded to a gas turbine. It is a complex machinery consisting of two sections: the first section includes two compression stages while the second comprises three compression stages.

In order to decrease the nitrogen temperature at the exit of the compression stage, leveling it at its input value before compression, a heat exchanger is positioned at the end of each stage. In these conditions, the compression is nearly equivalent to an isotherm process which requires less mechanical work for the compression.

Variables considered in the construction of the data matrix X employed in the PCA diagnoser consist of sensor and actuator (positioner) variables. Their tag names used in the refinery are summarized in Table 1 together with their description. As it can be observed, variables considered for the detection and isolation of faults are the nitrogen (N_2) flow through the multishaft compressor measured at different points, the commanded actuators signals, and their actual values used to regulate the position of the IGVs. All the variables included in the PCA data matrix X are process variables; no variables related to thermodynamic parameters

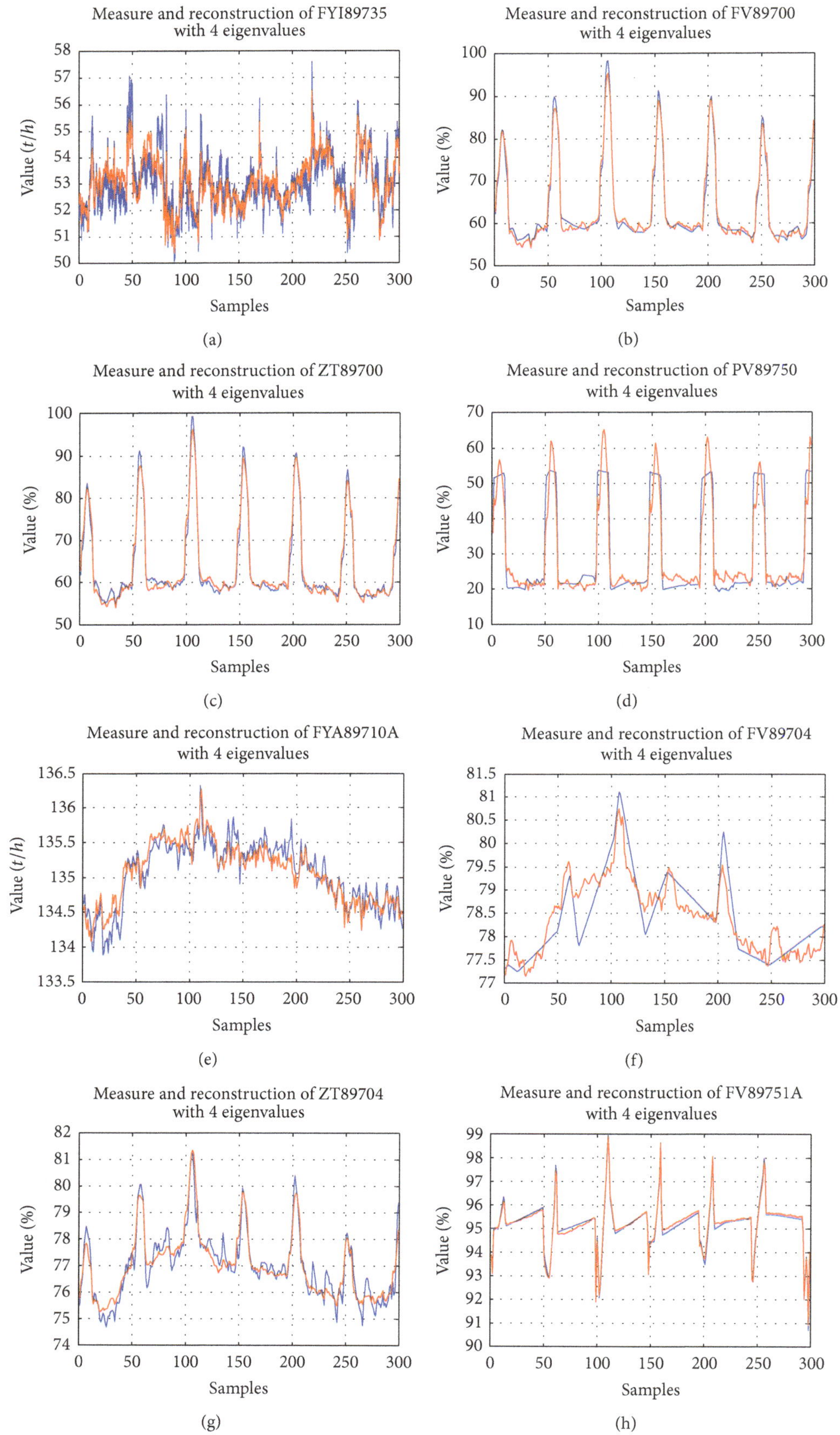

Figure 2: Continued.

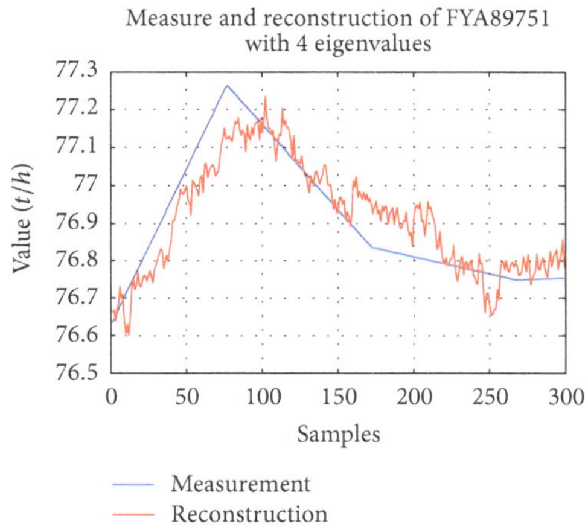

(i)

FIGURE 2: Figures (a)–(i) show the trends of the real variables (blue) and their reconstruction with PCA (red).

TABLE 1: Considered process variables.

Tag name	Description
FYI_89735	N_2 mass flow through the BLNC first section
FV_89700	Positioner of the Inlet Guide Vanes (IGV) relative to the first and second stages of BLNC
ZT_89700	Feedback of IGV position relative to the BLNC first and second stages
PV_89750	Measurement of the vent position at the entrance of first section of BLNC
FYA_89710A	N_2 mass flow through the second section of BLNC
FV_89704	Positioner of the IGV relative to the third stage of BLNC
ZT_89704	Feedback of IGV position relative to the third stage of BLNC
FV_89751A	Throttle valve position relative to inlet high pressure nitrogen gas (HNG)
FYA_89751	N_2 mass flow at the head of the high pressure column

were considered for the implementation of the FDI system in the case considered here since it has been verified that these variables were not particularly affected by the analyzed faults.

5. Fault Detection and Isolation System

The first step of PCA method concerns the selection of the Principal Component retained in the model. At this regard, for the computation of the matrix A (2) to be effective, the absence of faults in the system data is required. Furthermore, the absence of measurement noise has to be assumed. In fact, if the calculation of matrix A is performed from a noisy dataset, errors on variables reconstruction may influence the PCA model and, consequently, the overall fault identification process. For the application to the multishaft centrifugal compressor, to avoid measurement noise effects, PCA training data have been gathered just after instruments calibration. The resulting PCA model together with the application of adaptive thresholds guarantees good performances in terms of fault detection and isolation even in the presence of perturbations due to altered ambient conditions like, for example, the ones that can be daily experienced due to the alternation of day and night. At the same time, the succession of the seasons may call for the FDI system update, and to improve system sensibility, different PCA training dataset could be considered.

Accordingly, with the PCA model, the measurement space has been partitioned into two orthogonal spaces: the principal component subspace, which includes data variations according to the principal component model, and the residual subspace, which includes data variation not explained by the model. Applying the ANOVA procedure developed by the authors (see [20, 21]), the dimension of the PC subspace is set to four. In order to verify if the selected number of PCs is adequate to correctly explain the original system's variables, the signals of the original variables are compared with the reconstructed ones. Plots of Figure 2 show the measured and reconstructed signal. With the exception of Figure 2(f) (positioner of the IGV) and Figure 2(i) (N_2 mass flow at the head of the high pressure column) where slight mismatching can be noticed, all the reconstructions are in good agreement with measurements.

After having trained the model on system data in the absence of faults, that is, after having chosen the PCs that make up the *loading matrix P* used for the generation of the structured residuals (see (1)) the diagnoser has been tested on both single and multiple faults.

Faults that may possibly occur in the centrifugal compressor concern errors in the sensor readings and/or in the actuators. By inspection of historical data of the compressor at issue, the most common faults were found to be faults of the

TABLE 2: Most common compressor faults.

Tag name	Description	Time dependency
FYI_89735	Fault in the first stage N_2 mass flow sensor	Abrupt fault; incipient fault
ZT_89700	Fault in the first stage IGV positioner	Abrupt fault; incipient fault; intermittent fault
FYA_89710	Fault in the third stage N_2 mass flow sensor	Abrupt fault; incipient fault
ZT_89704	Fault in the third stage IGV positioner	Abrupt fault; incipient fault; intermittent fault
FYA_89751	Fault in the N_2 mass flow sensor relative to N_2 flow at the head of the high pressure column	Abrupt fault; incipient fault

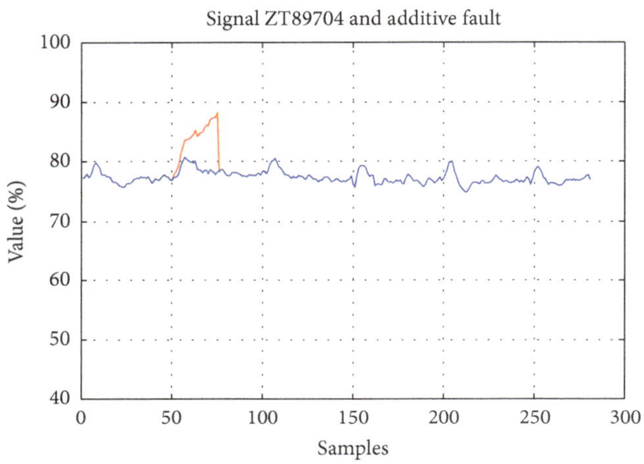

FIGURE 3: Third section IGV position (blue) and ramp additive fault (red).

actuators and mass flow sensors as specified in Table 2. To test the FDI system both single and multiple faults are simulated. The simulation is necessary to test the system on critical and not easy to detect faults such as drift faults and multiple faults.

5.1. Single Fault Case: Third Stage IGV Positioner Fault. The diagnoser has been tested on the detection and isolation of faults of the Inlet Guide Vanes (IGV) of the third stage of the compressor. An abrupt failure of the actuator, which caused its complete breakdown, was documented on the historical data at disposal; this failure was correctly detected by the Fault Diagnosis module, but since the detection in this case is quite trivial, we have chosen to test the diagnoser performances on an intermittent fault by modifying the real data by the addition of step or ramp variations. These kinds of faults may be likely caused by a temporary malfunction of the leverage used for IGV handling.

A drift on the IGV positioner has been simulated with the addition of a ramp signal up to 10% of the variable amplitude starting from the 50th sample as shown in Figure 3. When the drift of the actuator causes the measure to rise up to 5%–7% of

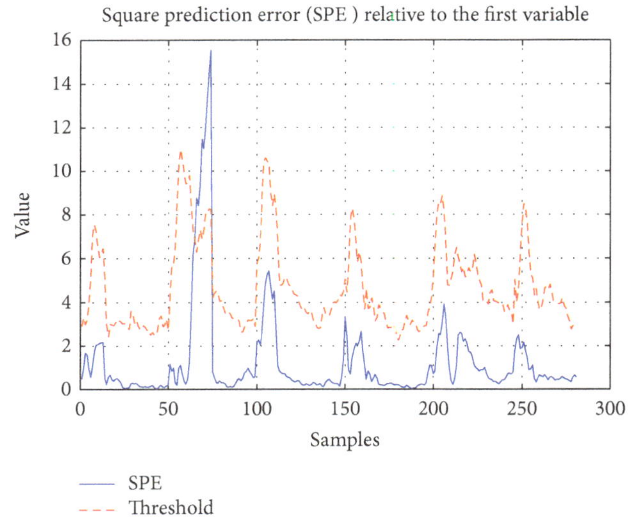

FIGURE 4: SPE relative to the N_2 mass flow sensor and its threshold.

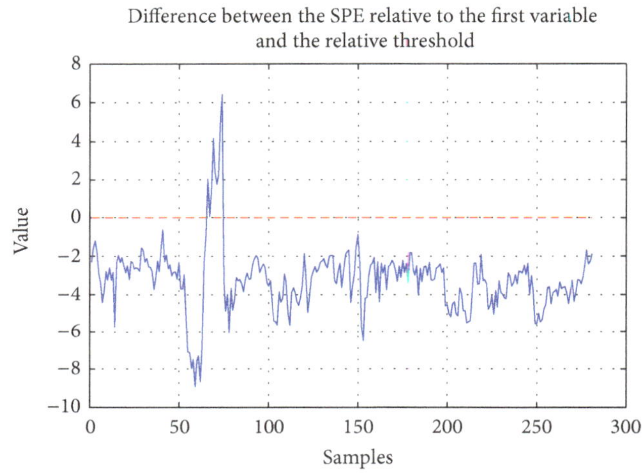

FIGURE 5: Differences between the SPE relative to N_2 mass flow sensor and its threshold.

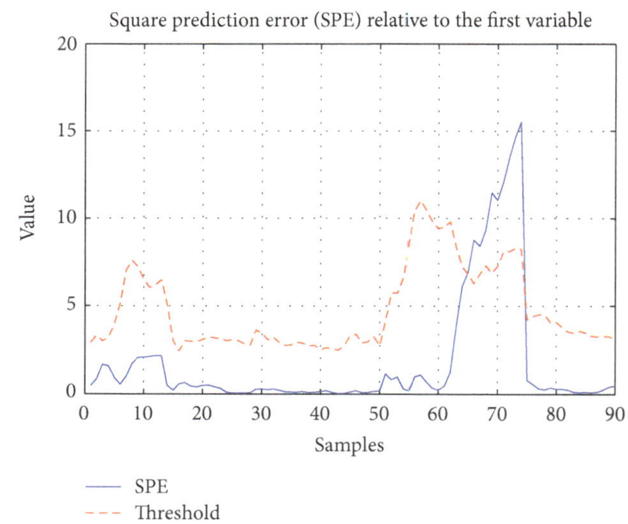

FIGURE 6: Zoom on threshold overcoming.

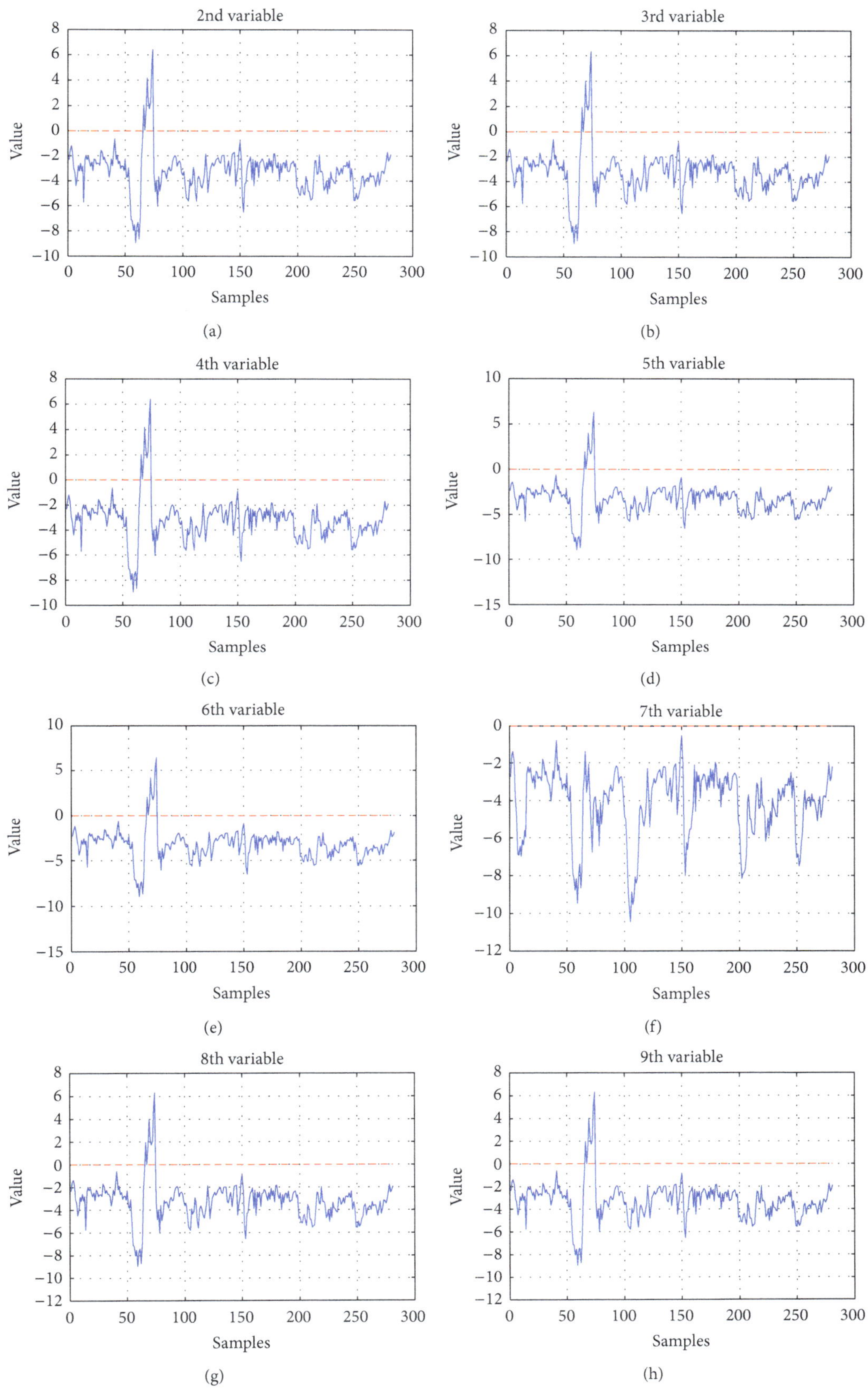

FIGURE 7: Difference between the SPE and its relative threshold: only the SPE relative to the 7th variable (f) stays under the threshold.

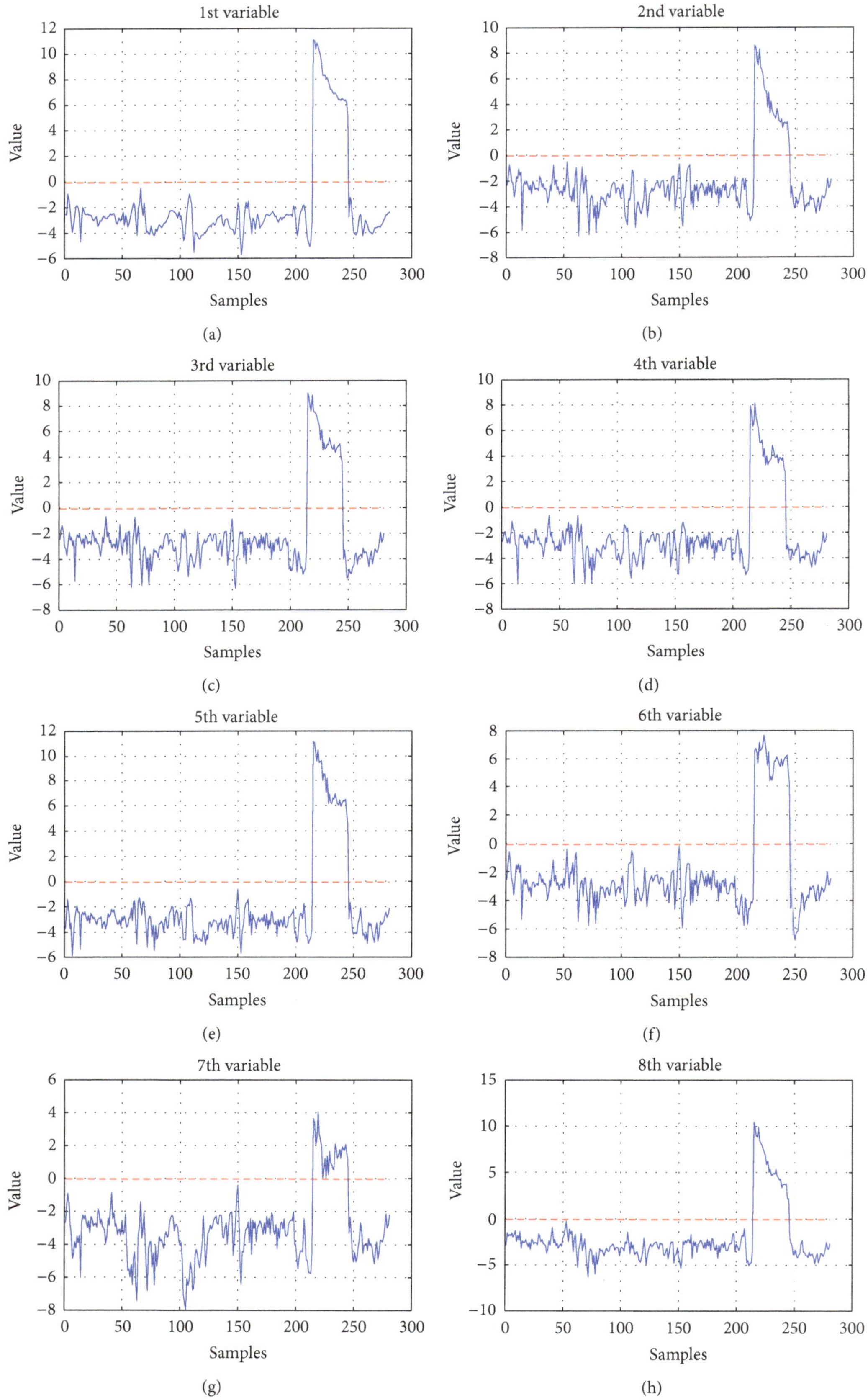

Figure 8: Continued.

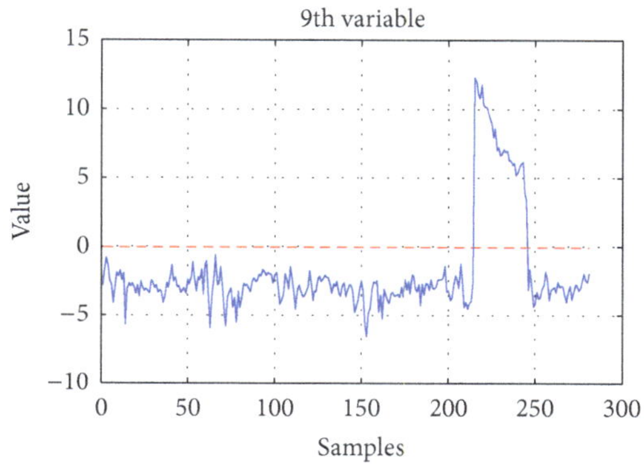

(i)

FIGURE 8: PCA detection results when considering single directions: all the SPEs exceed their thresholds.

its standard value, the square prediction error overcomes the assigned threshold. Figures 4, 5, and 6 show SPE computed on the first variable together with its difference from the adaptive threshold (see [11, 15]); it can be noticed that, at sample 62, SPE value exceeds the threshold. Since each SPE takes into account all the variables but the one or the ones associated with the direction(s) under inspection, it is necessary to check the temporal trends of all the other residuals. As a matter of fact, all the residuals are influenced by the IGV positioner fault with the exception of the residual associated with the fault variable.

This situation allows the correct isolatation of the fault. Figure 7 shows the deviations of the SPE from the threshold; it can be noticed that the difference remains negative for all the experiment sampling time only in correspondence of the 7th variable (ZT_89704).

These results show that the fault is detected at the 12th sample after its onset; given a sampling time of five minutes, the fault detection time can be computed in the order of 60 minutes. The behavior of the diagnoser in term of promptness is linked to the adopted large sampling time; given that actual process dynamics which are rather slow, the result can be considered fully satisfactory. Given the interest to investigate the performances of the proposed system on an operational dataset covering a large period of time (about two years), the adoption of a sampling time of five minutes was consequentially forced by the dataset at disposal. It is clear that the proposed FDI system, when implemented online, can process data at a faster sampling rate (typically one minute). No sensible limitations are imposed by the computational load of the proposed approach; the lower limit is determined at the I/O acquisition level: the DCS employed for controlling the machinery under study does not handle sampling period lower than 0.1 seconds; moreover, for many of the considered variables in Table 2, a sampling time not lower than one minute is generally set.

5.2. Multiple Faults Case: First and Third Stage IGV Positioners Fault. To check the validity of the proposed system on the detection of multiple faults, faults on the two IGV partitioners of the first and third section, respectively, have been simulated.

The simulated faults are supposed to be simultaneous, and they have been constructed with the following characteristics:

(i) Positioner ZT_89700: a trend of the signal is simulated at the time instants where the IGV position is opened around 60% of its total value;

(ii) Positioner ZT_89704: a bias of 10% of its average value is simulated.

In Figure 8, the differences between the single direction SPEs and the relative thresholds are shown. As it has to be expected, all the SPEs exceed the threshold.

The system correctly isolates the faults occurred indicating the presence of a multiple fault on the 3rd and 7th variables. Computed residuals concerning some of the variable pairs are reported in Figures 9 and 10. As expected the only SPE that remains under the threshold is the one that is associated with the variables in fault, that is, variable 3 and variable 7.

6. Conclusions

A Fault Diagnosis system for the detection and the isolation of expected faults of a rotary machine based on the Principle Components Analysis technique has been developed. The considered machine is a multishaft centrifugal compressor located in an integrated gasification and combined cycle of a refinery plant. The adoption of a model-free technique is justified by the fact that in the process industry, rich process data are available while, conversely, the development of a physical model is a demanding task that may not assure suitable results.

For the Practical implementation of the PCA, the choice of the number of principal components to be retained in the model has been based on an approach centered on the ANalysis Of Variance, ANOVA test, and for what concern the detection and the isolation issues, a structured residual approach has been applied and an adaptive threshold has

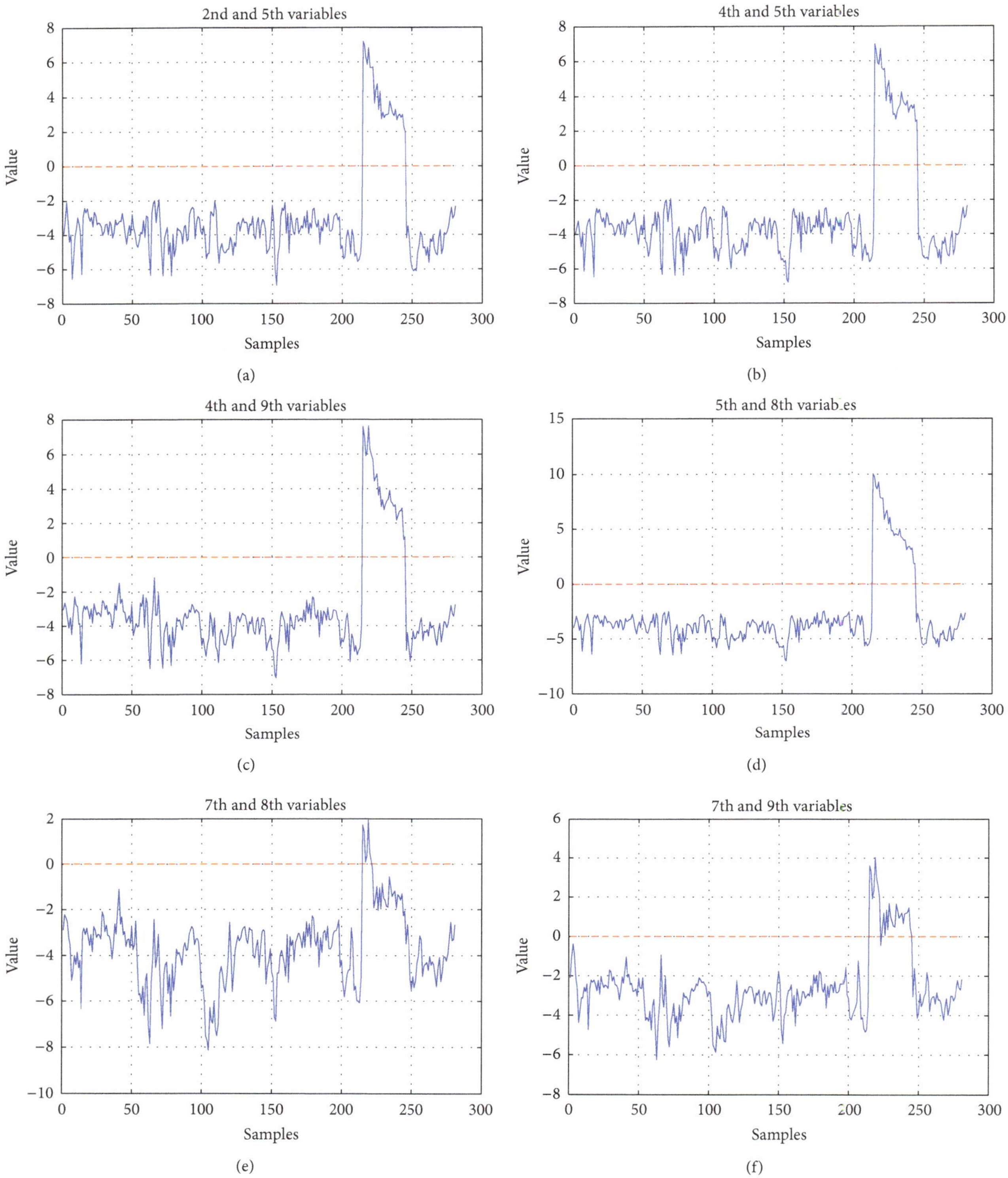

FIGURE 9: Some PCA detection results when considering directions pairs: the SPEs exceed their thresholds.

been adopted for their analysis that was well suited in dealing with the process variability.

The present method has been successfully applied in the development of a Fault Detection and Isolation (FDI) system of a multishaft centrifugal compressor located in a refinery plant; malfunctions on the compressor sensors and actuators have been examined and isolation of single and multiple faults as well as process faults have been successful.

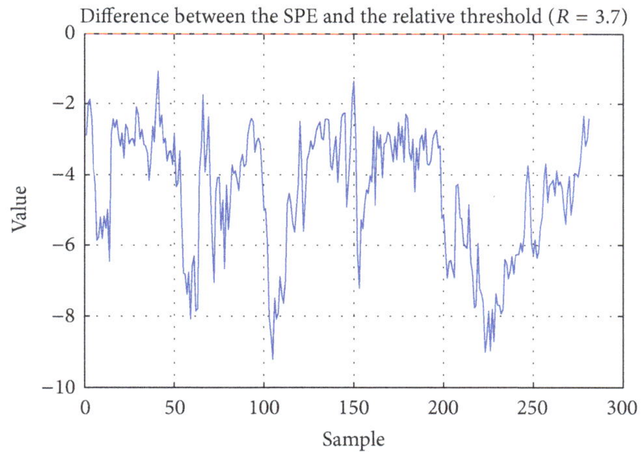

FIGURE 10: SPE calculated from the faulty variables pair (variable 3 and variable 7) remains under threshold.

References

[1] J. T. Gravdahl and O. Egeland, *Compressor Surge and Rotating Stall: Modeling and Control*, Springer, London, UK, 1999.

[2] B. De Jager, "Rotating stall and surge control: a survey," in *Proceedings of the 34th IEEE Conference on Decision and Control*, pp. 1857–1862, New Orleans, La, USA, 1995.

[3] D. C. Liaw, C. C. Song, and J. T. Huang, "Robust stabilization of a centrifugal compressor with spool dynamics," *IEEE Transactions on Control Systems Technology*, vol. 12, no. 6, pp. 966–972, 2004.

[4] J. T. Gravdahl and O. Egeland, "Centrifugal compressor surge and speed control," *IEEE Transactions on Control Systems Technology*, vol. 7, pp. 567–579, 1999.

[5] J. T. Gravdahl and O. Egeland, "Speed and surge control for a low order centrifugal compressor model," *Modeling, Identification and Control*, vol. 19, no. 1, pp. 13–29, 1998.

[6] T. B. Ferguson, *The Centrifugal Compressor Stage*, Butterworths, London, UK, 1963.

[7] T. Dalton and R. Patton, "Model-based fault diagnosis of a two-pump systems," in *Proceedings of the IFAC World Congress*, pp. 79–84, San Francisco, Calif, USA, 1996.

[8] S. Nold and R. Isermann, "Model-based fault detection for centrifugal pumps and AC drives," in *Proceedings of the 11th IMEKO World Congress*, Houston, Tex, USA, 1998.

[9] H. J. Liao and S. Z. Huang, "The fault diagnosis for centrifugal compressor based on time series analysis with neutral network," in *Proceedings of the 3rd International Conference on Advanced Computer Theory and Engineering (ICACTE '10)*, Chengdu, China, August 2010.

[10] M. Mugnaini, V. Quercioli, M. Catelani, R. Singuaroli, and A. Fort, "Characterization of centrifugal compressors' thermo-elements used in journal and thrust bearing temperature monitoring," in *Proceedings of the 19th IEEE Instrumentation and Measurement Technology Conference*, vol. 1, pp. 421–424, May 2002.

[11] P. Huang, Z. Pan, X. Qi, and J. Lei, "Bearing fault diagnosis based on EMD and PSD," in *World Congress on Intelligent Control and Automation*, pp. 1300–1305, 2010.

[12] S. Zhang, T. Asakura, X. Xu, and B. Xu, "Fault diagnosis system for rotary machines based on fuzzy neural networks," in *International Conference on Advanced Intelligent Mechatronics*, pp. 199–204, 2003.

[13] W. Q. Lim, D. H. Zhang, J. H. Zhou, and H. L. Chan, "Vibration-based fault diagnostic platform for rotary machines," in *Proceedings of the 36th Annual Conference on IEEE Industrial Electronics Society (IECON '10)*, pp. 1404–1409, 2010.

[14] J. Zhihao, J. Shanqwei, J. Wen, and W. Bangchum, "Rubbing fault diagnosis of rotary machinery based on wavelet and support vector machine," in *Proceedings of the 1st International Workshop on Database Technology and Applications*, pp. 287–290, 2009.

[15] R. Isermann, *Fault-Diagnosis Systems. An Introduction from Fault Detection to Fault Tolerance*, Springer, Berlin, Germany, 2006.

[16] B. S. Shin, C. J. Lee, G. Lee, and E. S. Yoon, "Application of fault diagnosis based on signed digraphs and PCA with linear fault boundary," in *International Conference on Control, Automation and Systems (ICCAS '07)*, pp. 984–987, October 2007.

[17] S. Bezergianni and A. Kalogianni, "Application of principal component analysis for monitoring and disturbance detection of a hydrotreating process," *Industrial & Engineering Chemistry Research*, vol. 47, no. 18, pp. 6972–6982, 2008.

[18] Y. Tharrault, G. Mourot, J. Ragot, and D. Maquin, "Fault detection and isolation with robust principal component analysis," *International Journal of Applied Mathematics and Computer Science*, vol. 18, no. 4, pp. 429–442, 2008.

[19] R. Isermann, *Fault-Diagnosis Applications. Model-Based Condition Monitoring: Actuators, Drives, Machinery, Plants Sensors, and Fault-Tolerant Systems*, Springer, Berlin, Germany, 2011.

[20] S. M. Zanoli, G. Astolfi, and L. Barboni, "Applications of fault diagnosis techniques for a multishaft centrifugal compressor," in *Proceedings of the 18th Mediterranean Conference on Control and Automation (MED '10)*, Marrakech, Morocco, 2010.

[21] S. M. Zanoli, G. Astolfi, and L. Barboni, "Applications of fault diagnosis techniques for a multishaft centrifugal compressor," in *Proceedings of the 9th European Workshop on Advanced Control and Diagnosis (ACD '11)*, Budapest, Hungary, 2011.

[22] R. N. Clark, *Fault-Diagnosis in Dynamic Systems. Theory and Application*, Prentice-Hall, Heslington York, UK, 1989.

Analysis of Cavitation Instabilities in a Four-Blade Inducer

O. Coutier-Delgosha, A. Dazin, G. Caignaert, and G. Bois

Arts et Metiers ParisTech/LML Laboratory, 8 boulevard Louis XIV, 59046 Lille Cedex, France

Correspondence should be addressed to O. Coutier-Delgosha, olivier.coutier@lille.ensam.fr

Academic Editor: Meinhard Taher Schobeiri

The cavitating behavior of a four-blade inducer tested in the LML laboratory large test facility is considered in the present paper. Experimental investigations based on unsteady pressure measurements and records from a six-component balance mounted on the inducer shaft are performed. Spectral analysis of the signals enables to detect several characteristic frequencies related to unbalanced two-phase flow patterns. The objective of the present paper is the understanding of the physical phenomena associated to these frequencies. Therefore, wavelet decomposition, flow visualizations, and direct analysis of the high-frequency force, moment, and pressure signals are applied. Results at nominal flow rate only are considered. Not only classical unbalanced cavitation patterns, but also unexpected flow organizations are discussed.

1. Introduction

Rocket engine turbopump inducers are designed to operate in cavitating conditions, because of the low pressure of LH2 and LOX in the tanks. However, development of cavitation must be controlled to avoid any reduction of the inducer performance. Attention is also focused on instabilities associated with cavitation, since they are responsible for unbalanced flow pattern that compromise the rotor equilibrium. Self-oscillation of the sheet cavities on the blades may also generate problematic pressure fluctuations at the inducer outlet.

The understanding of these different unsteady flow behaviors has not been completed, yet, although significant progress has been achieved in the last decades. Experimental investigations of the behavior of various inducers, characterized by a blade number varying usually between 2 and 5, have been conducted for example in Pisa University (Italy) in collaboration with Avio, in Arts et Metiers ParisTech and CREMHyG (France) in collaboration with SNECMA Moteurs, and in Osaka University (Japan) by Tsujimoto and colleagues in cooperation with the JAXA program. In these different studies, several recurrent unstable behaviors have been identified: (i) rotating cavitation is characterized by a nonsymmetrical two-phase flow pattern, which rotates faster or slower than the inducer, leading to so-called super-synchronous or subsynchronous regimes, respectively. Rotation speed in the inducer rotating frame is generally low, leading in the absolute frame to frequencies close to the inducer rotation one. Such phenomenon, which leads to significant radial forces, has been detected in nearly all tested inducers [1–3] (Tsujimoto et al. 1997). (ii) A surge mode oscillation characterized by a global pulsation of the vapor areas at low frequency: such oscillations have been investigated in simpler configurations of 2D Venturi profiles or 2D foil sections in cavitation tunnels and analyzed in details by many authors [4–7]. (iii) Rotating stall in backflow vortices: such behavior has been identified by Tsujimoto and colleagues [8] in the case of cavitation clouds extending upstream from the inducer inlet. It basically consists in a slow rotation of this detached cavitation, at frequency that depends on the number of vortices.

The occurrence of these different flow patterns strongly depends on the cavitation development in the machine. A typical sketch in the case of a four-blade inducer is given for example in Figure 1.

At cavitation inception, only a steady and balanced flow pattern with one short attached cavity on each blade is observed from flow visualizations. When cavitation parameter is slightly decreased, a steady and alternate cavitating

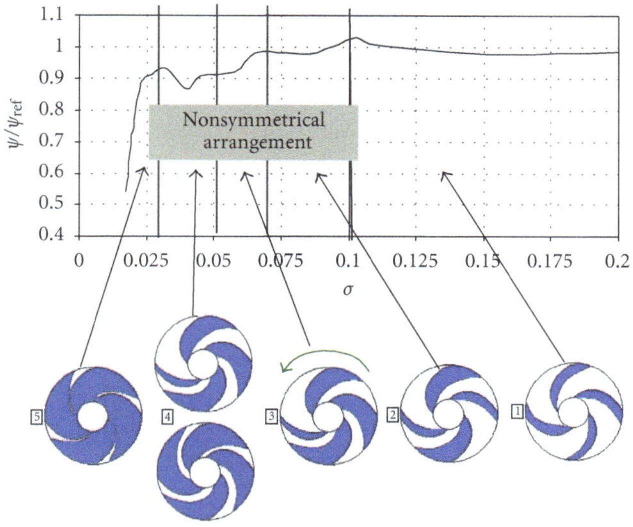

FIGURE 1: Sketches of cavitation patterns and performance evolution as the cavitation number decreases in a four-blade inducer [9].

FIGURE 2: Inducer geometry.

configuration appears with alternatively one short and one long cavity. For a lower cavitation parameter, just above breakdown, rotating cavitation can be identified: unbalanced attached cavities are observed in the different channels, their distribution rotating faster than the inducer. Finally near the breakdown of the inducer, a steady and balanced flow pattern with fully developed cavitation is observed.

More recently, supplementary unstable flow patterns have been identified in particular configurations: (i) higher-order rotating cavitation has been reported first by Fujii et al. [10], from Osaka University, in the case of a three-blade inducer. Frequency related to this phenomenon is 5 f_{ref} in the absolute frame. (ii) Higher-order cavitation surge: this axial instability has been found by the Osaka Team at frequency 5 f_{ref}, and also by researchers from Pisa University at frequencies 4.4 f_{ref} and 6.6 f_{ref} in the case of a two-blade inducer. In both cases, the phenomenon was detected at high flow coefficient.

These instabilities induce some strong radial forces that may perturb the rotor balance, and important pressure fluctuations in the lines. They must be quantified and controlled to avoid any major effect on the global pump behavior. Their detection is usually based, in all aforementioned references, on pressure transducers mounted on the inducer shroud at several angles in the same section. The analysis of the phase difference between the transducers enables to distinguish axial instabilities (characterized by zero-phase difference) from rotating cavitation (where a phase difference equal to the angle difference is obtained).

As can be seen, the physical characterization of the flow instabilities related to cavitation is still weak, which explains that only a few attempts have been performed to explain inception and mechanisms of rotating cavitation [11, 12]. Yoshida and colleagues state that inception of unstable behaviours is mainly governed by the sheet cavity length L_c. Sheet cavities longer than the cascade throat remain equal because they are fully constrained by the adjacent blade. In opposition, sheet cavities that do not reach the cascade throat ($0.8 < L_c/h < 1$) are characterized by a degree of freedom, which may lead to unstable cavitation such as synchronous nonsymmetrical flow patterns. However, no consensus is presently obtained concerning too the mechanisms that are responsible for the inception of unbalanced flow patterns as the ones that control the successive synchronous and non-synchronous regimes that are usually observed in inducers.

Conversely, several studies have been conducted to investigate the effects of the inducer design on its unsteady behavior: it has been found, for example, that the shape of the hub [13], of the leading edge [14], (Bakir et al. 2002), and even of the casing [15] have significant influence on cavitation instabilities. The effects of unequal spacing of the blades have been also investigated by numerical simulations [16]. All inducers have been usually manufactured with one to 4 blades in the inlet section.

In the present paper, the cavitating behaviour of a four-blade inducer (Figure 2) is investigated. Experiments have been performed in the large test facility of ENSAM Lille, with the collaboration of the CNES (French Space Agency) and SNECMA Moteurs. Although a large range of mass flow rates, inlet pressures, and rotation speeds have been tested, attention is focused presently on a few hydrodynamic conditions only, at nominal mass flow rate Q_n and reference rotation speed N_{ref}. The successive flow instabilities, which are encountered when the inlet pressure is progressively decreased, are analysed in details, in order to improve their characterization and physical understanding. Original force measurements performed with a balance mounted on the inducer shaft, associated with unsteady pressure measurements on the inducer shroud, enable to obtain data both in the absolute and inducer rotating frame, which enriches the analysis of the results.

Note that all results presented in the paper are dimensionless for reasons of confidentiality. So, the "+" indicates that each variable V^+ is V/V_0 where V_0 is a reference value which is not given.

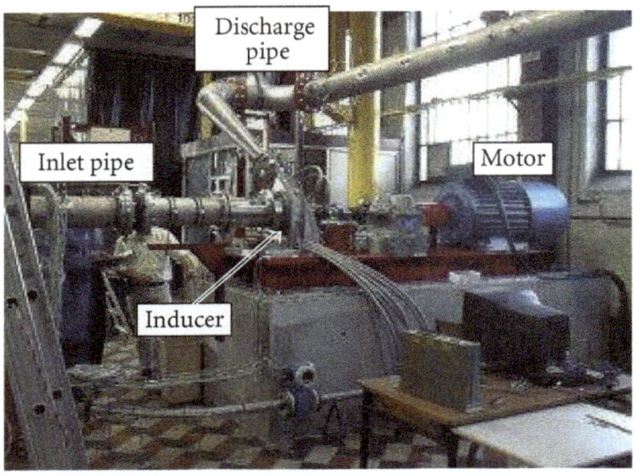

FIGURE 3: General view of the LML large test facility.

2. Experimental Setup

2.1. Test Facility. The LML laboratory large test facility devoted to the study of axial pumps in cavitating conditions has been used for the experiments. This two-stairs facility is equipped with a 200 kW motor that enables to reach an inducer rotation speed of 6000 rpm. The device upstairs (Figure 3) is mainly composed of the inducer to be tested and a stator downstream from the inducer that conducts the flow towards the discharge pipe. The inlet and discharge pipes are connected to two tanks. A free surface is maintained in the upstream one, so that the pressure can be controlled in the installation, from about 0.1 bars up to 16 bars. A part Q_s of the volume flow rate is taken at the inducer outlet for the axial equilibrium device, and then reinjected in the upstream tank. Downstairs are located two large resorption tanks used for the extraction of dissolved gas, a variable head loss device that enables to control the flow rate, and the vacuum pump used to decrease the pressure in the inlet tank. A heat exchanger is also connected to the facility to control the water temperature.

2.2. Acquisition Device. The inducer test section is equipped with a large variety of sensors devoted either to the characterization of steady flow properties, or to the analysis of the flow unsteady fluctuations.

Concerning steady flow properties, the torque M_x and the rotation speed N are measured with a Torquemaster TM213 torque meter, while a Rosemount AP6 pressure transducer (range 0–6 bars) is used for the inlet pressure P_i. A Rosemount DP7E22 differential pressure transducer (range 0–16 bars) connected between P_i and P_d is used to obtain directly the elevation ΔP. The main flow rate Q_m and the recirculating one Q_s are measured with Endress + Hauser electromagnetic flow meters. The motor shaft rotation speed is also controlled with a photoelectric cell.

Temperature measurements are performed on the upstream and downstream bearings of the inducer shaft, to control their increase at high rotation speed, and the water temperature is measured to regulate the heat exchanger operation so that a 25°C temperature is maintained in the flow. The torque M_x and the rotation speed N are acquired at frequency 100 Hz during 10 s, while all other signals are recorded at frequency 40 Hz during 140 s. From these records, mean and RMS values of the signals are calculated for each investigated flow condition. Note that the inlet and outlet pressures are measured in the inlet and delivery pipes more than one meter upstream and downstream from the inducer, so the calculated pressure head includes the head losses between the two pressure taps, including the one in the stator and in the downstream 90° bends that can be seen in Figure 4.

Absolute uncertainties on the mean flow characteristics are 0.5% of Q_{ref} for Q_m, 0.05% of Q_{ref} for Q_s, 0.25% of N_{ref} for N, 2% of the noncavitating torque value for M_x, and 0.2% of the measurement range for the upstream pressure and ΔP. Precision regarding the cavitation number and the head, torque, and flow rate coefficients depend on the flow conditions and will be indicated in the figures hereafter.

Unsteady flow properties are investigated with nine Kistler 701A piezo-electric pressure transducers (Figure 4): six of them are located in the upstream and delivery pipes, two other ones (P_1' and P_2') are located at the inlet of the inducer test section, and the last one P_3' is between the inducer and the stator. All transducers are mounted flush to the internal pipe wall. Four accelerometers are also installed on the inducer casing to measure the vibrations in the axial and radial directions. A six-component balance is mounted on the shaft to obtain the axial force F_x, the radial forces F_y and F_z, the torque M_x, and also the bending moments M_y and M_z. These 19 parameters are recorded simultaneously at frequency 2048 Hz with the LMS CADA-X code. An antialiasing filtering (Butterworth, 800 Hz, −50 dB/octave) is applied during acquisition.

Relative uncertainty on the pressure and vibration measurements is found to be close to 1% in noncavitating conditions and 2% in cavitating conditions. Precision on the forces and moments measured by the balance is estimated to 1% of the measurement ranges.

2.3. Experimental Process. Unsteady flow properties are investigated by varying continuously the inlet pressure from 2 bars down to the 25% drop of the inducer head. The duration of the pressure decrease is about 360 s, which may result in quasisteady flow conditions (this point is checked in [17]). During this process, the variable head loss device position is not modified, so the mass flow rate progressively decreases together with the inducer elevation. The present study focuses on the results obtained at nominal mass flow rate Q_n and reference rotation speed N_{ref}. However, data recorded at other rotation speeds are also used, because (i) visualizations have been performed at 0.6 N_{ref} and (ii) intensity of some fluctuations decreases with N, so they can be detected only at high values of N.

3. Methodology

Three methods have been applied for the analysis of the instabilities.

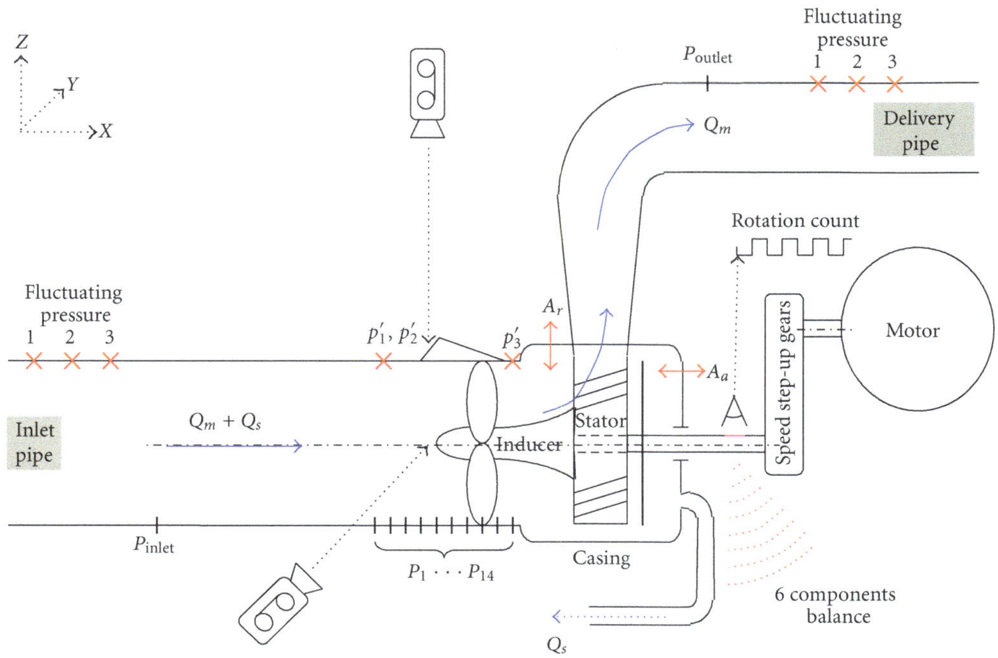

FIGURE 4: Scheme of the inducer test section including the acquisition equipment.

(i) Wavelet decomposition: this technique enables to identify on a signal the time of inception of the successive characteristic frequencies. It is based on a decomposition of the initial signal S into secondary signals S_i. Each signal S_i is obtained from a typical shape called "wavelet" and a discrete frequency corresponding to the period of the wavelet. For example, the secondary signal S_1 at frequency f_1 is composed of a succession of wavelets whose period is $1/f_1$, and magnitude varies according to the consistence of the initial signal with the shape of this wavelet. The residual, that is, S-S_1 is then decomposed into a secondary signal S_2 related to frequency f_2, and a new residual that will be decomposed again and so on. The initial signal is decomposed according to several discrete frequencies, until a residual of low magnitude is obtained. If high amplitude is observed locally on a secondary signal S_i, then it means that a fluctuation at frequency close to f_i is included in the initial signal. In the present study, decomposition into discrete wavelet (Daubechies no. 5) is applied to the pressure and force signals.

(ii) Study of the temporal signals: this method aims, through analysis of the force and pressure unsteady signals, to improve the understanding of the inception and development of the instabilities. The main objective consists in identifying transient unbalanced flow patterns, flow conditions at inception and/or vanishing of the instabilities, and phenomena related to the detected characteristic frequencies. The analysis focuses particularly on the phase difference between F_y and F_z, or P'_1 and P'_2 (mounted in the same cross-section with 90° difference between them). Attention is also paid to the mean values of F_y and F_z (the mean value of these two signals in noncavitating conditions has been set to zero).

In this analysis, low-pass filters have been applied in order to clearly distinguish some characteristic frequencies. Cut-off frequencies will be indicated systematically hereafter.

4. Wavelet Decomposition

The present inducer is characterized by a large number of nonsymmetrical flow patterns: so-called "synchronous" (unbalanced flow pattern rotating at inducer speed), "sub-synchronous," "super-synchronous" (unbalanced flow pattern rotating slower or faster than the inducer, resp.), and alternate (unbalanced flow pattern with identical opposite sheet cavities) regimes have been identified. An unknown fluctuation at frequency 0.9 f_{ref} in the absolute frame has been also detected.

These flow instabilities have been identified from the waterfall plots of F_y, F_z, P'_1, and P'_2, as indicated in Figure 5 for $N = N_{\text{ref}}$.

For example, super-synchronous rotating cavitation is clearly detected for $0.15 < \tau^+ < 0.22$. It is identified by the low frequency that is obtained in the rotating frame from the F_y signal, which is correlated with the peaks that can be seen from the analysis of P'_2, at frequency f_{RC} higher than the one of the inducer rotation. This pattern basically consists of large-scale fluctuations of the radial force components F_y and F_z (Figure 6), at frequency that progressively decreases from a few tens of Hertz down to zero when τ^+ is decreased. A 90° phase difference between F_y and F_z is observed, which suggests that these fluctuations may be associated to a rotating instability. Note that the values of the forces are not given in ordinate for confidentiality reasons: only the zero is indicated to facilitate the discussion.

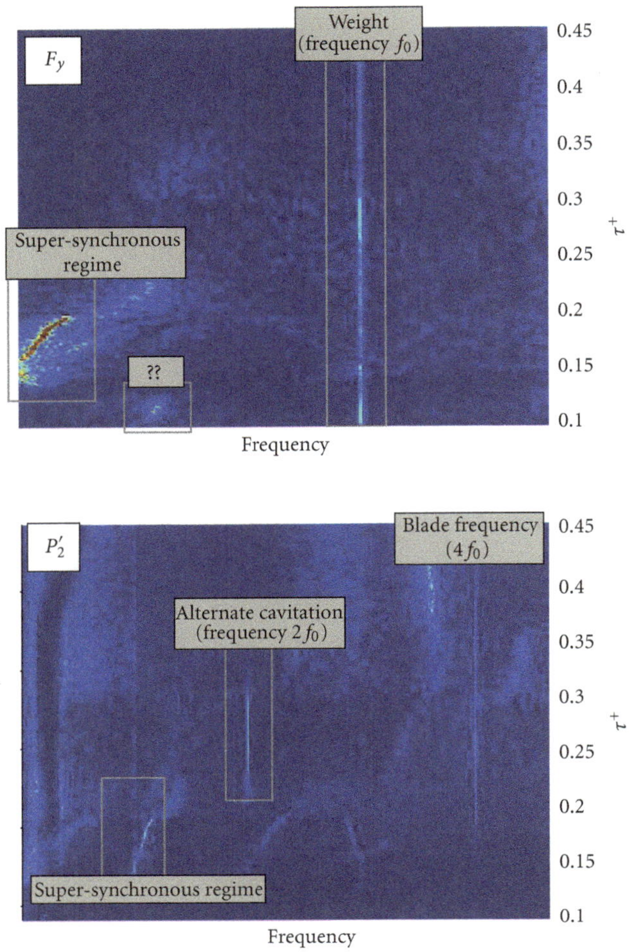

FIGURE 5: Waterfall plot for F_y and P_2' (Q_n, N_{ref}).

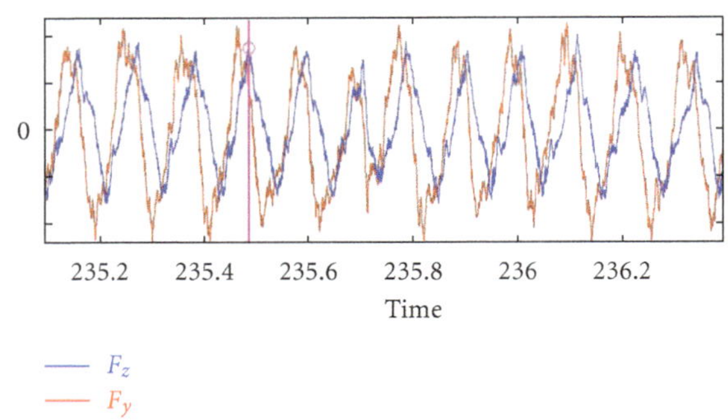

FIGURE 6: Time signals of F_y and F_z during the super-synchronous regime (Q_n, N_{ref}, cut-off frequency 400 Hz).

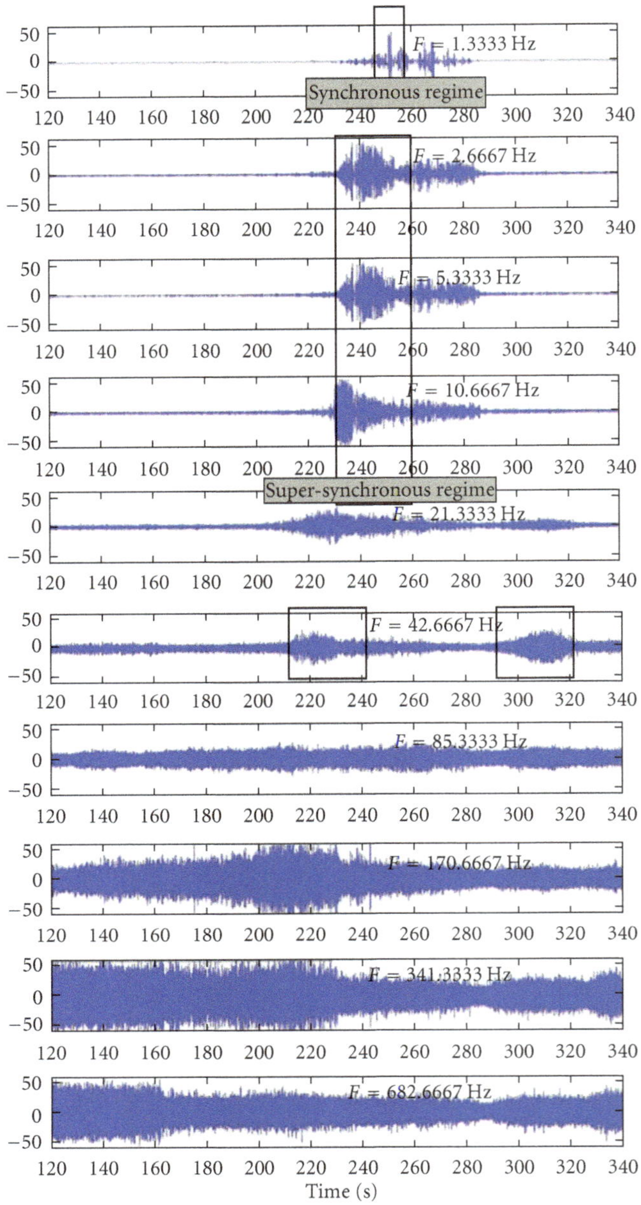

FIGURE 7: Wavelet decomposition of signal F_y.

For higher cavitation number ($0.22 < \tau^+ < 0.32$), a stable nonsymmetrical flow pattern is detected. It is mainly characterized by a clear frequency $2f_0$ on the spectrum analysis of P_1' and P_2' (see Figure 5), which is usually due to different sheet cavity sizes on successive blades, but identical sheet cavities on opposite blades. Since this arrangement is stable in the rotating frame, no frequency is detected in the waterfall plot of F_y or F_z.

In this section we focus on results at speed 1.2 N_{ref}, in order to have the clearest available fluctuations. Indeed, it has been shown by Coutier-Delgosha et al. (2009) that the rotation speed has some effects on the intensity of the radial loads, but does not modify the instability types.

The decomposition into discrete wavelet (Daubechies n°5) is applied to the F_y signal (Figure 7). Frequency f_1 is 682 Hz in the present case, and $f_{10} = 1.33$ Hz. The analysis is performed during the pressure decrease between $t = 120$ s and $t = 340$ s, that is from the inception of the first instability regime (alternate cavitation) to the performance breakdown. For high frequencies, high amplitude is nearly systematically obtained, whereas below 100 Hz, several points can be noticed.

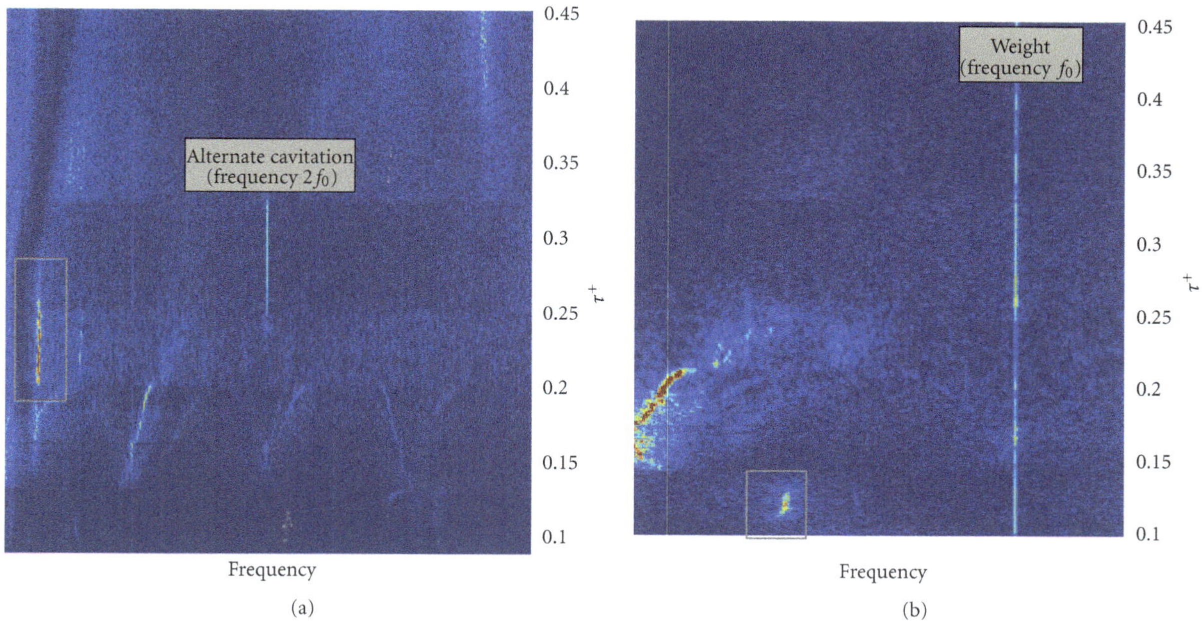

FIGURE 8: Waterfall plot of (a) P_2' and (b) F_y (Q_n, 1.2 N_{ref}).

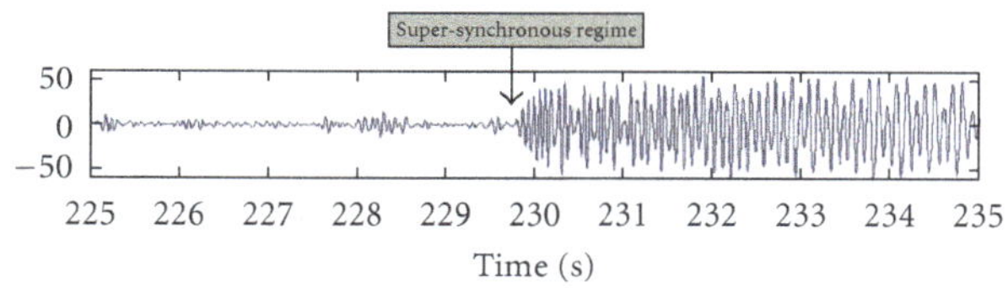

FIGURE 9: Wavelet decomposition for F_y. Zoom on the time interval $225\,\text{s} < t < 235\,\text{s}$ for $f = 10.66\,\text{Hz}$ (N_{ref}, Q_n).

(i) Close to 40 Hz, two sequences are characterized by a significant increase of the intensity: between $t = 210\,\text{s}$ and 230 s ($0.19 < \tau^+ < 0.23$), and between $t = 300\,\text{s}$ and 320 s ($0.075 < \tau^+ < 0.09$).

(ii) The inception of the super-synchronous instability can be observed clearly in the decomposition at frequency $f = 10.66\,\text{Hz}$, and also in the adjacent decompositions, with a lower intensity. It is observed that this characteristic frequency progressively decreases, until a short synchronous regime (which can be detected here at the lowest frequency) is obtained, for $253\,\text{s} < t < 255\,\text{s}$.

The same process is applied to the unsteady pressure P_2'. The main resulting information is a very clear unsteadiness characterized by a frequency close to 20/40 Hz for $210\,\text{s} < t < 230\,\text{s}$. Such frequency was already observed during the same time on the F_y signal, but not so clearly. Figure 7 shows the waterfall plot for P_2' at rotation speed 1.2 N_{ref}. A high-intensity signal at frequency close to 30 Hz is obtained, between the regime of alternate cavitation and the inception of the super-synchronous instability, that is, for $0.19 < \tau^+ < 0.23$. This corresponds to $210\,\text{s} < t < 230\,\text{s}$. It can be noticed that this frequency was much weaker at rotation speed N_{ref} (see Figure 5). Similarly, the second instability regime detected on the F_y wavelet decomposition (for $300\,\text{s} < t < 320\,\text{s}$) clearly appears on the waterfall plot at 1.2 N_{ref}, while it can hardly be detected at speed N_{ref} (cf. Figure 5).

It can be concluded that the wavelet decomposition does not enable any better understanding of the mechanisms related to the different instabilities. Conversely, it can be used to detect precisely their inception, as can be seen for example in Figure 9 in the case of the super-synchronous regime.

5. Analysis of the Temporal Signals

Several types of instabilities are encountered during the process of pressure decrease at 1.2 N_{ref} and Q_n. In the present paper, attention is focused on the regime of alternate cavitation.

This regime is obtained for $130\,\text{s} < t < 210\,\text{s}$, that is, for $0.23 < \tau^+ < 0.35$. It is characterized by sheet cavities of different sizes on two consecutive blades but identical on opposite ones. This nonsymmetrical arrangement results in a clear frequency $2f_0$ on the spectrum analysis of P_1' and P_2' (see Figure 8). Conversely, no evidence of this unbalanced flow pattern can be found in the waterfall plot of F_y or F_z, since it is stable in the inducer rotating frame.

Supplementary information can be obtained in the present case from the analysis of the temporal signals: if a low-pass filter with cut-off frequency 100 Hz is applied to the signals of F_y and F_z, in order to get rid of the high-frequency fluctuations, it becomes clear that mean values of F_y and F_z are significantly different during all the regime (Figure 10). Note that the values of the forces are not given in ordinate for

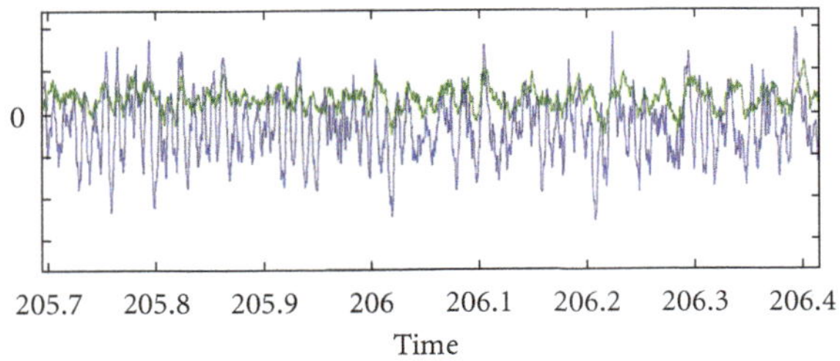

FIGURE 10: Temporal signals of F_y and F_z with low pass filter (cut-off frequency 100 Hz), Q_n, 1.2 N_{ref}.

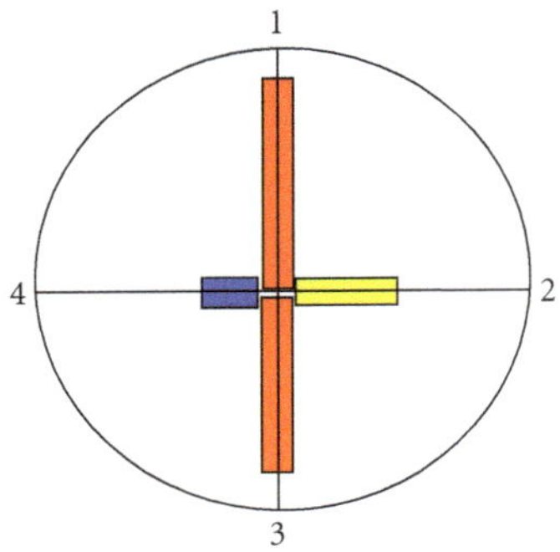

FIGURE 11: Plausible configuration for $0.23 < \tau^+ < 0.35$. Rectangles represent schematically the respective lengths of the sheet cavities on the four blades (1.2 N_{ref}, Q_n).

confidentiality reasons: only the zero is indicated to facilitate the discussion.

Although the values of the forces are not given in ordinate for confidentiality reasons, it is clear that F_y fluctuates around a negative average value $-F_{\text{mean}}$, while F_z is characterized by a positive mean value close to F_{mean}. It implies that the present nonsymmetrical flow pattern causes in addition a constant mean radial force, which does not rotate in the inducer rotating frame. It means that cavitation on opposite blades may not be fully identical. However, the $2f_0$ frequency on the pressure waterfall plots shows that two identical cavitation patterns separated by 180° are also present inside the inducer. The term "cavitation pattern" is voluntary used, since it may concern too sheet cavitation on the blade suction side as tip cavitation close to the inducer shroud.

The most plausible cavitation pattern is thus composed of two identical cavitation areas on two opposite blades (1 and 3), while the two other ones (2 and 4) are both of different size. In this configuration, a constant radial load is obtained in the inducer rotating frame because of the dissymmetry between blades 2 and 4. This load, represented schematically in Figure 11, may result (excepted in a particular orientation of axes y and z) in different mean values of the components F_y and F_z.

Another feature of the alternate cavitation regime can be remarked on the extended waterfall plots of F_y or F_z (see Figure 12). Frequency exactly equal to $3f_0$ is obtained for $0.23 < \tau^+ < 0.35$.

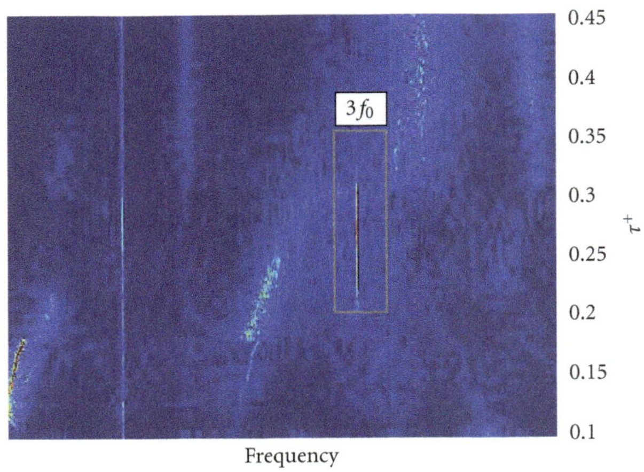

FIGURE 12: Extended waterfall plot of F_y (Q_n, N_{ref}).

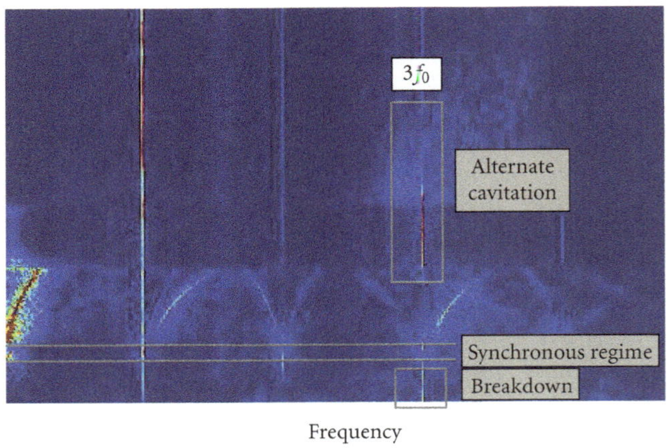

FIGURE 13: Extended waterfall plot of F_y (1.18 Q_n, N_{ref}).

The following points can be checked from the analysis of the waterfall plots obtained at various rotation speeds and mass flow rates.

(i) It is a $3f_0$ frequency, which changes with the inducer rotation speed.

(ii) It can be detected mainly above the nominal flow rate, that is, within the range of mass flow rate that leads to most of the regimes of radial loads. It is particularly strong at high mass flow rate (see Figure 13).

(iii) It is not specific to the alternate cavitation flow pattern: all regimes of developed and stable (i.e. nonrotating) cavitation result in the occurrence of the $3f_0$ frequency. It is observed for example in the case presented in Figure 13 (N_{ref}, 1.18 Q_n) that this frequency is obtained during three successive periods: (1) alternate cavitation, (2) short synchronous regime that occurs just after the super-synchronous one, and (3) regime of highly developed cavitation during the final performance breakdown.

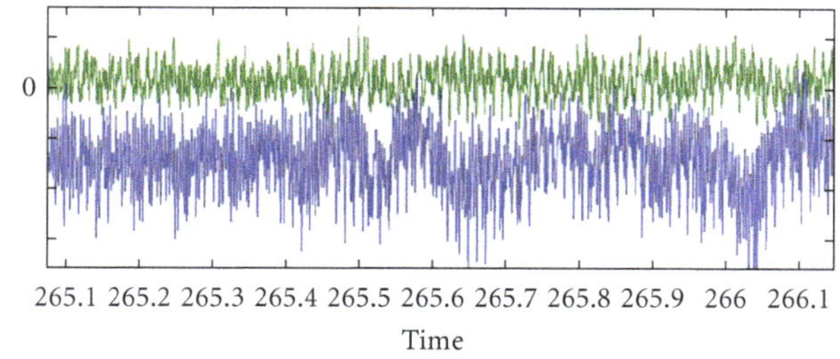

FIGURE 14: Temporal signals of F_y (in blue) and F_z (in green) (N_{ref}, 1.18 Q_n).

(iv) The $3f_0$ frequency can be detected even in noncavitating conditions, but strongly attenuated.

From these observations, no obvious explanation can be found regarding the occurrence of the $3f_0$ frequency. Conversely, some points enable to improve the understanding of this phenomenon. For example, no peak at frequency $3f_0$ is observed on the waterfall plots of the radial load intensity Fr, whatever the flow configuration is. It means that the radial load does not fluctuate; only its components do. Besides, the $3f_0$ frequency is clearly detected on the waterfall plots of the bending moments M_y and M_z, which suggests that this instability is related to a shaft bend stress. On the other hand, the frequency peak is very small on the waterfall plots of the axial load F_x, which eliminates any direct connection with an axial effort on the shaft.

Three different hypotheses can be postulated.

(i) A possible explanation is a rotating perturbation that would rotate at frequency $4f_0$ in the reference steady frame, which would generate a frequency $3f_0$ in the inducer rotating frame. Such phenomenon has been observed in experiments conducted in the Osaka University with a three-blade inducer, resulting in a $4f_0$ frequency in the inducer frame [10]. It is called high-order rotating cavitation by Tsujimoto and colleagues. Although the present detected frequency is not the same, this example shows that the hypothesis of a perturbation with high rotation speed is plausible. However, it is not the most probable one, because of the exact value $3f_0$ that is obtained.

(ii) It can be the result of a rotor/stator interaction at the inducer outlet. A method proposed by Brennen [18] enables to obtain the characteristic frequencies related to such interaction. In the present case, the stator in the discharge pipe is equipped with 13 blades, the inducer has 4 blades, so frequencies f_0, $2f_0$, $3f_0$, and $4f_0$ should appear on the plots with similar intensities. So, frequency $3f_0$ should not be predominant. Still it must be noticed that a frequency range 250 Hz–300 Hz has been previously identified as resonant frequencies of the installation. The $3f_0$ peak may thus be of higher intensity than the other ones if it belongs to this range. However this hypothesis is also rejected, because the $3f_0$ peak is of high intensity for all rotation speeds, from 0.6 N_{ref} up to 1.2 N_{ref}, while the value $3f_0$ is not systematically included into the range of resonant frequencies.

(iii) It can be related to a fluctuation of the radial load angle during the inducer rotation. More precisely, it can be anticipated that several fluctuations occurring during each inducer revolution may result in the measurement of the frequency $3f_0$. To investigate this point, a more in-depth analysis of the temporal signals of F_y and F_z has been conducted during the alternate cavitation regime for flow conditions N_{ref} and 1.18 Q_n, where frequency $3f_0$ is remarkably clear on the waterfall plots. Figure 14 presents the general aspect of the two signals (F_y in grey F_z in black), which confirms the difference of mean level that was mentioned previously.

A zoom on a shorter time period is performed in Figure 15(a), to obtain only 25 rotations of the inducer. Oscillations at frequency f_0 can be clearly observed on both signals. Two secondary oscillations of lower amplitude are also detected on the F_y signal, which may explain the presence of frequency $3f_0$ in the spectrum. In the case of F_z, these secondary oscillations do not appear so clearly. A second zoom is performed to focus on six rotations of the inducer only (see Figure 15(b)). Regular oscillations at frequencies f_0 and $3f_0$ are observed: the two secondary peaks are regularly positioned between the main oscillations. On this figure, secondary peaks can be also seen in the signal of component F_z (Figure 15(c)). It can be also noticed that the phase difference between F_y and F_z varies.

(i) The phase difference at low frequency f_0, which corresponds to the gap between the vertical full lines in Figures 15(b) and 15(c), is small. It equals 90°, since it is related to the inducer weight.

(ii) The phase difference between the secondary peaks is not the same. It can be estimated that it is close to 180°. It means that the phenomenon that generates the secondary oscillations results in fluctuations of F_y and F_z characterized by phase opposition. It must be mentioned that the analysis of various periods has enabled to draw this conclusion, while interpretation of Figures 15(b) and 15(c) is not so clear. On the basis of these observations, no definitive interpretation can be found for frequency $3f_0$.

However, a possible explanation, which could be validated only from high-speed visualizations, is given hereafter. It is based on a probable oscillation of the two sheet cavities of different sizes no. 2 and no. 4 denoted in Figure 11. These two areas may fluctuate around a mean size, and even reverse (i) because of the vertical pressure gradient due to the gravity within the inducer. The Froude number based on the inducer diameter and the axial velocity is about 0.2, which confirms that the gravity may have some influence on the flow during its passage in the machinery; (ii) or by effect of the two-phase structures which may be different at the top and at the

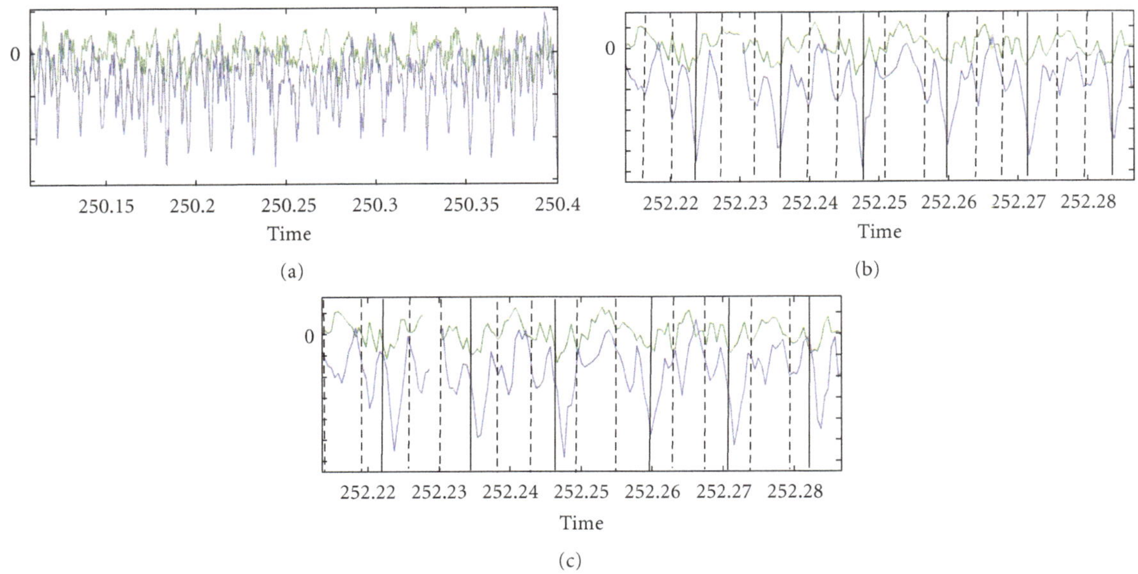

FIGURE 15: Zoom on the signals of F_y (in blue) and F_z (in green), 1.18 Q_n, N_{ref}, and low pass filter with cut-off frequency 400 Hz.

bottom of the inducer, which would be also an indirect effect of the gravity.

In this case, the oscillation of components F_y and F_z at frequency f_0, due to the inducer weight, may be coupled to this inversion of sheet cavities 2 and 4, which would occur twice during each rotation of the inducer, leading to a more complex signal. A scheme of the evolutions of F_y and F_z during a single rotation of the inducer is proposed in Figure 17. It is supposed here that the radial load related to the nonsymmetry between blades 2 and 4 is oriented according to the scheme given in Figure 16.

This particular orientation of F_y and F_z enables to explain the 180° phase difference between the two components in the secondary oscillations. Indeed, in such configuration, the inversion of the sheet cavities 2 and 4, which makes an inversion of the associated radial load, results in opposite variations of F_y and F_z. It must be noticed that the blade orientations drawn in Figure 17 represent schematically the direction of the radial load generated by nonidentical sheet cavities on opposite blades. It can be anticipated that these directions are close to the lines that would join the leading edge of the opposite blades.

It is observed in Figure 15 that the oscillations of F_y are of larger amplitude than those of F_z, which suggests that the angle between each component and the blades is not 45°. As indicated in Figure 16, the direction of F_y may be close to the direction of the radial load related to the nonsymmetrical flow pattern on blades 2 and 4. The expected evolutions of F_y and F_z during a single inducer rotation are drawn in Figure 17 on the basis of this hypothesis.

The eight schemes at the top of Figure 16 show the eight successive positions of sheet cavities on blades 2 and 4, during a revolution of the machinery, by step of 45°. It can be observed that two inversions of the small and big sheet

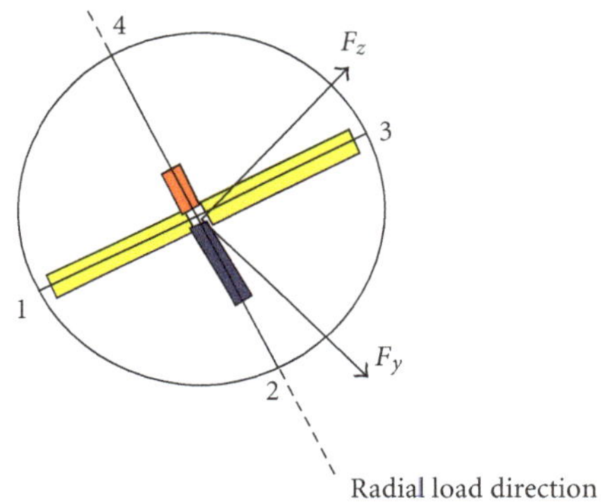

FIGURE 16: Expected orientation of F_y and F_z, and the blades (1.18 Q_n, N_{ref}).

cavities occur during one period, each time they cross the horizontal direction. Between two inversions, the difference of size between sheet cavities 2 and 4 first increases reaching a maximal value, and then decreases. It is maximal when the radial load is vertical, since in such configuration the pressure gradient, which controls the pulsation of the cavities according to our assumption, has the strongest effect on the sheet cavities located on blades 2 and 4.

On the central part of Figure 17, the curves in full line correspond to the evolutions of F_y and F_z without the radial load generated by the unbalanced flow pattern on blades 2 and 4: a sinusoidal evolution; due to the respective orientations of the gravity, F_y, and F_z, with a $\pi/2$ phase difference between the load components, is obtained in that case. This behaviour is modified by the fluctuations of the sheet cavities on blades 2 and 4, leading to the curves drawn in doted lines.

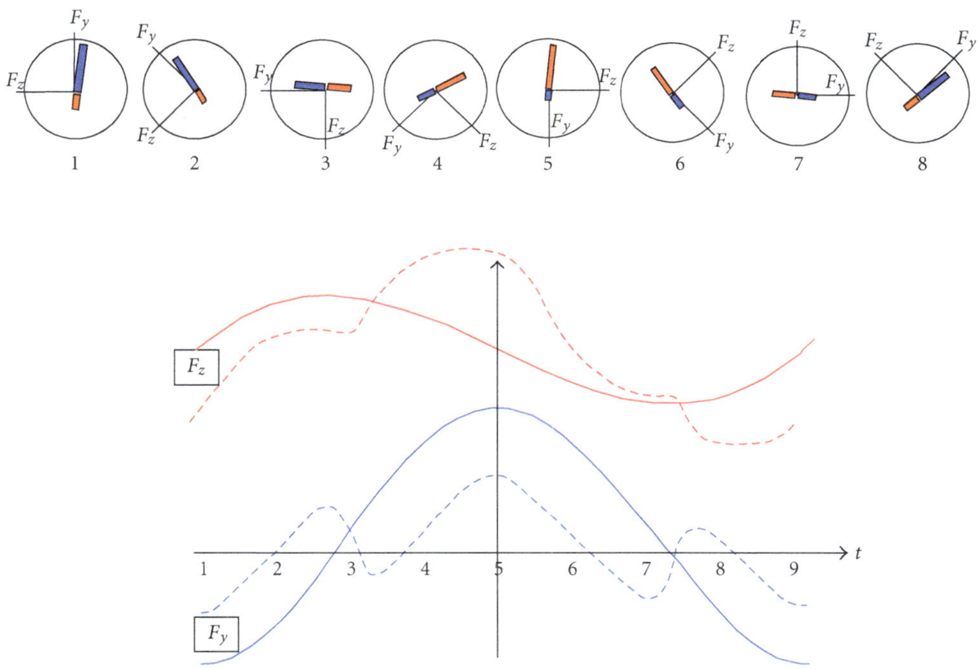

FIGURE 17: Scheme of the qualitative evolutions of F_y (blue curve) and F_z (red curve) during a single rotation of the inducer.

The evolution of F_y is composed of the following steps.

(i) It can be anticipated that the radial load will systematically be oriented from the small cavity towards the big one (i.e., from high-pressure areas towards low pressures). So, configuration 1 in Figure 17 is characterized by a component F_y related to the flow dissymmetry that is positive, and even almost maximal, since blades 2 and 4 are close to the vertical direction. Between steps 1 and 3, the radial load reaches its maximal value, and then it decreases, so the curves in full and dotted lines become closer. It must be reminded that all charts here focus on a qualitative behaviour, while the real evolutions of F_y and F_z may depend strongly on the relative values of the radial load related to the inducer weight and the one related to nonsymmetrical cavitation on blades 2 and 4.

(ii) Just after configuration 3, blades 2 and 4 cross the horizontal direction, so an inversion of the sheet cavity lengths is obtained. Consequently, the radial load due to the dissymmetry, and thus the sign of the F_y component related to this load, is also inverted. It results in a sudden decrease of F_y, under the curve in full line. Then, F_y reincreases progressively until step 5 is reached, since its direction becomes closer to the one of the gravity. In the same time, the radial load due to the dissymmetry also increases, but it is negative, so it slightly counterbalances the effect of the inducer weight.

(iii) Steps 6 to 8 lead to the same variations of F_y, but in the opposite direction: between steps 5 and 6, components related to the inducer weight and to the dissymmetry are both decreased. When the leading edges of blades 2 and 4 cross the horizontal direction (just after configuration 7), sheet cavities invert again, so the radial load that they generate also does, and F_y increases over the full line curve. Then, the effect of the weight decreases, leading to the decrease of F_y.

The curve for F_z is obtained by a similar process: from steps 1 to 3, the radial load due to sheet cavities on blades 2 and 4 is negative and decreases. After configuration 3, it inverts, so it becomes positive and increases more and more when the sheet cavities get closer to the vertical direction. After configuration 5, the radial load decreases again, and F_z also does and comes closer to the curve in full line. Just after step 7, sheet cavities invert again, and the F_z component related to the dissymmetry becomes negative and decreases down to its value at step 1.

This scenario generates for both components F_y and F_z three successive fluctuations during each rotation of the inducer, which is consistent with frequency $3f_0$ that is observed on the waterfall plots. In the configuration presented in Figure 17, the three pulsations of F_y occur at equal distance to each other, which explains the intense peak at frequency $3f_0$. Oscillations of component F_z are not so regular during one rotation, which is consistent with observations from Figure 15, and which also explains that the spectral line at $3f_0$ is weaker for F_z than for F_y.

6. Conclusion

The instabilities detected in a four-blade inducer have been discussed in the present paper by wavelet decomposition and analysis of the temporal signals of the two components F_y and F_z of the radial load on the shaft. It has been found that wavelet decomposition does not improve the understanding of the flow instabilities, while it enables a precise detection of inception and stop of each unbalanced flow pattern. An unexpected spectral line at $3f_0$, which occurs during alternate cavitation, has been discussed by direct analysis of the temporal signals of F_y and F_z. A possible scenario, based on the assumption that the gravity may have—directly or indirectly—some effect on the sheet cavities on the blades, is proposed. Only high-speed video would enable to confirm this hypothesis.

Nomenclature

f: Frequency (Hz)
f_0: Frequency of the inducer rotation (Hz)
F_x: Axial component of the force on the shaft (N)
F_y, F_z: Radial components of the force on the shaft (N)
N: Rotation speed (s^{-1})
N_{ref}: Reference rotation speed (s^{-1})
ΔP: Pressure elevation $P_d - P_i$ (Pa)
P_i: Absolute pressure in the inlet pipe (Pa)
P_d: Absolute pressure in the delivery pipe (Pa)
Pt_i: Total absolute pressure in the inlet pipe (Pa)
Pt_d: Total absolute pressure in the delivery pipe (Pa)
P_{vap}: Vapor pressure (Pa)
P'_1, P'_2: Fluctuating pressures at the inducer inlet (Pa)
P'_3: Fluctuating pressure at the inducer outlet (Pa)
M_x: Torque on the inducer shaft (Nm)
$M_{y,z}$: Bending moments on the shaft (Nm)
Q: Volume flow rate (m^3/s)
Q_n: Nominal flow rate (m^3/s)
Q_m: Main flow rate (in the delivery pipe) (m^3/s)
Q_s: Secondary flow rate (for axial equilibrium) (m^3/s)
r: Inducer tip radius (m)
t: Time (s)
ν: Outlet ratio hub radius/tip radius (—)
Ψ: Head coefficient $(Pt_d - Pt_i)/(\rho\ \omega^2\ r^2)$ (—)
χ: Torque coefficient $M_x/(\rho\ \omega^2\ r^5)$ (—)
ϕ: Flow rate coefficient $Q/(\pi\ \omega\ r^3\ (1 - \nu^2))$ (—)
τ: Cavitation number $(Pt_i - P_{\text{vap}})/(\rho\ \omega^2\ r^2)$ (—).

Acknowledgments

The present study was performed in the frame of contractual activity with SNECMA Moteurs and the CNES (French Space Agency). The technical staff of the LML laboratory was much involved in the development and the operation of the test facility. The authors wish to thank especially J. Choquet and P. Olivier for their collaboration.

References

[1] K. Kamijo, T. Shimura, and M. Watanabe, "An experimental investigation of cavitating inducer Instability," ASME Paper 77-WA/FW-14, 1977.

[2] B. Goirand, A. Mertz, F. Joussellin, and C. Rebattet, "Experimental investigation of radial loads induced by partial cavitation with liquid hydrogen inducer," in *Proceedings of the 3rd International Conference on Cavitation (ImechE '92)*, vol. C453/056, pp. 263–269, Cambridge, UK, 1992.

[3] J. de Bernardi, F. Rossellini, and A. Von Kaenel, "Experimental analysis of instabilities related to cavitation in turbopump inducer," in *Proceedings of the 1st International Symposium on Pump Noise and Vibrations*, pp. 91–99, Paris, France, 1993.

[4] R. A. Furness and S. P. Hutton, "Experimental and theoretical studies of two-dimensional fixed-type cavities," vol. 97, pp. 515–522, 1975.

[5] Q. Le, J. P. Franc, and J. M. Michel, "Partial cavities: global behavior and mean pressure distribution," *ASME Transactions Journal of Fluids Engineering*, vol. 115, no. 2, pp. 243–248, 1993.

[6] J. B. Leroux, O. Coutier-Delgosha, and J. A. Astolfi, "A joint experimental and numerical study of mechanisms associated to instability of partial cavitation on two-dimensional hydrofoil," *Physics of Fluids*, vol. 17, no. 5, article 052101, pp. 1–20, 2005.

[7] O. Coutier-Delgosha, B. Stutz, A. Vabre, and S. Legoupil, "Analysis of cavitating flow structure by experimental and numerical investigations," *Journal of Fluid Mechanics*, vol. 578, pp. 171–222, 2007.

[8] Y. Tsujimoto, "Cavitation instabilities in inducers," Tech. Rep. AVT-143 RTO AVT/VKI Lecture Series, von Karman Institute, Rhode-Saint-Genèse, Belgium, 2006.

[9] F. Joussellin, J. De Bernardi, B. Goirand, and Y. Delannoy, "Analyse par films ultrarapides de poches de cavitation sur l'Inducteur de la Turbopompe à Hydrogène d'un Moteur Fusée," in *5ème colloque de visualisation et de traitement d'images en Mécanique des Fluides*, Poitiers, France, 1992.

[10] A. Fujii, S. Azuma, Y. Yoshida, Y. Tsujimoto, H. Horiguchi, and S. Watanabe, "Higher order rotating cavitation in an inducer," *International Journal of Rotating Machinery*, vol. 10, no. 4, pp. 241–251, 2004.

[11] O. Coutier-Delgosha, Y. Courtot, F. Joussellin, and J. L. Reboud, "Numerical simulation of the unsteady cavitation behavior of an inducer blade cascade," *AIAA Journal*, vol. 42, no. 3, pp. 560–569, 2004.

[12] Y. Yoshida, Y. Sasao, K. Okita, S. Hasegawa, M. Shimagaki, and T. Ikohagi, "Influence of thermodynamic effect on synchronous rotating cavitation," *ASME Transactions Journal of Fluids Engineering*, vol. 129, no. 7, pp. 871–876, 2007.

[13] F. Bakir, S. Kouidri, R. Noguera, and R. Rey, "Experimental analysis of an axial inducer influence of the shape of the blade leading edge on the performances in cavitating regime," *ASME Transactions Journal of Fluids Engineering*, vol. 125, no. 2, pp. 293–301, 2003.

[14] Y. Yoshida, Y. Tsujimoto, D. Kataoka, H. Horiguchi, and F. Wahl, "Effects of alternate leading edge cutback on unsteady cavitation in 4-bladed inducers," *ASME Transactions Journal of Fluids Engineering*, vol. 123, no. 4, pp. 762–770, 2001.

[15] Y. D. Choi, J. Kurokawa, and H. Imamura, "Suppression of cavitation in inducers by J-Grooves," *ASME Transactions Journal of Fluids Engineering*, vol. 129, no. 1, pp. 15–22, 2007.

[16] H. Horiguchi, T. Takashina, and Y. Tsujimoto, "Theoretical analysis of cavitation in inducers with unequally spaced blades," *JSME International Journal, Series B*, vol. 49, no. 2, pp. 473–481, 2006.

[17] O. Coutier-Delgosha, G. Caignaert, G. Bois, and J. B. Leroux, "Influence of the blade number on inducer cavitating behavior," *Journal of Fluids Engineering*. In press.

[18] C. E. Brennen, "Pump vibration, concepts," in *Hydrodynamics of Pumps*, chapter 8, NREC and Oxford University Press, 1994.

10

Numerical Modeling of Unsteady Cavitating Flows around a Stationary Hydrofoil

Antoine Ducoin,[1] Biao Huang,[2] and Yin Lu Young[1]

[1] *Department of Naval Architecture and Marine Engineering, University of Michigan, 2600 Draper Drive, Ann Arbor, MI 48109, USA*
[2] *Department of Vehicle Engineering, School of Mechanical Engineering, Beijing Institute of Technology, Beijing 100081, China*

Correspondence should be addressed to Yin Lu Young, ylyoung@umich.edu

Academic Editor: Moustafa Abdel-Maksoud

The objective of this paper is to evaluate the predictive capability of three popular transport equation-based cavitation models for the simulations of partial sheet cavitation and unsteady sheet/cloud cavitating flows around a stationary NACA66 hydrofoil. The 2D calculations are performed by solving the Reynolds-averaged Navier-Stokes equation using the CFD solver CFX with the k-ω SST turbulence model. The local compressibility effect is considered using a local density correction for the turbulent eddy viscosity. The calculations are validated with experiments conducted in a cavitation tunnel at the French Naval Academy. The hydrofoil has a fixed angle of attack of $\alpha = 6^\circ$ with a Reynolds number of Re = 750,000 at different cavitation numbers σ. Without the density modification, over-prediction of the turbulent viscosity near the cavity closure reduces the cavity length and modifies the cavity shedding characteristics. The results show that it is important to capture both the mean and fluctuating values of the hydrodynamic coefficients because (1) the high amplitude of the fluctuations is critical to capturing the extremes of the loads to ensure structural safety and (2) the need to capture the frequency of the fluctuations, to avoid unwanted noise, vibrations, and accelerated fatigue issues.

1. Introduction

In liquid flows, cavitation occurs if the local pressure drops below the saturated vapor pressure, which leads to the formation of vaporous bubble(s) to relieve the negative pressure [1]. Cavitation is commonly observed when a hydraulic machine operates in high speeds or under off-design conditions, and is still a limiting factor in its design and operation. The occurrence of unsteady cavitation in marine propulsion devices such as hydrofoils, propellers, and waterjets can lead to problems such as pressure pulsations, sudden changes in loads, vibrations, noise, and erosion. Moreover, unsteady cavitation is usually the primary physical phenomenon behind performance alterations in a hydraulic machinery [2–4]. For this reason, it is crucial to be able to accurately predict the development and evolution of cavitation, and the resultant impact on the performance.

Phenomenologically, cavitation often involves complex interactions between turbulent flow structures and phase-change dynamics with large variations in fluid density and pressure fluctuations [5]. These physical mechanisms are still not well understood because of the complex, multiscale, multiphase phenomenon. Moreover, there are significant computational challenges associated with the accuracy, stability, efficiency, and robustness of numerical algorithms for the simulation of unsteady, turbulent cavitating flows.

In the numerical modeling of cavitating flows, the selection of cavitation models plays a major role on the prediction of the onset, growth, breakup, and collapse of cavitation bubbles. In recent years, significant efforts have been made in the development of cavitation models; examples of recent review articles can be found in [6–8]. Most cavitation models assume the flow to be homogenous and isothermal and apply either a barotropic equation of state or a transport equation to solve for the variation of the mixture density in the multiphase flow.

In barotropic cavitation models, the local mixture density (ρ_m) is assumed to depend only on the local pressure: $\rho_m = f(p)$. Examples of barotropic cavitation models include those proposed by [9, 10]. A recent experimental finding by [11] has shown that vorticity production is an important aspect of cavitating flows, especially in the cavity closure region. Specifically, this vorticity production is a consequence of the baroclinic torque:

$$\nabla \frac{1}{\rho_m} \times \nabla p. \quad (1)$$

Clearly, if a barotropic equation of state, $\rho_m = f(p)$, is used, the gradients of density and pressure are always parallel, which leads to zero baroclinic torque. Hence, barotropic cavitation models will not be able to properly predict the dynamics of cavitating flows [12], particularly for cases with unsteady cavitation because of the importance of cavitation in vorticity production.

The other popular approach to simulate cavitating flows is transport equation models (TEMs), which solves an additional transport equation for either the mass or volume fraction of vapor, with appropriate source/sink term(s) to regulate the mass transfer between the liquid and vapor phases. Different source/sink terms have been proposed, including the popular models presented in [6, 13–15]. Kubota et al. [13] coupled the Rayleigh-Plesset equation to the URANS equations to solve for the local void fraction based on an assumed bubble radius, and hence it is appropriate for modeling of unsteady cavitation. Merkle et al. [16] and Kunz et al. [14] employed the artificial compressibility method with a special preconditioning formulation to solve the multiphase URANS equations, comprised of the mixture volume, mixture momentum, and constituent volume fraction equations. Singhal et al. [15] developed a "full-cavitation model" based on the phase-change rate derived from a reduced form of the Rayleigh-Plesset equation for bubble dynamics and local flow conditions. The apparent advantage of the Singhal model comes from the convective character of the equation. Senocak and Shyy [17] compared different TEM models to develop an interfacial dynamics-based cavitation model (IDM), which allows direct interpretation of the empirical parameters in existing transport equation-based models. In the TEM approach, the mixture density is a function of the transport process. Consequently, gradients of the density and the pressure are not necessarily parallel, suggesting that TEM can accommodate baroclinic vorticity generation.

The objective of this study is to evaluate the predictive capability of three popular transport-based cavitation models for the simulation of quasisteady and unsteady cavitating flow around a stationary hydrofoil. The aim is to improve the understanding of the influence of multiphase flow on the turbulent closure model, and the interplay between vorticity fields, cavity dynamics, and hydrodynamic performance. The numerical models are first presented. Three popular transport-based cavitation models (named herein as the Kubota model, the Merkle model, and the Singhal model for simplicity) and the k-ω SST turbulence model are summarized, followed by presentation of the density correction model for the turbulent viscosity modification. The foil geometry, fluid mesh, and boundary conditions are presented, followed by a brief summary of the experimental setup of [18] used for validation of the computations. The results are then analyzed for an NACA 66 hydrofoil at a fixed angle of incidence ($\alpha = 6°$) and Reynolds number (Re = 750,000) for a case with quasisteady partial sheet cavitation ($\sigma = 1.49$), and a case with unsteady sheet/cloud cavitation ($\sigma = 1.00$). To study the influence of cavitation model and its interaction with turbulence modeling, results are compared for three different cavitation models and for different turbulent viscosity modifications. Finally, the variation of the predicted cavity length, lift and drag coefficients over a range cavitation numbers are compared with experimental measurements.

2. Transport-Based Cavitation Models

2.1. Conservation of Mass and Momentum. The URANS equations, in their conservative form, for a Newtonian fluid without body forces and heat transfers are presented below along with the mass transport equation in the Cartesian coordinates [19]:

$$\frac{\partial \rho_m}{\partial t} + \frac{\partial (\rho_m u_j)}{\partial x_j} = 0, \quad (2)$$

$$\frac{\partial (\rho_m u_i)}{\partial t} + \frac{\partial (\rho_m u_i u_j)}{\partial x_j} = -\frac{\partial p}{\partial x_i} + \frac{\partial}{\partial x_j}\left[(\mu_m + \mu_T)\left(\frac{\partial u_i}{\partial x_j} + \frac{\partial u_j}{\partial x_i} - \frac{2}{3}\frac{\partial u_k}{\partial x_k}\delta_{ij}\right)\right], \quad (3)$$

$$\frac{\partial \rho_l \alpha_l}{\partial t} + \frac{\partial (\rho_l \alpha_l u_j)}{\partial x_j} = \dot{m}^+ + \dot{m}^-, \quad (4)$$

$$\rho_m = \rho_l \alpha_l + \rho_v \alpha_v, \quad (5)$$

$$\mu_m = \mu_l \alpha_l + \mu_v \alpha_v, \quad (6)$$

where ρ_m is the mixture density, ρ_l is the liquid density, ρ_v is the vapor density, α_v is the vapor fraction, α_l is the liquid fraction, u is the velocity, p is the pressure, μ_m is the mixture laminar viscosity, μ_l and μ_v are, respectively, the liquid and vapor dynamic viscosities, and μ_T is the turbulent viscosity. The subscripts (i, j, and k) denote the directions of the Cartesian coordinates. The source term $\dot{m}^+$ and the sink term $\dot{m}^-$ in (4) represent the condensation and evaporation rates, respectively.

2.2. Kubota Model. The Kubota model [13] assumes a constant nuclei density in the fluid domain. The growth and collapse of the bubble clusters are governed by the simplified Rayleigh-Plesset equation for single-bubble dynamics [20]. The cavitation process is governed by the mass transfer equation given in (4), and the source and sink terms are defined as follows:

$$\dot{m}^{-} = -C_{k_{\rm dest}} \frac{3\alpha_{\rm nuc}(1-\alpha_v)\rho_v}{R_B}\left(\frac{2}{3}\frac{p_v - p}{\rho_l}\right)^{1/2}, \quad p < p_v, \quad (7)$$

$$\dot{m}^{+} = C_{k\_{\rm prod}} \frac{3\alpha_v\rho_v}{R_B}\left(\frac{2}{3}\frac{p - p_v}{\rho_l}\right)^{1/2}, \quad p > p_v, \quad (8)$$

where $\alpha_{\rm nuc}$ is the nucleation volume fraction, R_B is the bubble diameter, p_v is the saturated liquid vapor pressure, and p is the local fluid pressure. $C_{k\_{\rm dest}}$ is the rate constant for vapor generated from the liquid in a region where the local pressure is less than the vapor pressure. Conversely, $C_{k\_{\rm prod}}$ is the rate constant for reconversion of vapor back into liquid in regions where the local pressure exceeds the vapor pressure. As shown in (7) and (8), both the evaporation and condensation terms are assumed to be proportional to the square root of the difference between the local pressure and vapor pressure because the Kubota model was derived by assuming that the bubble dynamics are governed by the simplified Rayleigh-Plesset equation [20]. In this work, assumed values for the model constants are $\alpha_{\rm nuc} = 5 \times 10^{-4}$, $R_B = 1 \times 10^{-6}$ m, $C_{k\_{\rm dest}} = 50$, and $C_{k\_{\rm prod}} = 0.01$, which are the default values in CFX [5] and are used because of their supposed general applicability.

It should be noted that (7) and (8) are different from the original model proposed by [13], and instead follow the form presented in [21]. Since the evaporation rate is much higher than the condensation rate, different coefficients are needed for condensation and evaporation terms, as shown in (7) and (8). In addition, α_v is replaced by $\alpha_{\rm nuc}\ (1 - \alpha_v)$ in the vaporization term to account for the decrease of the nucleation site density with the increase of the vapor volume fraction.

2.3. Merkle Model. Several researchers have adopted the Merkle model proposed by [16] (e.g., see [12]), which has been presented in both the volume fraction form and the mass fraction form. It was derived primarily based on dimensional arguments for large-bubble clusters instead of individual bubbles. Consequently, the source and sink terms for the Merkle model shown in (9) are directly related to the pressure difference, $p - p_v$, instead of the square root of the pressure difference as in the Kubota model:

$$\dot{m}^{-} = \frac{C_{m\_{\rm dest}}\rho_l\rho_l {\rm MIN}(p - p_v, 0)(1-\alpha_v)}{(0.50\rho_l U_\infty^2)\rho_v\, t_\infty}, \quad p < p_v,$$

$$\dot{m}^{+} = \frac{C_{m\_{\rm prod}}\rho_l {\rm MAX}(p - p_v, 0)\alpha_v}{(0.50\rho_l U_\infty^2)t_\infty}, \quad p > p_v. \quad (9)$$

In this work, the empirical factors are set to be $C_{m\_{\rm dest}} = 1$ and $C_{m\_{\rm prod}} = 80$, which follows the constants used by Senocak and Shyy [17] for simulation of quasisteady cavitation around a NACA 66(MOD) hydrofoil. The mean flow time scale is defined as $t_\infty = c/U_\infty$ [12, 17].

2.4. Singhal Model. The Singhal mass transfer model proposed by [15] is originally known as the full-cavitation model. This model is also based on the reduced form of the Rayleigh-Plesset equation for bubble dynamics, but empirical constants have been added to take into account the interaction between water and vapor phases. In the derivation of the Singhal model, the bubble size is determined by the surface tension force, and the characteristic velocity, which reflects the effect of the local relative velocity between the liquid and vapor phases. Similar to the Kubota model, both evaporation and condensation terms are assumed to be proportional to the square root of the difference between the local pressure and vapor pressure. The Singhal model assumes that the mass transfer rate is related to the local turbulence kinetic energy, k. Following [8], the equations below represent a modified form of the original Singhal model, which consider the effects of noncondensable gas. The evaporation and condensation terms are defined as follows:

$$\dot{m}^{-} = C_{s\_{\rm dest}}\frac{\sqrt{k}}{\gamma}\rho_l\rho_v\left[\frac{2}{3}\frac{p_v - p}{\rho_l}\right]^{1/2}\left(1 - \frac{\alpha_v\rho_v}{\rho_m}\right), \quad p < p_v,$$

$$\dot{m}^{+} = C_{s\_{\rm prod}}\frac{\sqrt{k}}{\gamma}\rho_l\rho_l\left[\frac{2}{3}\frac{p - p_v}{\rho_l}\right]^{1/2}\frac{\alpha_v\rho_v}{\rho_m}, \quad p > p_v. \quad (10)$$

In this work, the empirical factors, $C_{s\_{\rm dest}}$ and $C_{s\_{\rm prod}}$ (resp., the destruction and the production coefficients for the Singhal model), have the following values: $C_{s\_{\rm dest}} = 0.02$ m/s and $C_{s\_{\rm prod}} = 0.01$ m/s, and the surface tension coefficient is assumed to be $\gamma = 0.0717$ N/m. f_v is the vapor mass fraction, defined as $f_v = \alpha_v/\rho_m$. It should be noted that in the present form of the mass transfer equations, the empirical factors $C_{s\_{\rm dest}}$ and $C_{s\_{\rm prod}}$ have to be expressed with m/s units to be consistent with the dimensions of $\dot{m}^+$and $\dot{m}^-$ in (4), which is in units of kg/(m^{3}s^1).

2.5. Turbulence Model and Local Compressibility Correction. The numerical results shown in this paper are performed using the commercial CFD code, CFX, to solve the URANS equations. The k-ω SST (shear stress transport) turbulence model is used, which combines the advantages of the original k-ε and k-ω models by using the k-ω model near the wall, and the k-ε model away from the wall [6]. The k-ω SST has been found to give good predictions of boundary layer detachment characteristics, and it has been validated for the prediction of subcavitating flow around the NACA66 hydrofoil shown in this paper [22].

It should be noted that the original URANS models were developed for fully incompressible single-phase flows and were not intended for flow problems involving compressible multiphase mixtures. As shown in [23], standard two-equation models (such as k-ε and k-ω models) overpredict the turbulent eddy viscosity in the rear part of the cavity,

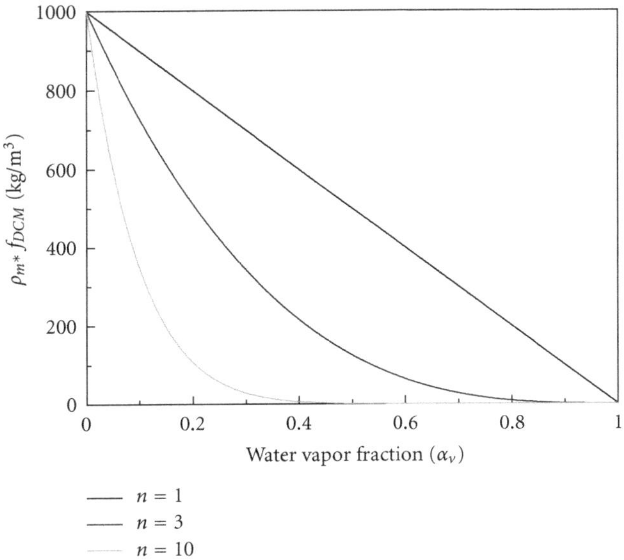

FIGURE 1: Local compressibility modification of the mixture density with different n values according to (11).

which results in overprediction of turbulent stresses, causing the reentrant jet in cavitating flows to lose momentum. Thus, the reentrant jet is not able to cut across the cavity sheet, which significantly modifies the cavity shedding behavior. For partial sheet cavitation in steady flow conditions, inability of the reentrant jet to reach the leading edge will lead to stabilization of the sheet cavity, which prevents the breaking and shedding of the sheet cavity. For unsteady flow conditions, inability of the reentrant jet to fully reach the leading edge will lead to frequent partial shedding of the cavity near the trailing edge, which tends to increase the cavity shedding frequency.

To improve numerical simulations by taking into account the influence of the local compressibility effect of two-phase mixtures on turbulent closure models, Coutier-Delgosha [23] proposed to reduce the mixture turbulent viscosity based on the local liquid volume fraction α_l by substituting μ_T in (3) with μ_{T_mod}:

$$f(n) = \frac{\rho_v + (1-\alpha_v)^n(\rho_l - \rho_v)}{\rho_v + (1-\alpha_v)(\rho_l - \rho_v)}, \qquad \mu_{T\_mod} = \mu_t f(n). \tag{11}$$

The variation of the modified effective density, $\rho_m f(n)$, with the vapor volume fraction, α_v, for different n values is shown in Figure 1.

In [23], they recommended $n = 10$ to better simulate the reentrant jet behavior and vapor shedding process. In [24], the same model was applied for the simulation of supercavitating flow around a hydrofoil, and obtained favorable agreement between numerical and experimental results for $n = 2$. The same modification has also been applied by [25] for the NACA 66 hydrofoil considered in this paper; they used $n = 3$ and obtained good agreements with experimental measurements.

3. Experimental Setup and Description

To evaluate the predictive capabilities of the three different cavitation models with different local compressibility corrections, numerical computations are compared with experimental measurements of the NACA 66 hydrofoil conducted at the cavitation tunnel at the Research Institute of the French Naval Academy. The test section has a cross-sectional area of $0.192\,\mathrm{m}^2$ and length of 1 m. The inflow velocity ranges between 3 m/s and 15 m/s, and the pressure in the test section ranges between 30 m bar to 3 bars. The tunnel inflow turbulence intensity, defined as $V_{\infty\mathrm{rms}}/V_\infty$ at the inlet of the test section, is about 2%. The foil has a uniform cross-section with an NACA 66 thickness distribution with a maximum thickness-to-chord ratio of 12%, and an NACA $a = 0.8$ camber distribution with a maximum camber-to-chord ratio of 2%. The chord length is $c = 0.15$ m and the span length is $s = 0.191$ m. The hydrofoil is made of stainless steel, and behaved like a 2D rigid hydrofoil even though it was mounted using a cantilevered setup with a small gap (1 mm) between the free end of the hydrofoil and the test section wall. Pressure measurements were carried out using seventeen flush-mounted piezo-resistive transducers (Keller AG 2 MI PAA100-075-010) with a maximum pressure of 10 bars. The transducers locations were aligned along the chord on the suction side of the hydrofoil at mid-span, starting from the foil leading edge at a reduced coordinate of $x/c = 0.1$ to the trailing edge at $x/c = 0.9$, with increments of 0.1 c. Lift and drag were measured using a resistive gauge hydrodynamic balance with a range up to 150 daN in lift and 15 daN in drag. Readers should refer to [18, 26] for additional details about the rigid hydrofoil experimental setup and results. The experimental results presented in this paper are taken from [18] for cases with partial sheet cavitation and from [26] for cases with unsteady sheet/cloud cavitation.

4. Numerical Setup and Description

To demonstrate and validate the numerical model, results are shown for the rigid rectangular NACA 66 hydrofoil described above. All the results shown in this paper corresponds to the hydrofoil fixed at $\alpha = 6°$ and subject to a nominal free stream velocity of $V_\infty = 5$ m/s, which yields a moderate-to-high Reynolds number of $\mathrm{Re} = V_\infty c/\nu_l = 0.75 \times 10^6$.

The density and dynamic viscosities of the liquid are taken to be $\rho_l = 999.19\,\mathrm{kg/m^3}$ and $\mu_l = \rho_l\nu_l = 1.139 \times 10^{-3}$ Pa·s, respectively, which correspond to fresh water at 25°C. The vapor density is $\rho_v = 0.02308\,\mathrm{kg/m^3}$ and the vapor viscosity is $\mu_v = 9.8626 \times 10^{-6}$ Pa·s. The vapor pressure of water at 25°C is $p_v = 3169$ Pa. The 2D fluid domain is shown in Figure 2, which corresponds to the height of the experimental test section at the French Naval Academy. The computational domain has an extent of about 5 c upstream and 10 c downstream of the foil to simulate near-infinite boundary conditions at the inlet and outlet. A no-slip boundary condition is imposed on the hydrofoil surface, and symmetry conditions are imposed on the top and bottom boundaries of the tunnel. The inlet velocity is set to be $V_\infty = 5$ m/s and the outlet pressure is set to vary

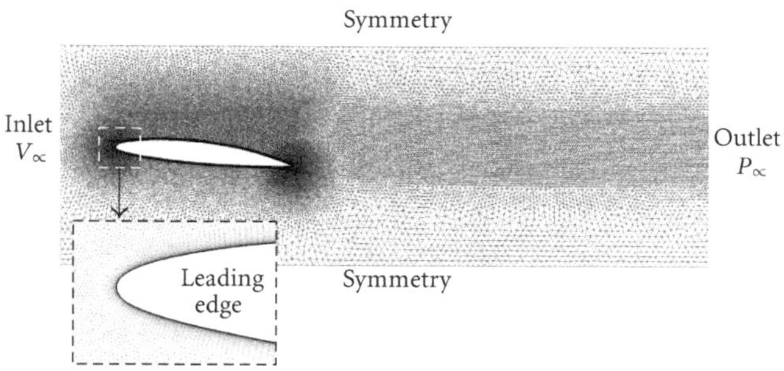

FIGURE 2: 2D fluid mesh and boundary conditions.

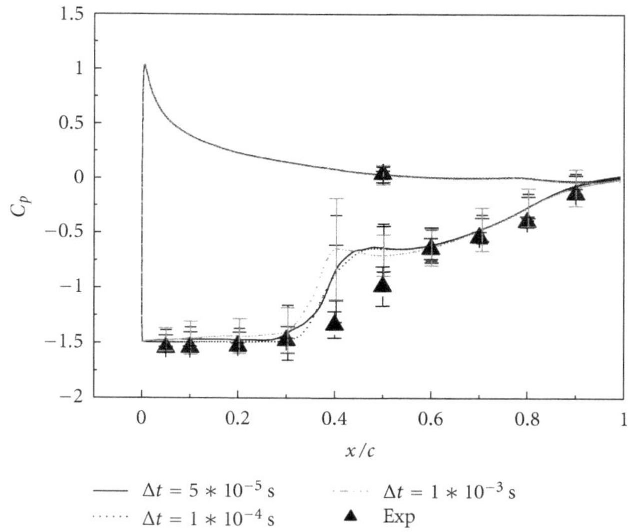

FIGURE 3: Comparisons of the measured versus predicted pressure coefficients obtained using different time step sizes, Δt. The lines represent the mean values and the error bars represent the level of fluctuations. $\sigma = 1.49$, Re = 750,000, $V_\infty = 5$ m/s, and $\alpha = 6°$. Merkle cavitation model, $n = 3$.

according to the cavitation number, defined as $\sigma = (p_\infty - p_v)/(0.5\rho_l V_\infty^2)$, where p_∞ is the tunnel pressure. A constant turbulent intensity of 2% is set at the inlet boundary and is equal to the experimentally measured turbulent intensity.

All cavitating runs have been initialized with a steady-state calculation using fully wetted models to avoid any vapor fraction at the initial time step. The tunnel pressure is then decreased progressively until the particular cavitation number is reached. Only 2D results are shown in this work. The 2D fluid mesh (shown in Figure 2) is composed of 120,000 elements with 50 structured elements across the foil boundary layer, which is selected to ensure $y^+ = yu_\tau/v_l = 1$, where y is the thickness of the first cell from the foil surface, and u_τ is the wall frictional velocity. The regions outside the boundary layer have been discretized with unstructured triangular elements. Mesh refinements are performed at the foil leading edge, trailing edge, and in the wake region. The time integration scheme is a second-order backward Euler algorithm, and the spatial derivatives are computed using a second-order upwind scheme. Mesh convergence and time discretization have been studied in detail for the case of the rigid NACA 66 hydrofoil at a fixed angle of attack of $\alpha = 6°$ in steady flow; see [22] for more details.

A temporal convergence study is carried out in the case of partial sheet cavitation, at an angle of attack of $\alpha = 6°$ with $\sigma = 1.49$. Figure 3 shows temporal convergence of the predicted surface pressure coefficients, $C_P = (P - P_\infty)/(0.5\rho V_\infty^2)$, along with the measured values. The error bars represent the maximum and minimum pressure fluctuations. The numerical results were generated using the Merkle model with $n = 3$. The largest time step size used, $\Delta t = 10^{-3}$ s, yielded a much stronger pressure gradient in the cavity closure region compared to the measured values. The two finer time step sizes yielded very similar results and compared better with the measured values. Hence, $\Delta t = 10^{-4}$ s is chosen for the computations from here on, which gives an average CFL number of CFL $= V_\infty \Delta t/\Delta x = 1$.

5. Results

5.1. Partial Sheet Cavitation. The influence of cavitation models and the effect of the turbulent viscosity are first analyzed for the case with quasisteady partial sheet cavitation ($\sigma = 1.49$). Figure 4 shows the predicted vapor volume fraction and turbulent kinetic energy obtained using the Kubota, Merkle, and Singhal cavitation models with different n values for the turbulent viscosity modification (see (11)). The vapor fraction contours for the three different cavitation models with $n = 1$ and $n = 3$ are, respectively, shown in the first and third rows in Figure 4; a vapor volume fraction of 0 (black) corresponds to pure water, whereas 1 (white) corresponds to pure vapor. The corresponding turbulent kinetic energy contours are shown in the second and fourth rows in Figure 4, where the lowest turbulent kinetic energy is shown in white, and the highest turbulent kinetic energy is shown in black.

For $n = 1$ (no turbulent viscosity modification), the three cavitation models yielded very different levels of turbulent kinetic energy in the cavity and at the cavity closure region, which in turn modified the vapor volume fraction and cavity shape, as shown in the first two rows of Figure 4. The Merkle model has the lowest level of turbulent kinetic energy in the vapor region, which leads to a longer sheet cavity compared to the other two models. Compared to the Merkle model, the Kubota model has a higher level of turbulent kinetic energy in the pure vapor region, and a shorter cavity length. It is observed that the production and destruction terms in the Merkle model (9) are related to the pressure difference, instead or the square root of the pressure difference for the Kubota and Singhal models (resp., (7), (8), and (10)). For an equivalent pressure level in subcavitating flow, the direct dependence of the production and destruction terms on the pressure difference leads to a longer cavity for the Merkle model. The Singhal model leads to the highest turbulent kinetic energy in the vapor region. This is because the vapor production and destruction terms for the Singhal model, as shown in (10), are directly related to the square root of

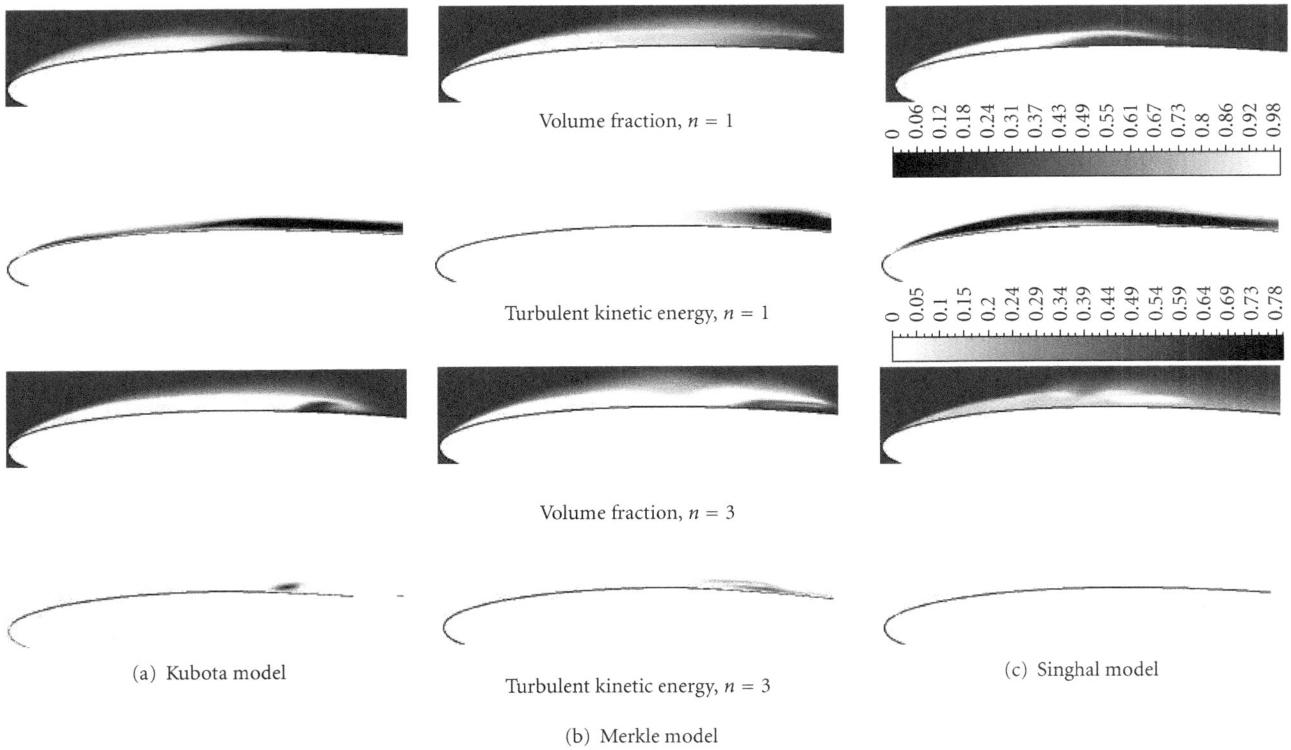

FIGURE 4: Comparisons of the vapor fraction and turbulence kinetic energy contours for the three different cavitation models with $n = 1$ and $n = 3$ for $\sigma = 1.49$, Re = 750,000, $V_\infty = 5$ m/s, and $\alpha = 6°$.

the local turbulent kinetic energy, which creates a nonlinear feedback mechanism that increases the pressure gradient in the cavity closure region, and hence tends to destabilize the cavity.

For $n = 3$, the results for all three models in the bottom two rows of Figure 4 show a general decrease in the turbulent kinetic energy in the vapor region, and slightly more unsteady behavior at the cavity trailing edge, compared to $n = 1$. The results show that the density correction to the turbulent viscosity has different degrees of impact on the three cavitation models. The Merkle model predicts similar cavity lengths with and without the density modification, although the decrease of the turbulent kinetic energy allows the reentrant jet at the cavity trailing edge to move upstream, and hence increases the flow unsteadiness and leads to partial shedding of the cavity. The Kubota model shows a noticeable increase in cavity length and more unsteadiness at the cavity trailing edge with the modification. The Singhal model appears to be the most sensitive to the turbulent viscosity modification. Again, this is because of the dependence on the square root of the local turbulent kinetic energy in the production and destruction terms. When $n = 3$, the turbulent kinetic energy is completely suppressed in the cavity region, which allowed the cavity to grow longer but also lead to unsteady partial shedding of the vapor structure near the cavity trailing edge where the pressure gradient is high.

The increase of unsteadiness with increasing n values for (11) is due to the development of a reentrant jet as a result of the reduced turbulent eddy viscosity level at the cavity closure. Figure 5 shows the predicted vapor fraction contours and the velocity vectors at the cavity closure region for $\sigma = 1.49$ obtained using the Merkle model with $n = 1$ (no modification), $n = 3$, and $n = 10$. With $n = 1$ (Figure 5(a)), the turbulent eddy viscosity in the cavity closure region is too high, which prevented the upstream motion of the reentrant jet, and hence the cavity is fully stable. With $n = 3$ (Figure 5(b)), the reduction of the turbulent viscosity in the cavity closure region allows the reentrant jet to develop and move upstream, which lead to partial shedding of the cavity. With $n = 10$ (Figures 5(c) and 5(d)), the turbulent eddy viscosity is reduced to near zero in the pure vapor region, which allowed the reentrant jet to reach the cavity leading edge. Subsequently, the cavity detached from the leading edge (Figure 5(d)) and then shed downstream as a cloud cavity. This complete breakup of the cavity from the leading edge and unsteady transformation to cloud cavitation obtained using $n = 10$ are different from experimental observations by [18], whom reported partial sheet cavitation with behavior similar to that obtained using $n = 3$ at $\sigma = 1.49$.

As suggested in [7, 27], the baroclinic torque created by the mixture density and pressure gradients in the cavitating

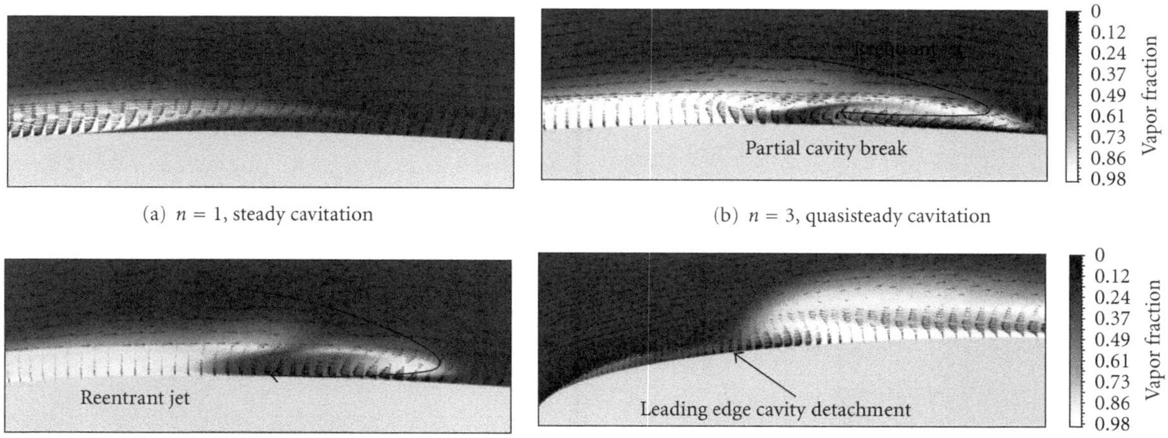

(a) $n = 1$, steady cavitation

(b) $n = 3$, quasisteady cavitation

(c) $n = 10$, unsteady cavitation, t_1

(d) $n = 10$, unsteady cavitation, $t_1 + 0.03$ s

FIGURE 5: Comparison of the vapor fraction contours and velocity vectors for different n values with the Merkle model at $\sigma = 1.49$, Re = 750,000, $V_\infty = 5$ m/s, and $\alpha = 6°$.

region are responsible for the alteration of the vorticity field, which is evident via the vorticity transport equation:

$$\frac{\partial \boldsymbol{\omega}}{\partial t} + \nabla \times (\boldsymbol{\omega} \times \mathbf{u}) = \frac{\nabla \rho_m \times \nabla p}{{\rho_m}^2} + (\upsilon_m + \upsilon_t)\nabla^2 \boldsymbol{\omega}, \quad (12)$$

where $\boldsymbol{\omega}$ and $\mathbf{u}$ are, respectively, the fluid vorticity and velocity vectors. υ_m is the laminar kinematic viscosity and υ_t is the turbulent kinematic viscosity.

The first term on the left-hand side (LHS) of (12) represents the rate of change of vorticity at a fixed point. The second term on the LHS represents the convection as well as the stretching and tilting of the vorticity due to velocity gradients. The first term on the right-hand side (RHS) of (12) is called the baroclinic torque, which is nonzero only in a variable-density flow field. The baroclinic torque corresponds to the generation of vorticity from unequal acceleration caused by unaligned pressure and density gradients, for example, the generation of a shear layer caused by the faster acceleration of the lighter fluid (vapor) relative to the heavier fluid (water). The second term on the right-hand side of (12) represents the laminar and viscous diffusion of the vorticity.

To examine the influence of cavitation on the production/alteration of vorticity around the case with quasisteady partial sheet cavitation, the strengths of $(\nabla \times (\boldsymbol{\omega} \times \mathbf{u}))$ and $(\nabla \rho_m \times \nabla p/{\rho_m}^2)$ are compared in Figure 6 for the different cavitation models with different n values. As expected, the baroclinic term modifies the vorticity field in regions with high density and pressure gradients, that is, along the liquid-vapor interface and near the cavity closure. The strength of the baroclinic torque increases with n, particularly near the cavity closure region because the reduction of the turbulent eddy viscosity tends to promote the unsteady shedding of the cavities. Consequently, the vorticity field is modified, as evident via the contour of the vorticity transport/stretching term. The baroclinic term is found to be the lowest for the Merkle model, which shows a smoother cavity with consequently lower density and pressure gradients. The baroclinic term is higher for the Singhal and Kubota cavitation models, where the magnitude of the baroclinic torque is comparable to the vorticity transport/stretching term near the cavity closure, which will tend to increase the flow unsteadiness, as observed earlier in Figure 4.

Figures 7, 8, and 9 show the comparison of predicted wall pressure coefficients, obtained at $\sigma = 1.49$, for the Kubota, Merkle, and Singhal cavitation models, respectively, along with the measured values from [18]. On each of the three figures, the mean and fluctuations of the predicted pressure coefficients obtained using $n = 1$, 3, and 10 are shown. The nearly constant pressure region with $C_P = -1.49$ corresponds to the leading edge, partial sheet cavity on the suction side, which is followed by a sudden pressure jump at the cavity closure due to transition back to subcavitating flow. In general, the pressure fluctuations increased, particularly at the cavity closure region, with increasing n values because of the decrease of turbulent eddy viscosity and the additional vorticity generation/modification by the baroclinic torque. The Kubota cavitation model (Figure 7) has very steady pressure behavior without modification ($n = 1$), which showed a cavitating region up to $X/c = 0.2$. The results are approximately the same for $n = 3$ and 10, which showed a constant pressure region up to $X/c = 0.3$, and larger pressure fluctuations at the cavity closure. The Merkle model (Figure 8), on the other hand, showed negligible difference in the predicted pressure coefficients between the different n values except for the increase in pressure fluctuations at the cavity closure for $n = 10$. Moreover, the predicted pressure distributions by the Merkle model are in good general agreement with experimental measurements. The Singhal model (Figure 9) shows the largest difference between the different n values because of the explicit dependence on the turbulent kinetic energy, which changes with the turbulent eddy viscosity as well as vorticity generation/distortion by the baroclinic torque. Consequently, as n increases, the cavity length (constant pressure region) predicted by the Singhal model decreases. A general unsteadiness is observed along the entire suction side pressure distribution. Moreover, the predicted pressure

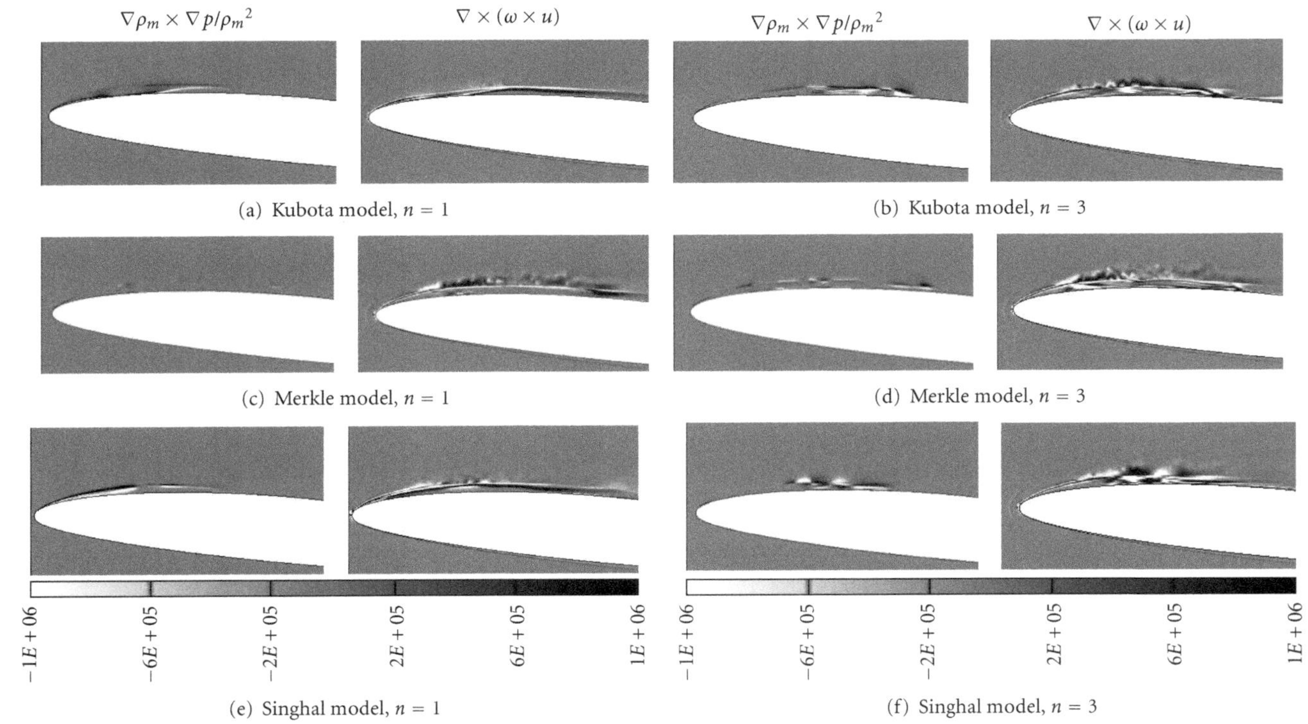

FIGURE 6: Comparisons of the baroclinic torque and vorticity transport/stretching term in (12) for the different cavitation models with $n = 1$ and $n = 3$ for $\sigma = 1.49$, Re = 750,000, $V_\infty = 5$ m/s, and $\alpha = 6°$.

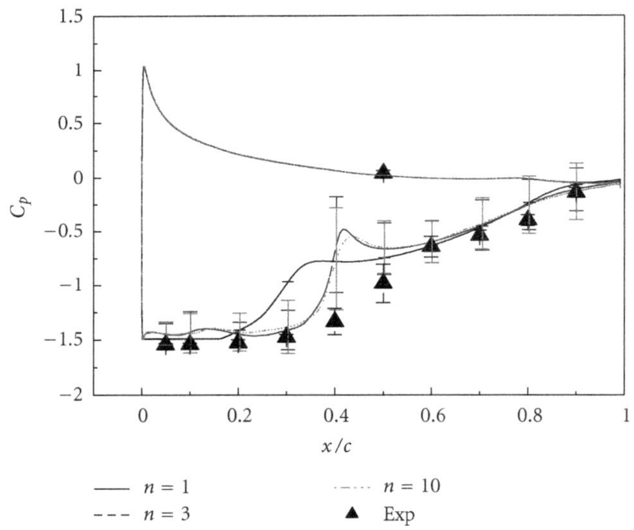

FIGURE 7: Computed versus measured pressure coefficients for different n values, Kubota cavitation model, $\sigma = 1.49$, Re = 750,000, $V_\infty = 5$ m/s, and $\alpha = 6°$. The lines represent the mean values and the error bars represent the level of fluctuations.

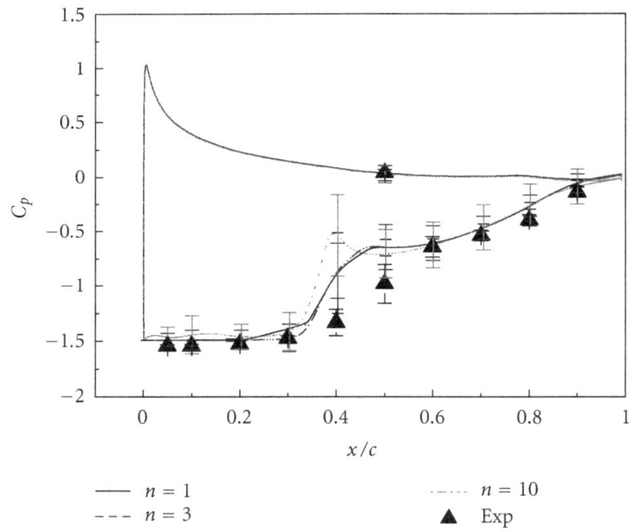

FIGURE 8: Computed versus measured pressure coefficients for different n values, Merkle cavitation model, $\sigma = 1.49$, Re = 750,000, $V_\infty = 5$ m/s, and $\alpha = 6°$. The lines represent the mean values and the error bars represent the level of fluctuations.

coefficients show greater discrepancy with experimental measurements compared to the Kubota and Merkle models. In general, the results obtained with the Merkle cavitation model with $n = 3$ matched better with experiments.

5.2. Unsteady Sheet/Cloud Cavitation. To evaluate the predictability of the three different cavitation models for unsteady sheet/cloud cavitation, results are shown in this section for the case with $\sigma = 1.00$, Re = 750,000, and $\alpha = 6°$.

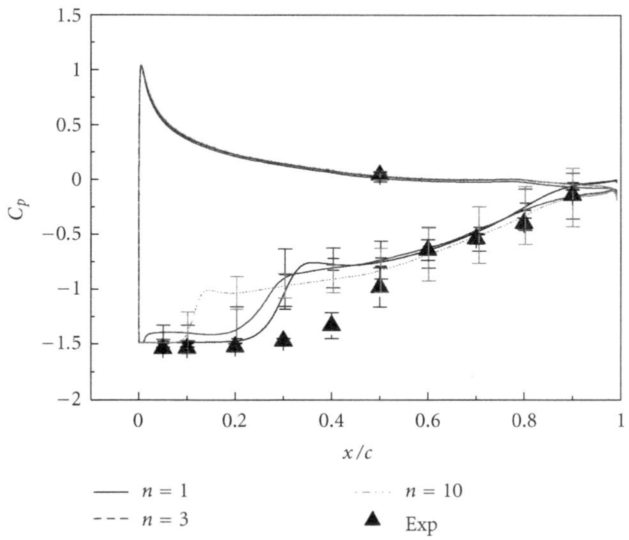

FIGURE 9: Computed versus measured pressure coefficients for different n values, Singhal cavitation model, $\sigma = 1.49$, Re = 750,000, $V_\infty = 5$ m/s, and $\alpha = 6°$. The lines represent the mean values and the error bars represent the level of fluctuations.

It should be noted that the Kubota model did not converge for this case with $\Delta t = 0.0001$ s. The authors believe that it is due to the high value of C_{dest} in (7), which makes the calculation unstable. Hence, results are shown in this case for the Merkle and Singhal models only.

At $\sigma = 1.00$, Re = 750,000, and $\alpha = 6°$, unsteady sheet/cloud cavitation was reported in [26]. Figures 10 and 11 show the evolutions of the vapor fraction, turbulent kinetic energy, and vorticity for $n = 1$ and $n = 3$ with the Merkle and Singhal cavitation models, respectively. Globally, the results show significant reduction in the turbulent kinetic energy in the cavity region, particularly at the cavity closure, and longer cavity with increasing n values because of the reduction of the turbulent eddy viscosity. For the Merkle model (Figure 10), it is observed that the level of vorticity is closely related to the vapor fraction contour and the distribution of turbulent kinetic energy. For $n = 3$, when the cavity develops, the decrease of turbulent kinetic energy lead to an increase of vorticity near the wall (Figure 10(b)), which allowed the reentrant jet to form and move upstream (t_4) while allowing the cavity to expand to the foil trailing edge; subsequently, the cavity breaks and lifts off at the at the leading edge (t_6) to form a large cavitation cloud, which then convects downstream. For the Merkle model, the cavity shedding frequency at $n = 3$ is approximately 3.5 Hz, which agrees with the measured value, as shown in Table 1. On the contrary, for $n = 1$, the cavity partially breaks around the midchord (see the vapor fraction in Figure 10(a), at t_4), which limited the maximum cavity length to $L_c/c = 0.8$, and increased the shedding frequency to about 4.0 Hz, see Table 1. Moreover, because of the presence of cloud cavitation near the foil trailing edge (Figure 10(a) at t_6 and t_8), the overprediction of turbulent viscosity without the density

TABLE 1: Comparison of the measures and predicted cavity shedding frequencies for the Merkle and Singhal cavitation models with $n = 1$ and $n = 3$ for the case of unsteady sheet/cloud cavitation. $\sigma = 1$, Re = 750,000, $V_\infty = 5$ m/s, and $\alpha = 6°$.

Models	Merkle		Singhal		Experiments
	$n = 1$	$n = 3$	$n = 1$	$n = 3$	
Shedding frequency (Hz)	4.0	3.5	22.5	20.5	3.5

correction ($n = 1$) suppresses the formation of the counter-clockwise trailing edge vortex, which changes the vorticity patterns around the foil, as well as the convection of the cloud in the wake. When the modification is applied ($n = 3$), the level of turbulent kinetic energy in the cloud is lower, which allows the counter-clockwise trailing edge vortex to form and interact with the cloud cavity (Figure 10(b) at t_6 and t_8).

As shown in Figure 11, the Singhal cavitation model is more sensitive to the turbulent viscosity modification. Without the modification ($n = 1$), the level of turbulent kinetic energy is very high near the cavity closure (see Figure 11(a)), which limited the length of the sheet cavity to approximately $X/c = 0.6$ (t_4), and the cavity partially breaks at $X/c = 0.4$ (t_6), followed by successive shedding of smaller vapor structures (t_8). The cavity appears to be more unstable and transient than that predicted by the Merkle cavitation model, as shown by the scattered vorticity distribution near the wall (see Figure 11(a)). When the modification is applied ($n = 3$), the Singhal model also predicted an increase in the cavity length up to $L_c/c = 1$, although general cavity behavior is much more unstable than that predicted by the Merkle model. The cavity breaks around $X/c = 0.2$. However, it does not shed as one big cavity; instead, the aft end of the cavity is broken into multiple little cloud cavities, as shown in Figure 11(b). Therefore, the shedding frequencies predicted by the Singhal model (about $f = 22.5$ Hz for $n = 1$ and $f = 20.5$ Hz for $n = 3$) are much higher than those by the Merkle model and did not agree with experimental measurements, as summarized in Table 1. Again, the turbulent viscosity modification allowed the trailing edge vortex to develop earlier, which interacted with the cloud cavity as it sheds into the wake.

Figure 12 shows the baroclinic torque and the vorticity transport/stretching terms, at $t = t_6$ for both the Merkle and Singhal models. For both models, the baroclinic torque concentrates along the regions where the cavity breaks, although the magnitudes seem to be small compared to the vorticity transport/stretching term. Nevertheless, the influence of the cavity on the vorticity distribution is clearly evident by the significant modification of the vorticity transport/stretching term (Figure 12) and vorticity contours (Figures 10 and 11). Hence, the interdependency between cavitation generation/production and vorticity generation/modification due to the baroclinic torque is not negligible, particularly in terms of its effect on the cavity stability and shedding frequency.

Based on available experimental data presented in [26], an analysis of suction side loading can be calculated by

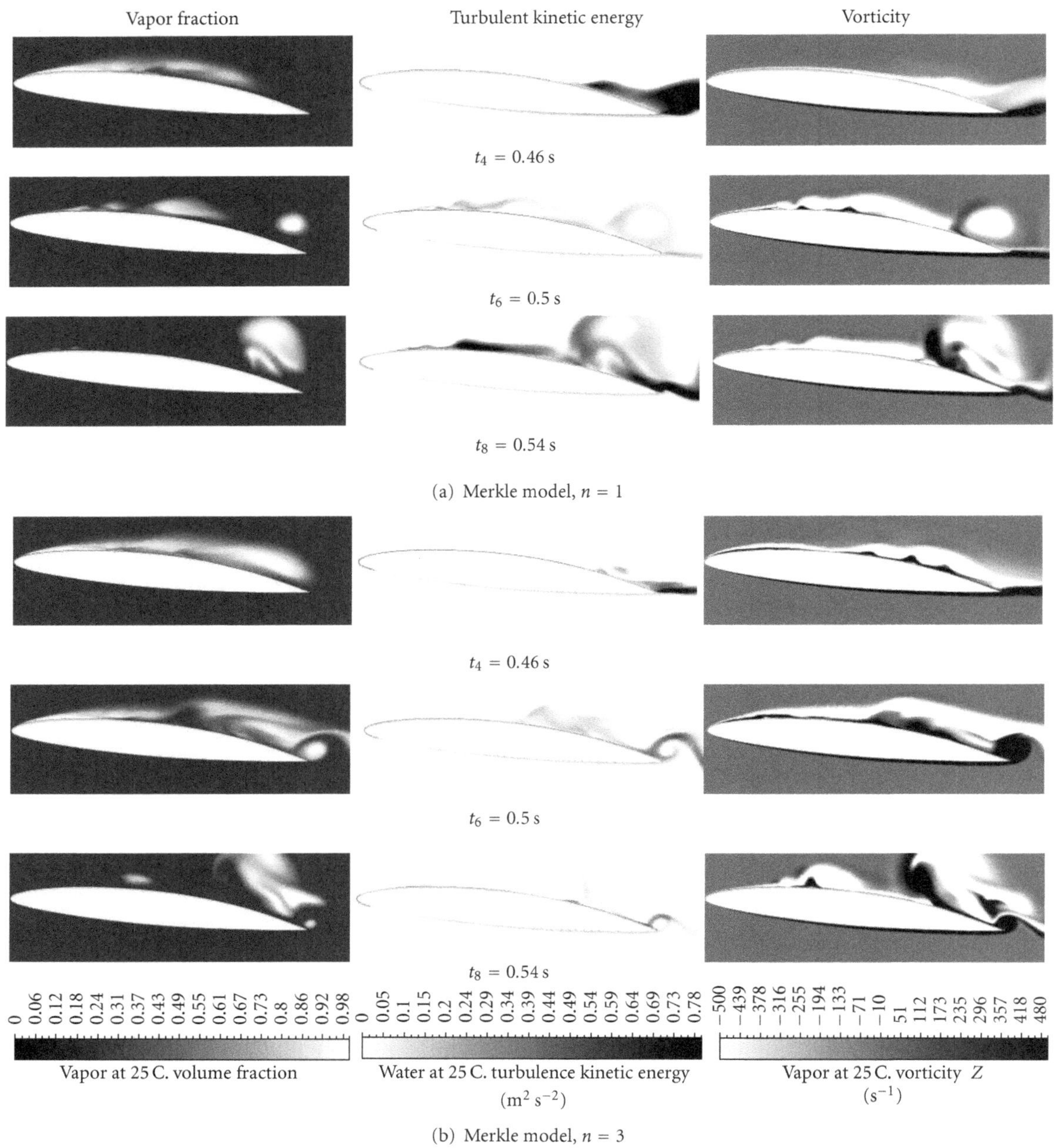

FIGURE 10: Comparison of the vapor fraction (left), turbulent kinetic energy (middle), and vorticity (right) contours for the Merkle cavitation models with (a) $n = 1$ and (b) $n = 3$, in the case of unsteady sheet/cloud cavitation with $\sigma = 1.00$, Re = 750,000, $V_\infty = 5$ m/s, and $\alpha = 6°$.

summing the pressure coefficients along the suction side. The approximate suction side lift coefficient can be written as:

$$C_l^+(t) = \sum_{i=1}^{10} C_P\left(\frac{x_i}{c,t}\right)\Delta\left(\frac{x_i}{c}\right), \quad (13)$$

where $C_P(x_i/c,t)$ is the pressure coefficient at location x_i/c, and $\Delta(x_i/c) = 0.1$ is the nondimensional distance between two consecutive pressure transducers. A total of 10 pressure transducers were placed along the suction side at the mid-span of the foil [22]. For comparison purposes, the same procedure is applied to calculate the numerically derived suction side lift coefficients. It should be noted that since the flow is mostly attached along the pressure side, the changes to the total lift coefficient should be dominated by the suction side dynamics, which is represented by the suction side lift coefficient.

As shown in Figure 13, a fair comparison is observed between the experimental measurements and numerical computations obtained using the Merkle model with $n = 3$. The peaks generally correspond to the time instance when the maximum cavity length is observed (t_4 in Figure 10(b)),

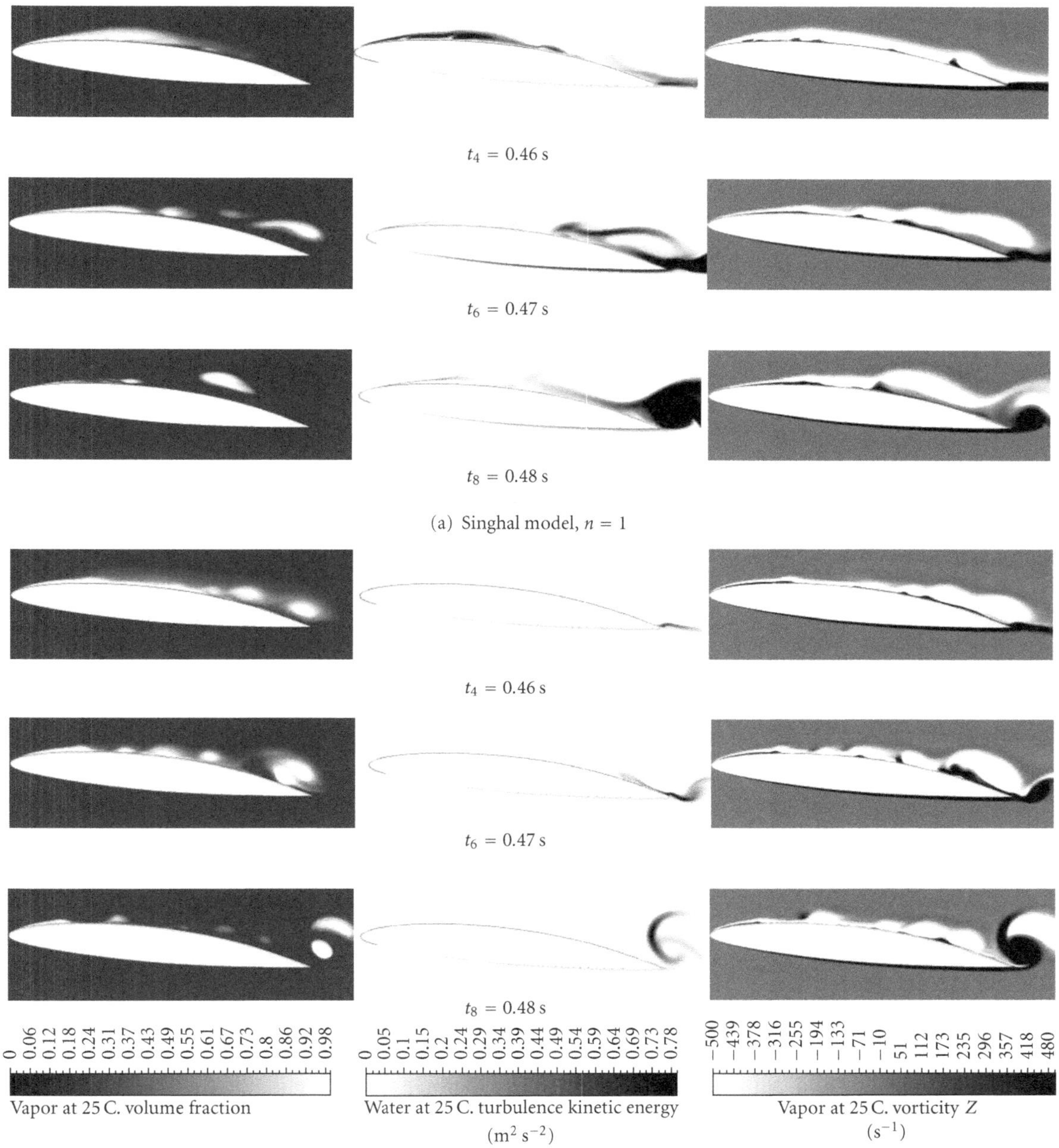

(b) Singhal model, $n = 3$

FIGURE 11: Comparison of the vapor fraction (left), turbulent kinetic energy (middle), and vorticity (right) contours for the Singhal cavitation models with (a) $n = 1$ and (b) $n = 3$, in the case of unsteady sheet/cloud cavitation with $\sigma = 1.00$, Re = 750,000, $V_\infty = 5$ m/s, and $\alpha = 6°$.

and the unloading phase corresponds to the shedding of the cavity. The results show that the rate of the unloading phase predicted by the Merkle model is faster than the measured value. This difference could be due to inaccurate prediction of the convection/destruction of vapor in the cloud, which is by definition very transient and sensitive to the production and destruction rate of vapor, and the interaction between the cavity and the vorticity field through the baroclinic torque. Nevertheless, the shedding frequency predicted by the Merkle model with $n = 3$ matched well with the measured frequency of $f = 3.5$ Hz. As shown in Table 1, the cavity shedding frequency is higher for $n = 1$ because the cavity is shorter and undergoes partial shedding at its trailing edge. Hence, it is important to take into account the influence of mixture density variation in the calculation of the turbulent viscosity in and near the cavity.

As shown in Figure 14, the Singhal model with $n = 3$ shows a much higher shedding frequency, $f = 20.5$ Hz,

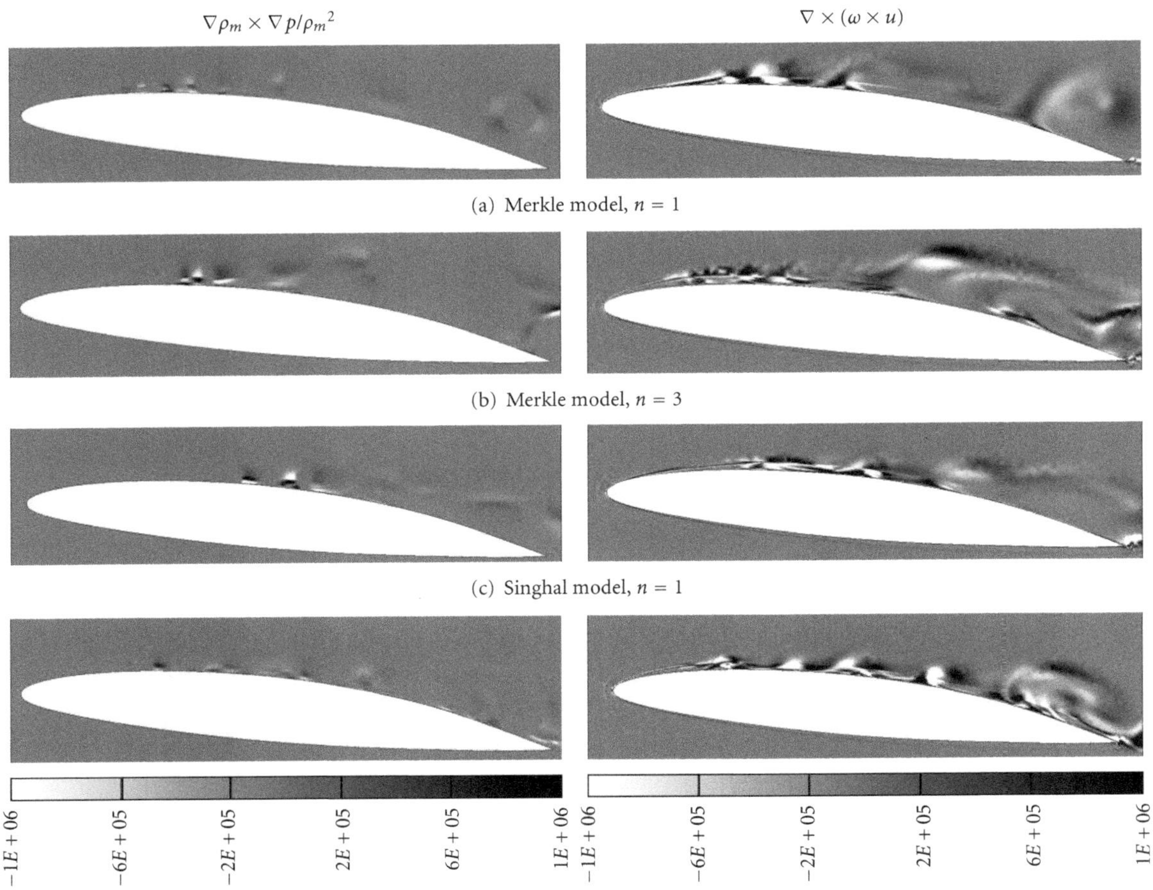

(d) Singhal model, $n = 3$

FIGURE 12: Comparison of the baroclinic torque (left) and vorticity transport/stretching term (right) in (12) at t_6 for the Merkle and Singhal models with $n = 1$ and 3 for $\sigma = 1$, Re = 750,000, $V_\infty = 5$ m/s, and $\alpha = 6°$.

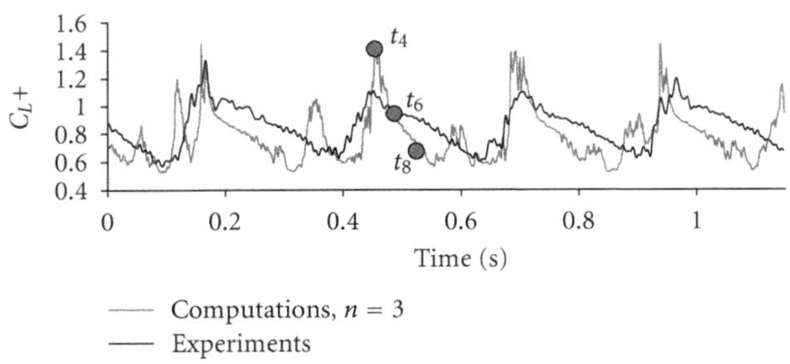

FIGURE 13: Comparisons of the measured and the predicted suction side lift coefficients obtained using the Merkle cavitation model with $n = 3$. $\sigma = 1$, Re = 750,000, $V_\infty = 5$ m/s, and $\alpha = 6°$.

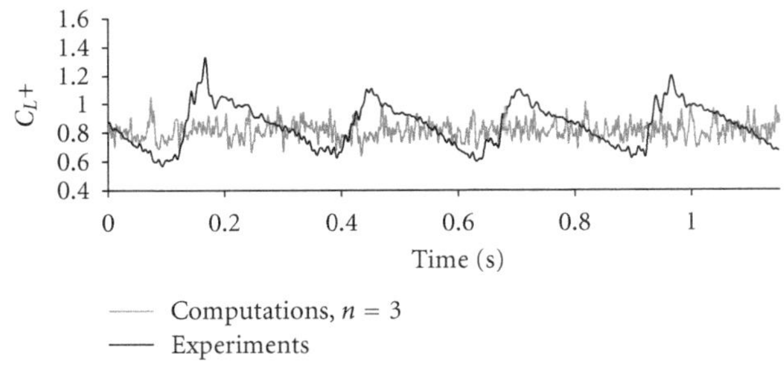

FIGURE 14: Comparisons of the measured and the predicted suction side lift coefficients obtained using the Singhal cavitation model with $n = 3$. $\sigma = 1$, Re = 750,000, $V_\infty = 5$ m/s, and $\alpha = 6°$.

than the measured frequency of 3.5 Hz. The high frequency predicted by the Singhal model is because of the continuous presence of vapor along the hydrofoil surface (see Figure 11(b)), which significantly perturbs the wall pressure distributions and leads to the generation of vorticity through the baroclinic torque (see Figure 12(d)). As a consequence, the predicted suction side lift coefficient shows a much higher shedding frequency and much lower amplitude compared to the measured values. The results shown in Table 1 confirms that the shedding frequency is higher for the Singhal model if $n = 1$, without the turbulent viscosity modification.

It should be noted that the importance of capturing and comparing both the mean and fluctuating values is evident in Figures 13 and 14. As shown in the figures, although the mean suction side lift coefficient predicted by both the

Merkle and Singhal models are approximately the same, and agreed well with experimental measurements, the amplitude and frequency of the fluctuations predicted by the Singhal model are very different from the Merkle model, which compared much better with experimental measurements.

Figure 15 shows the comparisons between the predicted vapor fraction contours obtained using the Merkle cavitation model with $n = 3$ and photographs from the experiments presented in [26] for $\sigma = 1$, Re = 750,000, and $\alpha = 6°$. The white vapor fraction represents the pure vapor and the black represents the pure water. The predicted flow streamlines are also shown to facilitate visualization of the flow dynamics. Similar to the comparison of the suction side loading, the numerical predictions obtained using the Merkle model with $n = 3$ are in good agreement with the experiments. The cavity expanded up to the foil trailing edge at $X/c = 1$ (t_1 to t_4), followed by breakup near the foil leading edge due to the reentrant jet at t_5. An interaction between the bubbly cavity mixture and the trailing edge vortex is observed at t_6 and t_7, followed by complete convection of the cloud cavity into the wake at t_8.

5.3. Comparisons of the Measured and Predicted Hydrodynamic Load Coefficients and Cavity Lengths. Figures 16, 17, and 18 show the measured and predicted normalized cavity lengths (L_c/c), lift coefficient (C_L), and drag coefficient (C_D) over a range of cavitation numbers (σ). The numerical results for all three models are obtained using $n = 3$, which compared the best with the experimental measurements, as shown in the previous section.

Figure 16 shows the predicted maximum and minimum cavity lengths, along with the measured maximum cavity lengths. Cavitation inception is observed at $\sigma \approx 2.4$ for both computations and experiments. For $1.5 < \sigma < 2.0$, the minimum and maximum values are equal, which suggest stable cavitation. It should be noted that all three cavitation models give very similar results for stable cavitation, and all agree well with the experimental measurements (star symbols). For $\sigma < 1.5$, the cavity becomes unstable, as indicated by the diverging maximum and minimum values of the cavity length, and differences could be observed between the three cavitation models. The results suggest that the cavity breaks directly at the leading edge for the Kubota model, as indicated by the near-zero value of the minimum length represented by the open circles in Figure 16 for $\sigma < 1.5$. However, experimental observations suggest the existence of partial sheet cavitation between $1.3 < \sigma < 1.6$, which is characterized by a partial break of the cavity with a stable vapor region at the cavity leading edge [18]. As shown in Figure 16, the Singhal and Merkle models agree well with the experimental observations, but not the Kubota model. As the cavitation number further decreases, the break of the cavity moves closer to the foil leading edge. At $\sigma = 1.00$, unsteady sheet/cloud cavitation occurs, as shown in the numerical and experimental results reported in the previous section. It should be noted that the measured maximum cavity lengths agree well with the numerical values for all three cavitation models considered. However, as evident via the results shown in the previous section and in Table 1, even if the minimal and maximal values of the cavity length agree well between the Singhal and the Merkle cavitation models, the shedding frequencies are very different. It should be noted that the Kubota model did not converge due to excessive fluctuations for $\sigma < 1.2$, and hence results are only shown for $\sigma > 1.2$ in Figures 16–18.

Figure 17 compares the measured and predicted lift coefficients ($C_L = F/(0.5\rho V_\infty^2 c)$, where F is the lift force) for different cavitation numbers. The lift coefficient increases slightly with the development of small, leading edge, and partial sheet cavitation ($1.3 < \sigma < 1.6$). The increase in lift is attributed to the suction effect created by the small leading edge cavitation bubble caused by the high pressure gradient. As the cavity grows with decreasing cavitation number (i.e., $\sigma < 1.2$, $L_{C\max}/c > 0.8$), the lift coefficient decreases. Both the slight increase in the lift coefficient for $1.3 < \sigma < 1.6$ and subsequent drop in lift coefficient for $\sigma < 1.2$ are captured by the Merkle and Singhal models. The numerical predictions show large fluctuations in the lift coefficients for $\sigma < 1.2$, as indicated by the error bars, due to the unsteady cavity dynamics. The numerical results show good agreement with the experimental measurements for $\sigma > 1.6$, in the quasisteady cavitation and fully wetted flow regimes. The predicted lift coefficients by all three models are slightly less than the measured values for $1.3 < \sigma < 1.6$, in the partial sheet cavitation flow regime. In that range, the differences between the three cavitation models are relatively small, with the Merkle model showing the most favorable, and the Singhal model showing the least favorable comparisons. The Kubota model also predicts much higher fluctuations compared to the other two models in the unsteady cavitation flow regime, and the fluctuations grow with decreasing cavitation number.

Comparisons of the predicted and measured drag coefficients ($C_D = D/(0.5\rho V_\infty^2 c)$, where D is the drag force) are shown in Figure 18. All three cavitation models underpredicted the drag coefficients, but all three trends agree well with the experiments. It should be noted that the underprediction of the drag may be attributed to 3D effects, where downwash and wall boundary layer effects tend to increase the drag. The results show that the development of cavitation significantly increases the drag, particularly for unsteady cavitation. The combination of decreases in lift and increases in drag leads to a decrease in the efficiency, which is a classical issue for cavitation. It should be noted that the magnitudes of the fluctuations in the lift and drag coefficients are higher for the Merkle model than the Singhal model at $\sigma = 1.00$, which is consistent with the results shown in Figures 13 and 14.

6. Conclusions

The predictive capability of three popular cavitation models is investigated for the prediction of cavitation flow around a NACA 66 hydrofoil at a fixed angle of incidence ($\alpha = 6°$) at Re = 750,000, for a case with partial sheet cavitation ($\sigma = 1.49$), and a case with unsteady sheet/cloud cavitation ($\sigma = 1.00$). Comparisons of the predicted and measured

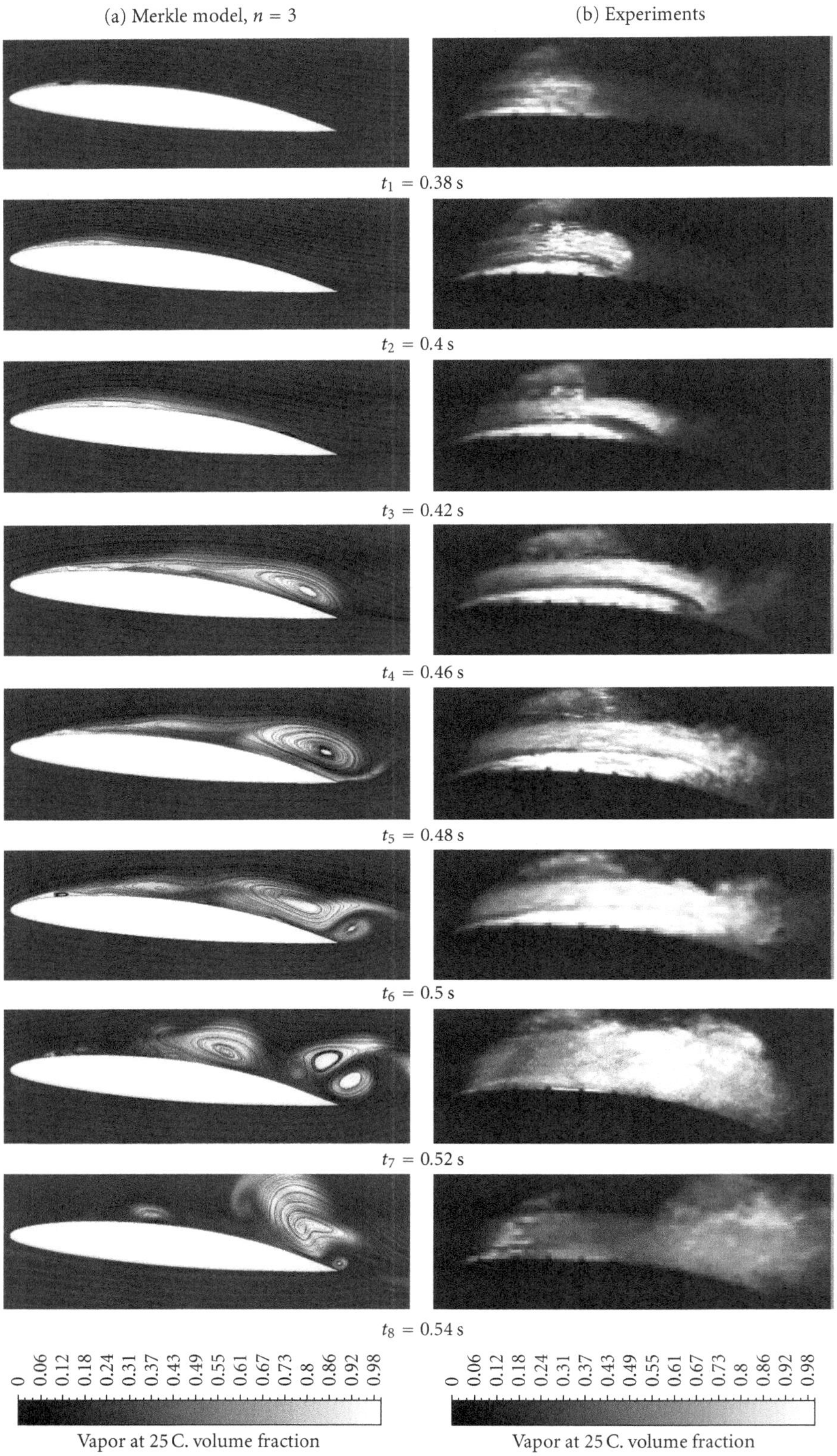

Figure 15: Comparisons of the predicted vapor fraction contours and flow streamlines predicted using the Merkle cavitation model with $n = 3$ on the left with experimental observations on the right. $\sigma = 1.00$, Re = 750,000, $V_\infty = 5$ m/s, and $\alpha = 6^\circ$.

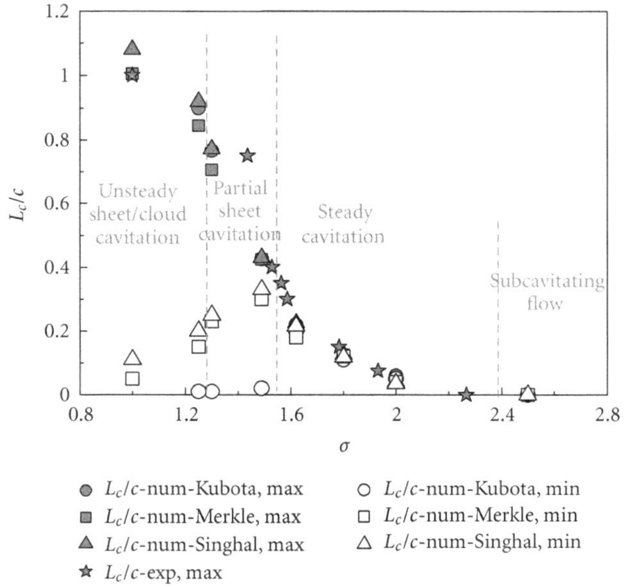

FIGURE 16: Comparison of the predicted and measured normalized maximum and minimum cavity lengths (L_c/c) at various cavitation numbers. Re = 750,000, V_∞ = 5 m/s, $\alpha = 6°$, and $n = 3$.

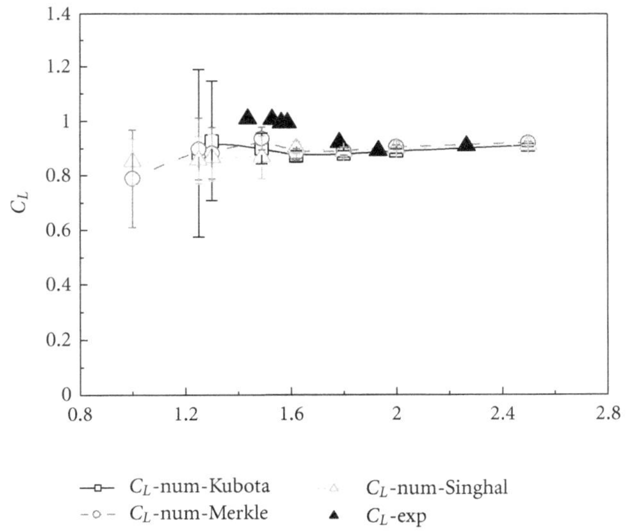

FIGURE 17: Comparisons of the measured and predicted lift coefficients (C_L) for various cavitation numbers. Re = 750,000, V_∞ = 5 m/s, $\alpha = 6°$, and $n = 3$.

cavity lengths and hydrodynamic coefficients over a range of cavitation numbers are also shown. In addition to the application of different cavitation models, different levels of density correction to the turbulent viscosity are applied to investigate the interaction between cavitation and turbulent flows. The aim of this density correction is to take into account the local compressibility effects in the cavitating region, especially near the cavity closure, where standard turbulence models tend to overpredict the turbulent viscosity, which will modify the cavity shedding dynamics.

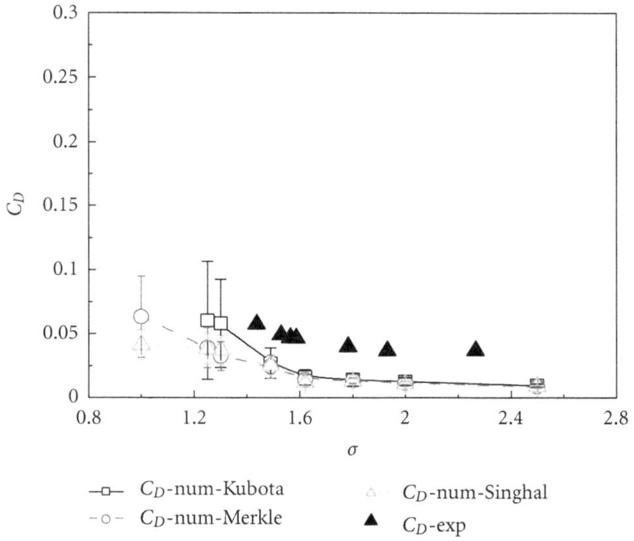

FIGURE 18: Comparison of the measured and predicted drag coefficients (C_D) for various cavitation numbers. Re = 750,000, V_∞ = 5 m/s, $\alpha = 6°$, and $n = 3$.

The general conclusions of the paper are as follows:

(i) the density modification helps to reduce the turbulent viscosity in the cavitating region, which decreases the turbulent kinetic energy in the cavity, as well as around the cavity closure. With the reduction of the turbulent viscosity, the reentrant jet is allowed to develop underneath the cavity and move upstream while allowing the cavity to expand toward the foil trailing edge. Consequently, the cavities tend to be slightly longer and more unsteady with the density modification. This lead to a better comparison with experimental measurements and observations for both the quasisteady cavitation case and the unsteady sheet/cloud cavitation case;

(ii) the density correction to the turbulent viscosity has varying degrees of impact on the different cavitation models. The Singhal model is the most sensitive to the turbulent viscosity modification, which already overpredicted the general unsteadiness due to the explicit dependence of the source and sink terms on the turbulent kinetic energy, the addition of the density correction will further increase the overall unsteadiness, and hence lead to greater discrepancies compared to experimental measurements. On the other hand, the Merkle model seems to be the least sensitive to the turbulent viscosity modification, and tends to be more stable than the Singhal and Kubota models, particularly for unsteady cavitating flows. Addition of the density correction for the Merkle model helps to avoid overprediction of the turbulent eddy viscosity, which allows the reentrant jet to properly develop, and hence leads to improved comparisons between predictions and measurements;

(iii) the baroclinic torque is responsible for the generation of additional vorticity due to shear flow created by the unequal acceleration between the lighter vapor and heavier liquid phases. Hence, in regions with high, unaligned density and pressure gradients, for example, along the liquid-vapor interface and the cavity closure region, the vorticity field is modified by the cavitation, which in turn changes the cavitation dynamics. The interdependency between the cavity and vorticity dynamics through the baroclinic torque is not negligible, particularly for unsteady cavitating flows. The magnitude of the baroclinic torque tends to increase with the power factor n in the density correction for the turbulent viscosity (11), which in turn increases the flow unsteadiness, particularly near the cavity closure. New detailed experimental measurements of the mean and fluctuating velocity fields in the cavitating region are needed to investigate the validity of the density correction, and the importance of the baroclinic torque on the cavitating response of hydrofoils;

(iv) the results show that while predictions by all three cavitation models compare well with experimental measurements in the quasisteady partial sheet cavitation range, the predicted cavity shedding frequencies and load fluctuations predicted by the three cavitation models are very different in the unsteady sheet/cloud cavitation range. Moreover, the numerical stabilities of the cavitation models are very sensitive to the dependency of the condensation and evaporation rates on the power to the pressure difference term (1.0 in the Merkel model and 0.5 in the Kubota and Singhal models), which is probably attributed to the interdependency between the cavitation and vorticity dynamics through the baroclinic torque. The Singhal model is different from the Kubota and Merkel models in that it explicitly depends on the square root of the turbulent kinetic energy (see (10)), which increases the coupling effect between the vorticity distribution in the turbulent flow field and the cavity dynamics, and hence tends to destabilize the cavity. Consequently, the predicted cavity shedding frequencies was much higher than the measured values, and the amplitude of the load fluctuation was much lower than the measured values. Physically, as noted in [28], transition to turbulence will sweep away attached cavities. Hence, explicit dependence of the sink (vaporization) term on the turbulent kinetic energy does not agree with experimental observations presented in [28], nor with experimental measurements shown in this paper. Nevertheless, additional detailed experimental measurements of the mean and fluctuating velocity fields, as well as the time histories of the vapor fraction distribution in the cavity is needed to truly quantify the validity of the various cavitation models;

(v) in general, the Merkel model with a density correction of $n = 3$ shows the best agreement with experimental measurements and observations for both steady and unsteady cavitating flows, particularly in terms of the cavity shedding frequency. The predicted shedding for unsteady sheet/cloud cavitation obtained using the Merkle model with $n = 3$ matches with the measured value of 3.5 Hz. The Kubota model is not able to converge for the case of unsteady sheet/cloud cavitation, and the Singhal model predicts much higher cavity shedding frequency and much lower amplitude of fluctuations compared to the measured values;

(vi) the analysis showed that the destruction and production coefficients in the Singhal model are not dimensionless parameters, but instead should have units of velocity to be consistent with the units of the mass transport equation;

(vii) the results demonstrate the importance of capturing both the mean and fluctuating values of the hydrodynamic coefficients and cavity lengths for the case of unsteady sheet/cloud cavitation because (1) the fluctuations are in the same order as the mean values, and hence knowledge of the fluctuations is critical to capturing the extremes of the loads to ensure structural safety; (2) the need to capture the frequency of the fluctuations, which are critical to avoid unwanted noise, vibrations, and accelerated fatigue issues. Moreover, it is also critical to compare both the amplitude and frequencies of fluctuations in addition to the mean values when evaluating the accuracy of numerical models.

This paper helps to improve the understanding of the predictive capability of the three popular cavitation models. The impact of the density correction term on the different cavitation models and its influence on the cavitation dynamics, vorticity field, and the force coefficients have been investigated. Additional research is needed to better understand the influence of the local compressibility and baroclinic torque on the cavitation and wake dynamics, as well as turbulent vorticity fields, particularly for cases with spatially/temporally varying inflow, and for cases with rigid and/or elastic body motions. Such research is important because accurate prediction of the cavity and load fluctuations is critical when analyzing the hydroelastic stability, fatigue, noise, vibration, and erosion characteristics of hydraulic machineries.

Acknowledgments

The authors gratefully knowledge the Office of Naval Research (ONR) and Dr. Ki-Han Kim (program manager) for their financial support through Grant no. N0014-09-1-1204.

References

[1] G. K. Batchelor, *An Introduction to Fluid Mechanics*, Cambridge University Press, 1967.
[2] C. E. Brennen, *Cavitation and Bubble Dynamics*, Oxford Engineering & Sciences Series 44, Oxford University Press, New York, NY, USA, 1995.
[3] R. T. Knapp, J. W. Daily, and F. G. Hammitt, *Cavitation*, McGraw-Hill, New York, NY, USA, 1970.
[4] D. Joseph, "Cavitation in a flowing liquid," *Physical Review E*, vol. 51, no. 3, pp. R1649–R1650, 1995.
[5] C.-C. Tseng and W. Shyy, "Modeling for isothermal and cryogenic cavitation," *International Journal of Heat and Mass Transfer*, vol. 53, no. 1–3, pp. 513–525, 2010.
[6] F. R. Menter, *Improved Two-Equation k-Turbulence Models for Aerodynamic Flows*, NASA Technical Memorandum, 1992, 103975.
[7] S. Kim and S. Brewton, "A multiphase approach to turbulent cavitating flows," in *Proceedings of the 27th Symposium on Naval Hydrodynamics*, Seoul, Korea, 2008.
[8] M. Morgut, E. Nobile, and I. Biluš, "Comparison of mass transfer models for the numerical prediction of sheet cavitation around a hydrofoil," *International Journal of Multiphase Flow*, vol. 37, no. 6, pp. 620–626, 2011.

[9] Y. Delannoy and J. L. Kueny, "Two phase flow approach in unsteady cavitation modelling," in *Proceedings of the Spring Meeting of the Fluids Engineering Division*, pp. 153–158, June 1990.

[10] Y. Chen and S. D. Heister, "Modeling hydrodynamic nonequilibrium in cavitating flows," *Journal of Fluids Engineering*, vol. 118, no. 1, pp. 172–178, 1996.

[11] S. Gopalan and J. Katz, "Flow structure and modeling issues in the closure region of attached cavitation," *Physics of Fluids*, vol. 12, no. 4, pp. 895–911, 2000.

[12] I. Senocak and W. Shyy, "Evaluation of cavitation models for Navier-Stokes computations," in *Proceedings of the ASME 2002 Fluids Engineering Division Summer Meeting Montreal (FEDSM '02)*, pp. 395–401, Quebec, Canada, July 2002.

[13] A. Kubota, H. Kato, and H. Yamaguchi, "A new modelling of cavitating flows: a numerical study of unsteady cavitation on a hydrofoil section," *Journal of Fluid Mechanics*, vol. 240, no. 1, pp. 59–96, 1992.

[14] R. F. Kunz, D. A. Boger, D. R. Stinebring et al., "A preconditioned Navier-Stokes method for two-phase flows with application to cavitation prediction," *Computers & Fluids*, vol. 29, no. 8, pp. 849–875, 2000.

[15] A. K. Singhal, M. M. Athavale, H. Li, and Y. Jiang, "Mathematical basis and validation of the full cavitation model," *Journal of Fluids Engineering*, vol. 124, no. 3, pp. 617–624, 2002.

[16] C. L. Merkle, J. Z. Feng, and P. E. O. Buelow, "Computational modeling of the dynamics of sheet cavitation," in *Proceedings of the 3rd International Symposium on Cavitation*, Grenoble, France, 1998.

[17] I. Senocak and W. Shyy, "A pressure-based method for turbulent cavitating flow computations," *Journal of Computational Physics*, vol. 176, no. 2, pp. 363–383, 2002.

[18] J. B. Leroux, J. A. Astolfi, and J.-Y. Billard, "An experimental study of unsteady partial cavitation," *Journal of Fluids Engineering*, vol. 126, no. 1, pp. 94–101, 2004.

[19] B. E. Launder and D. B. Spalding, "The numerical computation of turbulent flows," *Computer Methods in Applied Mechanics and Engineering*, vol. 3, no. 2, pp. 269–289, 1974.

[20] C. E. Brennen, *Fundamentals of Multiphase Flow*, Cambridge University Press, 2005.

[21] P. Zwart, A. Gerber, and T. Belamri, "A two-phase flow model for predicting cavitation dynamics," in *5th International Conference on Multiphase Flow*, Yokohama, Japan, 2004.

[22] A. Ducoin, J. A. Astolfi, F. Deniset, and J.-F. Sigrist, "Computational and experimental investigation of flow over a transient pitching hydrofoil," *European Journal of Mechanics B*, vol. 28, no. 6, pp. 728–743, 2009.

[23] O. Coutier-Delgosha, R. Fortes-Patella, and J. L. Reboud, "Evaluation of the turbulence model influence on the numerical simulations of unsteady cavitation," *Journal of Fluids Engineering*, vol. 125, no. 1, pp. 38–45, 2003.

[24] X. Li, G. Wang, Z. Yu, and W. Shyy, "Multiphase fluid dynamics and transport processes of low capillary number cavitating flows," *Acta Mechanica Sinica*, vol. 25, no. 2, pp. 161–172, 2009.

[25] J. H. Seo and S. K. Lele, "Numerical investigation of cloud cavitation and cavitation noise on a hydrofoil section," in *Proceedings of The 7th International Symposium on Cavitation (CAV '09)*, Ann Arbor, Mich, USA, August 2009.

[26] J. B. Leroux, O. Coutier-Delgosha, and J. A. Astolfi, "A joint experimental and numerical study of mechanisms associated to instability of partial cavitation on two-dimensional hydrofoil," *Physics of Fluids*, vol. 17, no. 5, Article ID 052101, 20 pages, 2005.

[27] K. R. Laberteaux and S. L. Ceccio, "Partial cavity flows. Part 1. Cavities forming on models without spanwise variation," *Journal of Fluid Mechanics*, vol. 431, pp. 1–41, 2001.

[28] J. P. Franc and J. M. Michel, "Unsteady attached cavitation on an oscillating hydrofoil," *Journal of Fluid Mechanics*, vol. 193, pp. 171–189, 1988.

11

Computational Fluid Dynamic Analysis of a Vibrating Turbine Blade

Osama N. Alshroof,[1] Gareth L. Forbes,[2] Nader Sawalhi,[3] Robert B. Randall,[1] and Guan H. Yeoh[1,4]

[1] *School of Mechanical and Manufacturing Engineering, The University of New South Wales, Kensington, Sydney, NSW 2052, Australia*
[2] *Department of Mechanical Engineering, Curtin University, Bentley, Perth, WA 6102, Australia*
[3] *Mechanical Engineering Department, Prince Mohammad Bin Fahd University (PMU), AlKhobar 31952, Saudi Arabia*
[4] *Australian Nuclear Science Technology Organisation (ANSTO), Locked Bag 2001, Kirrawee DC, NSW 2233, Australia*

Correspondence should be addressed to Osama N. Alshroof, o.alshroof@unswalumni.com

Academic Editor: Zhaoye Qin

This study presents the numerical fluid-structure interaction (FSI) modelling of a vibrating turbine blade using the commercial software ANSYS-12.1. The study has two major aims: (i) discussion of the current state of the art of modelling FSI in gas turbine engines and (ii) development of a "tuned" one-way FSI model of a vibrating turbine blade to investigate the correlation between the pressure at the turbine casing surface and the vibrating blade motion. Firstly, the feasibility of the complete FSI coupled two-way, three-dimensional modelling of a turbine blade undergoing vibration using current commercial software is discussed. Various modelling simplifications, which reduce the full coupling between the fluid and structural domains, are then presented. The one-way FSI model of the vibrating turbine blade is introduced, which has the computational efficiency of a moving boundary CFD model. This one-way FSI model includes the corrected motion of the vibrating turbine blade under given engine flow conditions. This one-way FSI model is used to interrogate the pressure around a vibrating gas turbine blade. The results obtained show that the pressure distribution at the casing surface does not differ significantly, in its general form, from the pressure at the vibrating rotor blade tip.

1. Introduction

Blade vibration is an inherent characteristic in the operation of gas turbines, due to the passing of the rotor blades through the trailing wakes from the upstream stator blade stage. This unavoidable blade vibration needs to be taken into account in the design of these engines and as such the measurement of blade vibration is needed. Due to the high operating temperatures, pressures and the rotating environment within a turbine engine, the need for noncontact measuring techniques is apparent. A recent research program has revealed that the gas turbine internal casing pressure, and resulting casing vibration, can be used to determine some characteristics of the internal blade vibration [1, 2]. This work, however, relied on the assumptions that the pressure at the casing surface was not significantly different from that at the blade tip and that the pressure profile developed around the rotor blades followed the blade oscillatory motion.

Investigation of the pressure at the casing surface and at the tip of a vibrating rotor blade is undertaken within this paper through a "tuned" three-dimensional one-way fluid-structure interaction (FSI) model of a vibrating turbine blade stage using the commercial numerical code ANSYS (ANSYS Mechanical and ANSYS CFX). This paper also introduces and discusses the limitations on numerical modelling of complex structures, such as gas turbine engines, undergoing FSI, and the various simplifications currently employed to model this behaviour.

2. When Is FSI Modelling Required?

Most often problems which involve the interaction between fluids and structures can quite sensibly be modelled as uncoupled problems within their separate domains. In these uncoupled problems it is assumed that one domain is driven by the other, with the driven domain having no feedback effect on the driving domain. This results in what is called one-way fluid-structure interaction [3]. The choice of which is the driving and driven domain, respectively, is made from holistic knowledge of the problem being investigated. For instance, the acoustic excitation of air being driven by a speaker can, most often, sensibly be modelled without the need to look at the feedback effects of the driven air on the structural deflection of the load speaker surface. That is therefore an example of a classic FSI problem where the coupling effect of the interaction between the fluid and structural domains only needs to be taken into account in one direction, namely, the effect of the structure on the fluid. Another example, conversely, where the fluid is the driving domain and the structure is being driven, with little feedback effect between the domains, is the dynamic wind loading on many civil structures. The dynamic nature of the flow, and gusting of winds, provides variable dynamic driving loads on the structures' surfaces. However, the motion and displacement of the structure rarely have any influence on the wind loads which drive this motion. Thus the structure can be analysed separately, or uncoupled from the wind, under the given loads of interest. This, however, is not always the case for wind loads on civil structures, as the infamous Tacoma Narrows Bridge collapse is thought to be a result of FSI [4]. The Tacoma Narrows Bridge is believed to have collapsed due to the dynamic motion of the bridge under wind loads which introduced aeroelastic flutter in the bridge deck. The motion of the bridge caused by the wind created structural deflections which further enhanced the wind forces driving its motion, so that the air and the structure were interacting in a feedback system. It is when conditions such as these arise that modelling of the two-way FSI needs to be undertaken. This is often the case in situations where the amplitude of the structural deflection is large and/or the structural system is highly compliant.

Analytical modelling of complex structures has limited scope, so that numerical analysis is often required. The need for numerical solutions to most complex engineering problems has driven the development and wide application of Finite Element Analysis (FEA) and Computational Fluid Dynamics (CFD) numerical simulation in the respective structural and fluid domains. One way to numerically model coupled FSI problems is to couple these two computational domains together with independent solvers for each domain, passing boundary conditions from each domain to the other at each computational step [5]. This setup of independent numerical solvers for each domain is the so-called partitioned solver technique; this is in contrast to a monolithic FSI solver procedure where the coupled boundary conditions are solved together in both domains at each time step [6]. It should be noted, however, that a monolithic solver still often requires iteration of the obtained solution at each time step if the system is nonlinear, which is the case for strongly coupled domains.

Although two-way fluid-structure coupling is starting to be more widely available in commercial CFD/FEA packages, it still has not reached a level of integration and computational efficiency to enable it to be used for the analysis of complex structure and flow interactions without significant computational and time expense.

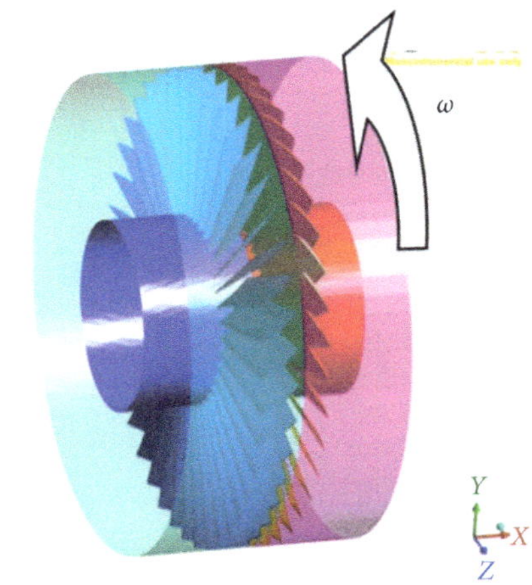

FIGURE 1: Schematic view of gas turbine stage, showing the hub, the external casing, and the blades (40 stator blades and 40 rotor blades).

3. Current FSI Modelling of Gas Turbines

Fluid flow and structural motion within a gas turbine represent a highly complex system to be numerically modelled. Gas turbines consist of multiple rotor and stator blade stages, as shown in Figure 1, which depicts a single stator and rotor blade stage, rotating relative to each other and producing complex flow and forces around both the stator and rotor blades. Full investigation of the flow and structural motion in such an engine requires a three-dimensional two-way coupled FSI model to be developed. Full three-dimensional two-way coupled modelling of a gas turbine stage with rotor blade vibration is theoretically possible with today's commercial numerical packages; however, the computational expense and setting up of such models are prohibitive and do not appear to be within feasible limits for a number of years yet. The limitations of commercial software on the two-way FSI modelling have also been noted by other authors [7, 8]. For instance, a three-dimensional single stator and rotor blade sector with two-way coupling on the rotor blade surface using ANSYS Workbench 12.1 may take months for a converged solution. As another example, the model within this paper required approximately 4 weeks to obtain a converged solution. Two-way coupling, on average, can take an order of magnitude longer in computational time when compared to a one-way model. One-way modelling is no more onerous than a pure CFD model, except for the more complicated flow that can develop due to the structural motion and the transient solution method which is often required.

Adding to this computational time is the significant increase in computational expense which arises when modelling a turbine stage, as the entire stage often needs to

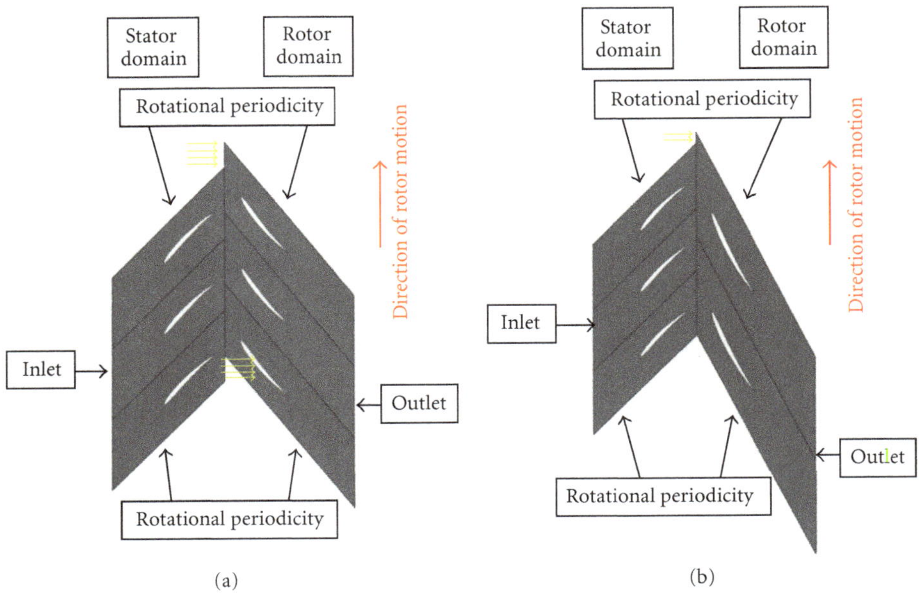

FIGURE 2: Modelling of rotor/stator domains requires the lowest common multiple of rotor/stator blades to be modelled for continuity of flow out of one domain and into another. (a) Integer pitch ratio, all flow out of stator domain flows back into rotor domain (b) noninteger pitch ratio, has unequal flow from stator to rotor domains.

be included in the model. In order to obtain a complete analysis, the number of blade sectors required to be modelled should be the lowest common ratio of the number of rotor to stator blades. The need to model this increased domain can be explained with the aid of Figure 2. In Figure 2(a) a two-dimensional blade cascade is shown for a stage with equal numbers of rotor and stator blade numbers (note that three sectors have been shown for clarity). This allows, with appropriate boundary conditions, a single sector to be modelled. Flow from the stator side of the rotor/stator junction can flow into the rotor side (highlighted in yellow). In Figure 2(b) a noninteger pitch ratio is shown. In this case, continuity of the flow from the stator to the rotor is not provided. Therefore in order to fully model a turbine stage an equal width stator and rotor domain need to be created. In the case of most real turbine stage designs, noninteger pitch numbers are desirable which can result in a large computational domain. For instance, a stator of 13 blades and a rotor of 17 blades would require the entire stage to be modelled for the continuity of the flow between the stator/rotor. However, the most recent version of ANSYS includes multirow methods, which reduce the computational domain needed to allow for noninteger pitch ratios without the need for modelling the whole domain [8]. It is unclear yet whether these multirow methods result in any loss of accuracy.

In addition to often requiring a large computational domain for complete analysis of gas turbine engines, the unsteady nature of the flow within the turbine stage also requires the use of a transient solution procedure. For instance, in reference to Figure 1, the rotor stage moves relative to the stator stage and causes oscillating flow and forces on the rotor blades. Therefore large computational domains and transient solution procedures have meant that a number of simplifying FSI modelling techniques for gas turbine engines have arisen.

To the best of the authors' knowledge, a complete three-dimensional transient two-way FSI of a turbine engine including blade vibration has not been achieved. Some notable simplified models which have been developed recently are

(i) two-way transient three-dimensional FSI model of an axial hydro turbine; this does not include the preceding stator stage, so it only considers static deformation and does not include blade vibration [7];

(ii) pseudo two-dimensional two-way transient FSI model of a gas turbine rotor blade vibration [9];

(iii) one-way transient three-dimensional FSI model of a gas turbine blade undergoing vibration [10];

(iv) significant numerical and experimental investigation of the one-way transient three-dimensional FSI model of a turbine blade undergoing vibration and the effect on blade flutter and casing tip gap [11].

The use of a harmonic balance approach to the solution of the fluid dynamics is by far the most efficient for steady state harmonic response forms [12]. Harmonic balance methods, however, do not seem to be available in commercial software packages at this stage. The increase in efficiency is reduced as full transient flow or motion should be modelled. Neglecting the full transient flow effects may not produce the actual blade motion, but as the dominant pressure pulses on the turbine blades during operation are from harmonic orders of the rotational speed, the dominant motion effects should still be present.

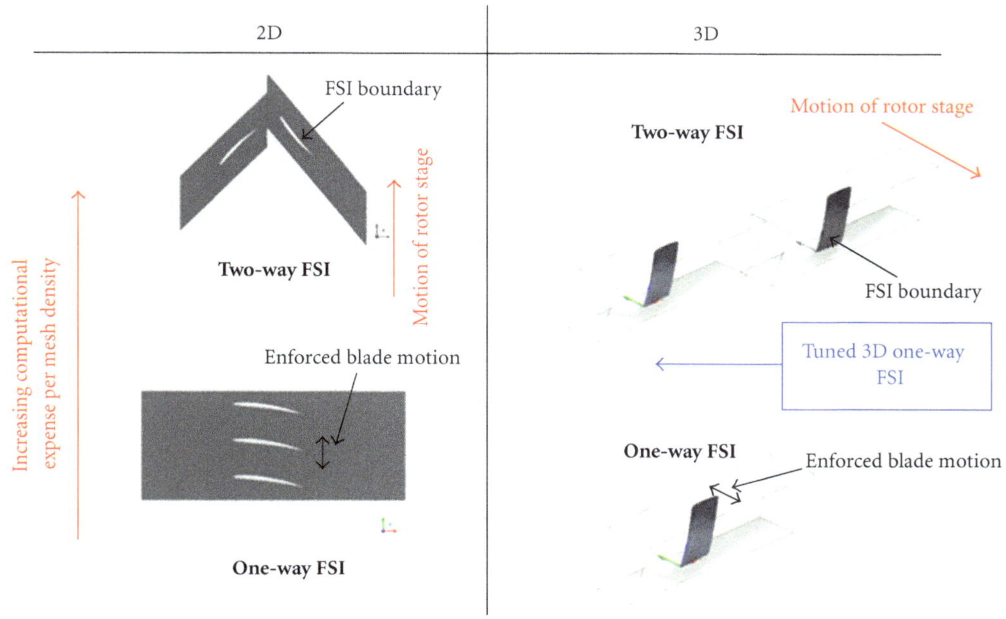

FIGURE 3: Schematic of different FSI numerical modelling procedures for gas turbines. The simplified tuned model used within this work lies between 3D One-way FSI and 2D Two-way FSI in computational expense for the same mesh density.

Figure 3 shows schematically the increase in computational expense for different FSI modelling approaches, with each employing simplifying assumptions from the full three-dimensional two-way FSI model. One-way FSI models give the option of reduced complexity by enforcing the vibration on the blade surface and solving the moving boundary condition CFD model of a fluid sector with a vibrating rotor blade. However, these models do not take into account the preceding stator blade row effects. Two-dimensional models give an obvious reduction in complexity by limiting the computational domain, but this eliminates the three-dimensional behaviour of the flow. This simplification does not allow resolution of the complex three-dimensional flow behaviour in the gap between the tip of the blade and casing wall which is the aim of this paper.

The model presented within this paper is a "tuned" three-dimensional one-way FSI model which enforces the rotor blade motion but also includes the preceding stator blade domain and "tunes" the blade motion to take into account the dynamic pressure on the rotor blade due to the stator passing period. This model was developed to give the computational efficiency of a one-way model and to include the effects of the two-way interaction of the blade motion and the driving pressure fluctuations on the rotor blade. Once this model is developed, the investigation of the pressure distribution at the rotor blade tip and the casing surface is undertaken.

4. Rotor Blade Vibration

As stated earlier, vibration of gas turbine blades is an inherent characteristic of their use and is driven predominantly by the periodic traversing of high and low pressure regions from the preceding stator blade row. This process can be explained, with the aid of Figure 1, by conceptualising how rotor blades would need to pass through the flow of the stator rows as the turbine rotates. This blade vibration has been enforced on the rotor blade in the following numerical modelling, forcing the rotor blade section to move transverse to the turbine's angle of attack.

A three-dimensional gas turbine was modelled as shown in Figure 1, with 40 blades on a 100 mm diameter hub rotating with angular velocity, ω, of 500 rad/s. The axial gap between the stator blades and the rotor blades is 10 mm and various radial gaps between the rotor blade tips and casing surface, of 0.5 mm, 1 mm, and 1.5 mm, were tested. In practice the smallest gap possible between the blade tip and casing surface is desirable so the sensitivity to the size of this gap was investigated. An equal number of stator and rotor blades were used to reduce the computational domain (as explained in Section 3), such that the enforced motion on the rotor blades was given the stator passing frequency of 20000 rad/s with the maximum blade tip peak to peak amplitude of 0.15 mm. In order to reduce the computational time needed to solve this problem, a three-dimensional wedge encompassing one stator blade and another three-dimensional wedge encompassing one rotor blade, as shown in Figure 4(a), were simulated using the boundary conditions outlined in the next section.

5. Boundary Conditions

The following boundary conditions were employed to obtain the time-dependent solution (Table 1).

It should also be noted that the rotor domain is rotating at 500 rad/s, as shown in Figure 1, whereas the casing surface is defined as a stationary nonslip wall.

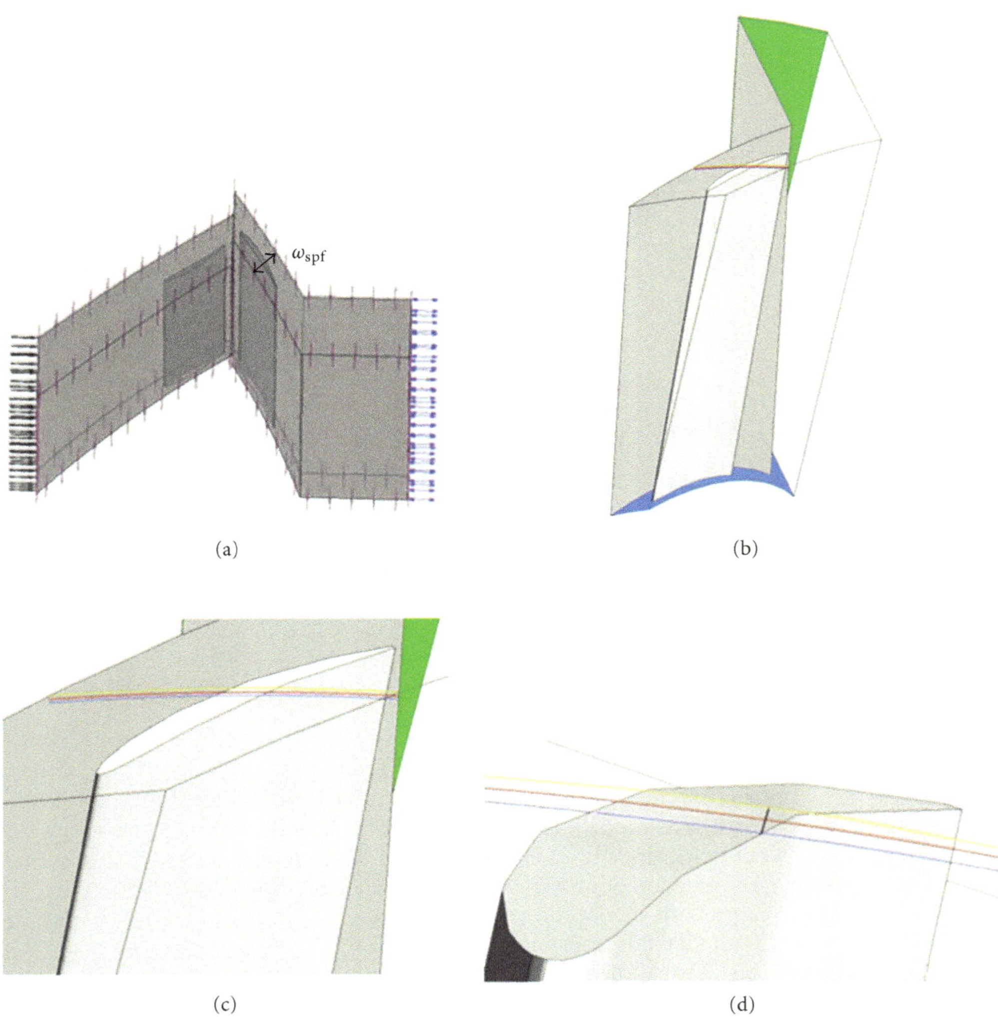

FIGURE 4: (a) Three-dimensional blade model showing blade vibration direction. (b) Three-dimensional model showing polylines where the pressure is observed. (c) Zoomed polylines. (d) Polylines and vertical line across blade tip/casing gap where the pressure is observed.

TABLE 1: Boundary conditions for the FSI model as shown in Figure 2.

Boundary	Description
Inlet	Inlet velocity = $[44.38i - 61.08j + 0k]$ m/s, in which i, j, and k are the unit vectors for x, y, and z coordinates as shown in Figure 3.
Outlet	Average relative static pressure = 0.
Casing surface	Nonslip wall
Hub	Nonslip wall
Stator blade	Nonslip wall
Rotor blade	Nonslip wall with forced vibration added to represent the response to forces from the periodic traversing of high and low pressure regions from the preceding stator blade row. The displacement of the rotor blade is presented as a sinusoidal displacement as a function of time and position in the z-direction as follows: $\delta(t,z) = 45\left(-A \times \sin\left(20000t - t_f\right)\right) \times \left(\lvert z\rvert - R_0\right)^2$, in which there are 40 stator blades, $t_f = 1.2868$ s is a phase offset calculated from the applied pressure, and $R_0 = 0.1$ m is the radius of the hub
The two sides of the wedge	Defined using rotational periodicity
The interface between the two domains	Defined as transient rotor stator [13]

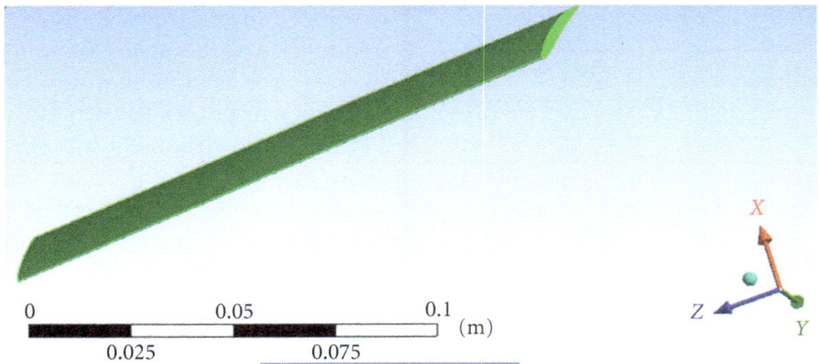

Bounding box	
Length X	2.4327e-002 m
Length Y	6.5411e-002 m
Length Z	0.15351 m

FIGURE 5: 150 mm long rotor blade model.

6. Rotor Blade Structural Analysis and Property Settings

The forces present on the rotor blade were first calculated under the flow conditions it would experience without any blade vibration. It is the resulting motion of the rotor blade, calculated as a result of these forces, which is used in the later enforced motion of the CFD model. The response of the rotor blade under these forces is discussed in this section. The material properties of the blade were altered to obtain the desired maximum blade tip displacement of 0.15 mm. What is undertaken within this section is a classic decoupled structural analysis under the incoming flow conditions.

A judicious choice of damping is also made to suppress higher modes of vibration for two reasons. Firstly, the first mode of vibration is often the most dominant and thus the most important to consider. Secondly, limiting the motion to only one mode of vibration allows for simpler motion to be observed when investigating the flow between the blade tip and casing surface (this avoids the necessity of low-pass filtering the calculated results).

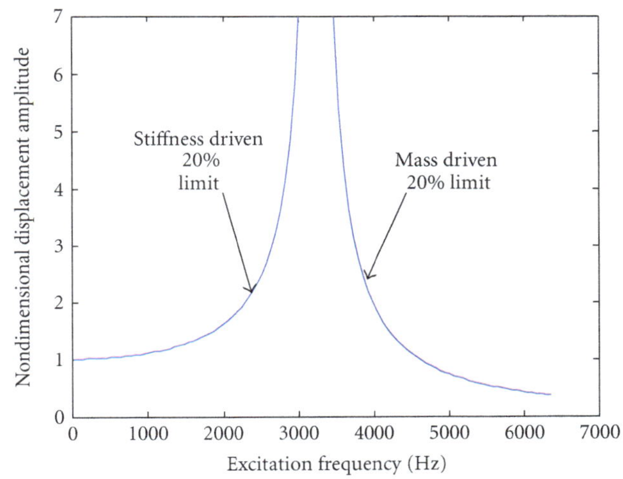

FIGURE 6: Nondimensional amplitude for the first mode with different driving frequencies. Shown are the chosen stiffness-driven limit 20% below the natural frequency and the mass-driven limit 20% above.

6.1. Blade Geometry. The blade under consideration is shown in Figure 5; it has a length of 150 mm and a chord length of 50 mm with an aerodynamic profile of NACA4506.

6.2. Natural Frequencies and Mode Shapes. The aim of the initial finite element model of one rotor blade is to tune the structural properties of the blade (mass density ρ, modulus of elasticity E) so the first natural frequency of the blade is placed roughly at $-/+$ 20% from the excitation frequency (blade pass frequency) as shown in Figure 6. Normally the separation has to be even larger than 20% to ensure that the system is not excessively excited. For the purpose of this study the 20% limit was selected to drive the system close to its first natural frequency and obtain reasonable blade deflections.

To simulate the connection of the blade to the rotor, the rotor blade FEA model was fixed at one end. Thus the model in its simplest configuration represents a cantilever beam with an aerofoil cross-section. The analytical expression to calculate the natural frequencies of a cantilever beam is given in (1) [14]. The analytical expression of (1) allows observation of the main properties which control the behaviour of the blade natural frequency and thus how it reacts to external excitations:

$$\omega_i = (\beta_i l)^2 \sqrt{\frac{EI}{\rho A l^4}}, \tag{1}$$

where, i: natural frequency Index ($i = 1, 2, 3, \ldots$); ω_i: natural frequency (rad/s); $\beta_i l$: boundary condition constants; namely, $\beta_1 l = 1.875104$, $\beta_2 l = 4.694091$, $\beta_3 l = 7.854757$, and $\beta_4 l = 10.995541$; E: Young's modulus of elasticity (Pa); I: second moment of area of the beam (blade) cross-section about the y-axis (m^4); ρ: mass density (kg/m^3); A: cross-sectional area (m^2); l: length (m).

TABLE 2: The first four natural frequencies of one rotor blade.

Mode	Frequency (Hz)
(1)	3831.2
(2)	23112
(3)	25322
(4)	48583

For a rotor with 40 blades, and a speed of 500 rad/s (79.58 Hz, 4774.68 rpm), the blade pass frequency, BPF (rotor speed times the number of blades) would be 20,000 rad/s (3183.1 Hz). The selection of the first natural frequency at 20% from the BPF means that the 1st natural frequency would be either 2546.5 (20% below) for a stiffness controlled (driven) system or 3819.7 Hz (20% above) for a mass controlled (driven) system. This also means that the vibration response will be in phase with the exciting force for the stiffness controlled system (180 degrees out of phase if mass was controlled). Depending on the engine of interest and the dominant engine order, both stiffness and mass controlled blade vibrations can be found in turbine engines.

To achieve a first natural frequency around 3820 Hz, the mass density of the material (ρ) and the Young's modulus of elasticity (E) were adjusted. The modulus of elasticity per mass density, which is often referred to as the *specific stiffness* or *specific modulus* needs to be optimized. In aerospace applications the specific modulus is required to be high indicating a minimum structure weight (hollow structure) with minimum deflections (higher modulus of elasticity or stiffness). To achieve a high specific stiffness structure with a first natural frequency around 3820 Hz, the mass density (ρ) was selected as 9 kg/m^3, while the modulus of elasticity (E) was increased to 3×10^{11} Pa. The first four natural frequencies of the blade are listed in Table 2, while corresponding mode shapes of the first three natural frequencies as shown in Figure 7. The most important of these is the first mode (pure bending). The second mode (mixed torsion and bending) is roughly 6-times higher, which agrees with rough calculations $(\beta_2 l/\beta_1 l)^2$, based on (1).

6.3. Damping. Modal damping, proportional to the mass and/or stiffness matrix, was added to the finite element model. To suppress the second and higher modes and retain the first mode, the damping factor (ψ), defined in (2) [15], was adjusted, in the material properties, to achieve this purpose:

$$\psi = \frac{2\zeta_i}{\omega_i}, \quad (2)$$

where, ζ_i is the damping ratio of mode i; ω_i is the natural frequency of mode i.

For a 2% damping in the first mode, (ψ) has been calculated to be 1.67×10^{-6} s. This value of (ψ) will correspond to a damping ratio of around 24% (very heavily damped) at the second mode. This gets even higher for higher modes, thus suppressing their effect. The suppression of the response of higher frequency modes was employed to avoid numerical problems in the simulation. The natural frequencies were outside the range directly excited by blade passage, and responses would be filtered to exclude higher frequency responses in any case, as has been done in Figures 14 and 15. Additionally the blade motion is related to the phase modulation of the blade pass signal, and this has been arranged to be close to the first mode, so that higher order modes are well outside the frequency band demodulated.

6.4. Deriving Blade Forced Vibration Equation (Amplitude and Phase). The rotor blade will be excited at the blade pass frequency (3183.1 Hz). The forced vibration equation for a cantilever beam along its length z and its change in time t, $\delta(t,z)$, is given in (3) [14]. The direction axial direction z for the blade of interest is shown in Figure 5. For use in this model the deflection of the blade along its length z has been simplified to a quadratic equation, as shown in (4). Application of (4) will result in a blade deflection shape which will look similar to the deflection shape shown in the top image in Figure 7. This deflection of the rotor blade is zero at the rotor hub and maximum at the blade tip whilst oscillating around its equilibrium position in time. The spatial distribution of the deflection is given by the "mode shape equation" and the osculation in time is given by the "time varying" section highlighted in (4):

$$\begin{aligned}\delta(t,z) &= \underbrace{C_1}_{\text{Constant}} \\ &\quad \times \overbrace{[\sin\beta_1 z - \sinh\beta_1 z - \alpha_1(\cos\beta_1 z - \cosh\beta_1 z)]}^{\text{Mode shape equation}} \\ &\quad \times \overbrace{(-0.00015 \times \sin(500 \times 40 \times t - 1.2868))}^{\text{Time varying}},\end{aligned} \quad (3)$$

where,

$$\alpha_1 = \left(\frac{\sin\beta_1 l + \sinh\beta_1 l}{\cos\beta_1 l + \cosh\beta_1 l}\right), \qquad \beta_1 l = 1.875104,$$

$$\delta(t,z) = \underbrace{\frac{1}{(0.25-0.1)^2}}_{\text{Scaling factor}} \times \overbrace{(z-0.1)^2}^{\text{Mode shape equation}} \times \overbrace{\left(\underbrace{-0.00015}_{\text{Max deflection}} \times \sin\left(\underbrace{500 \times 40}_{\text{BPF}} \times t \underbrace{-1.2868}_{\text{Phase}}\right)\right)}^{\text{Time varying}}. \quad (4)$$

The terms given in (4) are described as follows.

(i) t: simulation time steps defined in the CFD solver.

(ii) z: rotor blade longitudinal axis. Note that z values are defined as $0.1 \leq z \leq 0.25$. The limits of the blade z given such that zero deflection is obtained at

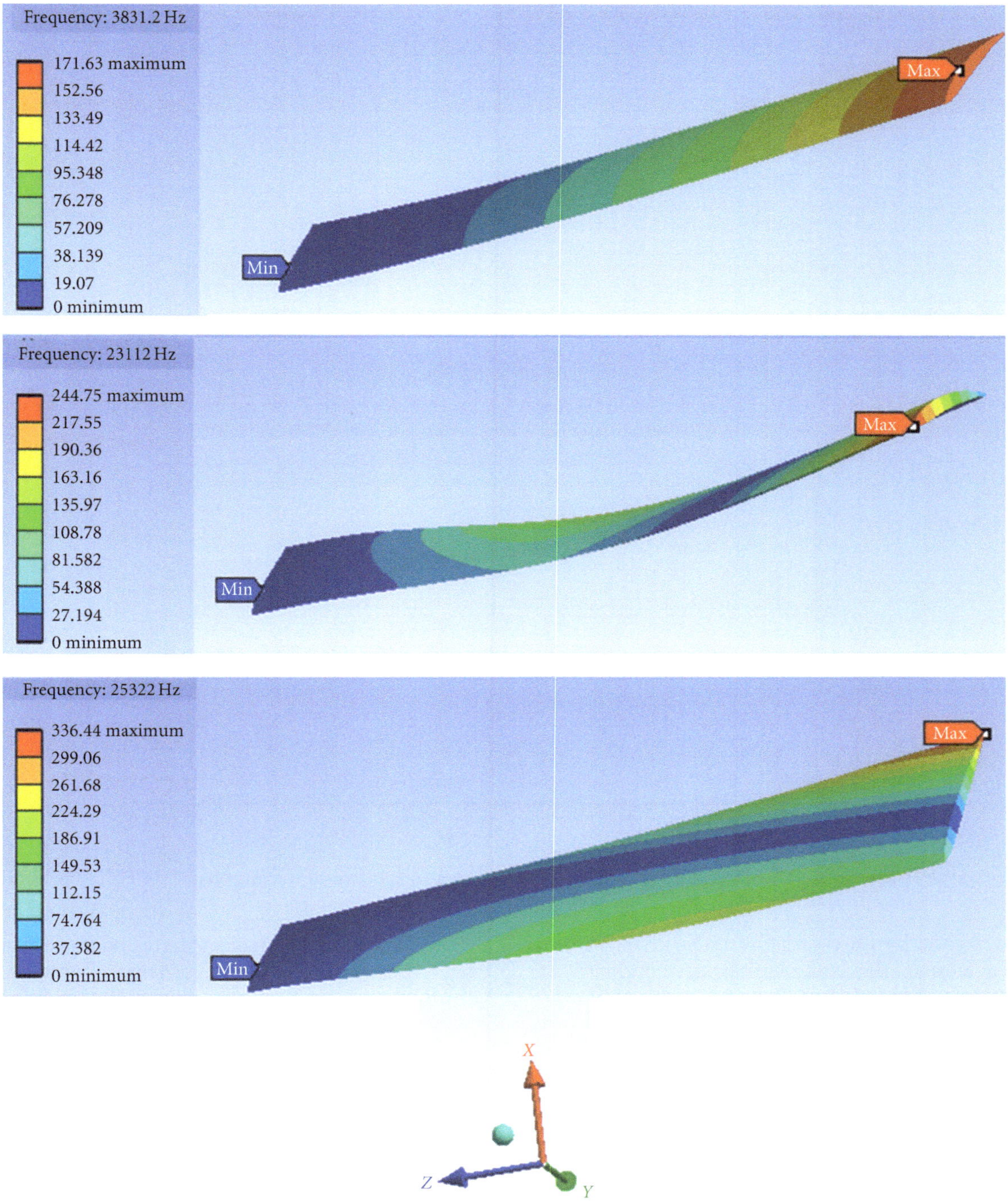

FIGURE 7: The first 3 mode shapes of the blade: top (pure bending), middle, a mixture of bending and torsion, and bottom, pure torsion.

the attachment to the rotor (0.1 m from the centre of the rotor) and maximum deflection at the tip (0.25 m). The scaling factor was incorporated to give a maximum deflection of 0.15 mm.

(iii) The rotor blade is already deflected under the average static pressure, that is, the blade starting position assumes that it has already deflected under the static pressure.

(iv) The phase of (3) was derived from running the CFD model on its own and observing the averaged pressure on one-third of the free end of the blade. The period for the pressure variation is 3183.1 Hz (BPF). The phase angle was derived as 1.2 rad and was used in the forced vibration equation to ensure that the response (deflection) is in phase with the pressure (excitation). However, as the model was run with the forced equation, it is the deflection of the blade that

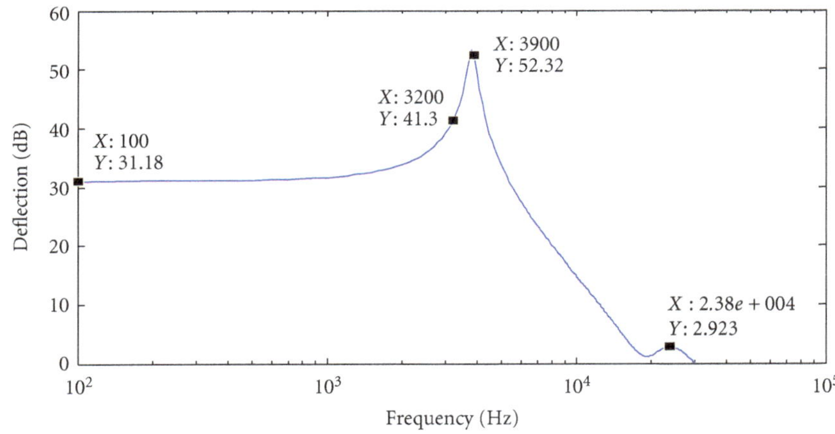

Figure 8: Harmonic response total maximum deflection Decibels (re 1.3×10^{-6}) for a 1000 Pa pressure acting on one surface of the blade.

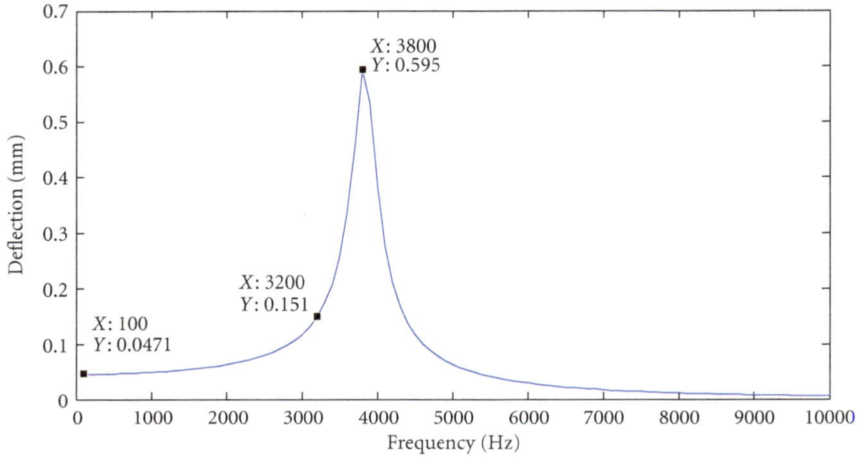

Figure 9: Harmonic response linear total maximum deflection for a 1000 Pa pressure acting on one surface of the blade.

creates the pressure variation rather than the other way around (this is the limiting assumption of any one-way FSI model).

(v) Amplitude of the blade vibration was arbitrarily selected as ±0.15 mm. This value corresponds to the blade response under a tip pressure of around 1000 Pa. The harmonic response (total deflection) under a constant 1000 Pa pressure applied to one surface is shown in Figure 8 (100–30000 Hz, dB—log scale) and zoomed in Figure 9 (100–5000 Hz linear). Figure 8, which displays the blade response over a wide range, shows that the dominant natural frequency in the model comes from the first natural frequency. In Figure 9, the maximum blade defection at the tip, when the blade is excited at the BPF, is shown to be 0.15 mm. The static displacement is 0.05 mm, while the resonant deflection will be about 0.6 mm.

7. CFD Solution Method

The air flow is considered to be compressible. A transient solver with a second-order backward Euler differencing scheme is used for the transient term and the SST turbulence model with a high-resolution scheme is employed for the advection terms in the transport equations governing mass, momentum, and energy. At each time step, internal iterations are carried out until the residuals for mass, momentum, and energy residuals fall below the criterion 10^{-6}. A time step of 1×10^{-6} s is used.

7.1. Details of the Grid. As the blade deforms, the fluid mesh needs to follow the displacement of the blade; therefore the mesh has to be expanded and compressed. As shown in Figure 10, hexahedral elements are created around the blade as well as for the rest of the computational domain. In order to adequately resolve the complex flow around the blade, an increased number of elements were placed around the surface of the blade. A total of 1.7 million elements was generated for the whole domain.

8. Investigation of the Pressure between the Blade Tip and Casing Surface

Now that the computational model has been described and set up, the pressure between the vibrating rotor blade tip

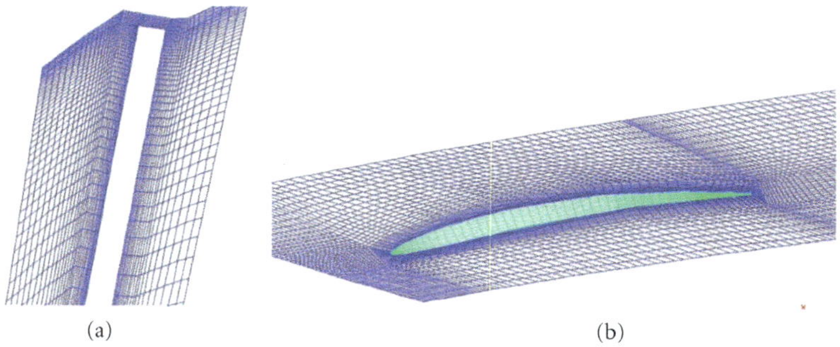

FIGURE 10: View of a structured grid at (a) horizontal plane crossing the blade tip and (b) cross-section plane located at the blade centre.

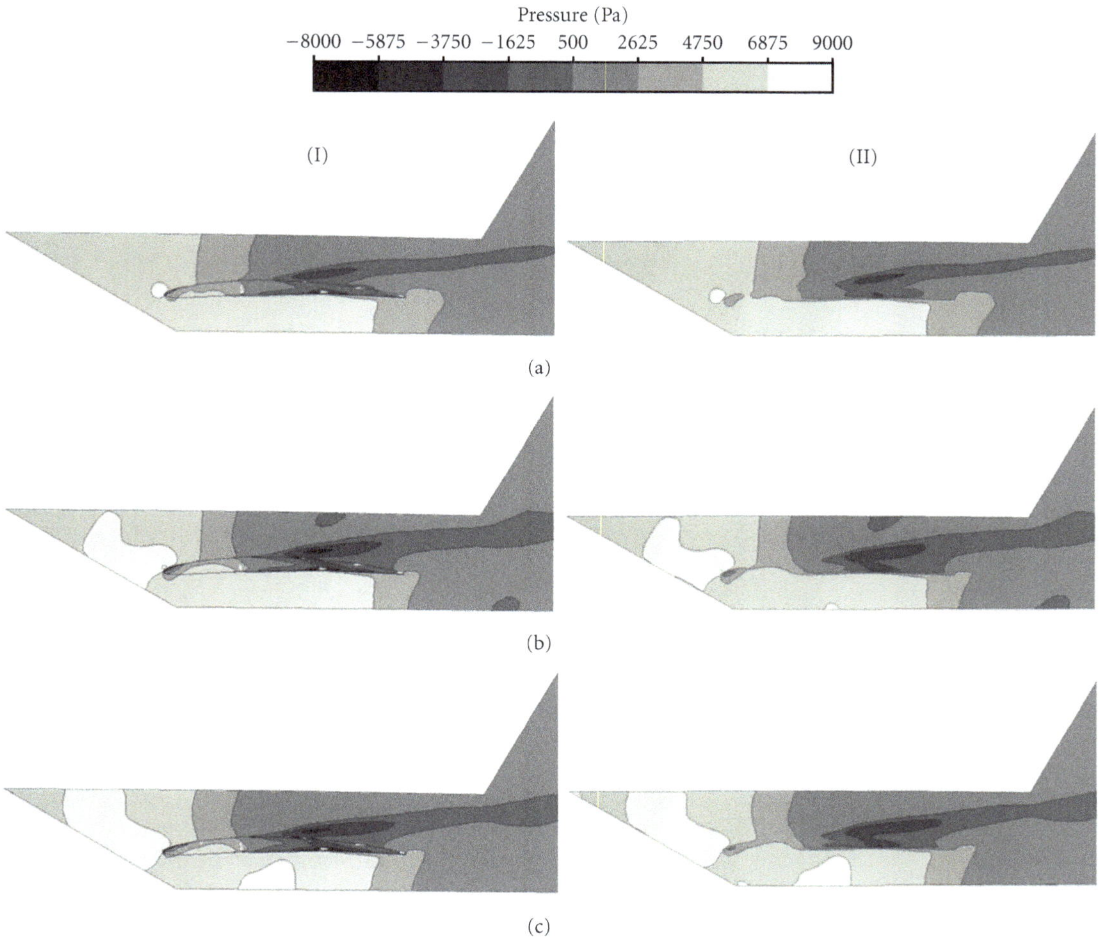

FIGURE 11: Pressure contours of surface located at (I) blade tip radius and at (II) the casing surface, for the case of 0.5 mm gap, when the blade is at (a) maximum deflection, (b) mean deflection, and (c) minimum deflection.

and the casing surface will now be investigated for a range of gaps between the blade tip and casing surface, namely, 0.5, 1, and 1.5 mm. The pressure between the blade tip and casing surface is being investigated to find if there is any significant difference between the pressure at the blade tip and that at the casing surface and whether the pressure profile around the blade follows its motion as it is oscillating around its equilibrium position.

During rotor rotation and the subsequent enforced rotor blade oscillation, a high pressure region, as expected, is maintained underneath the blade aerofoil and a low pressure region is maintained above the blade aerofoil. This is also observed at the blade tip; see Figure 11(I) which shows the pressure contours on a surface located at a radius equal to the blade tip, for the case of 0.5 mm gap, at maximum-minimum-mean values of blade deflection. Interestingly,

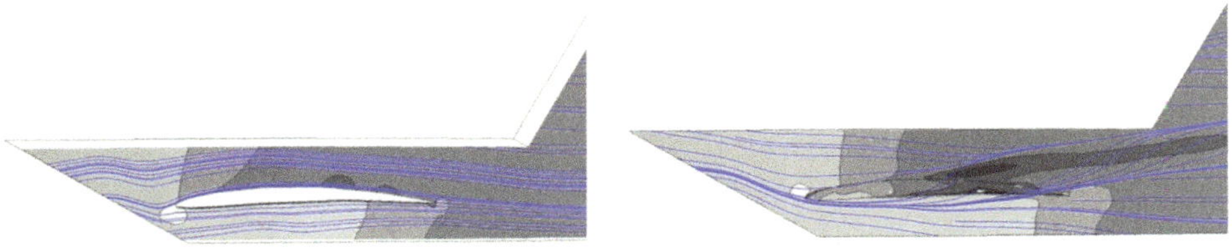

FIGURE 12: Pressure contours of surface located at (a) midway between blade tip and hub and (b) blade tip radios, for the case of 0.5 mm gap between the blade tip and casing surface.

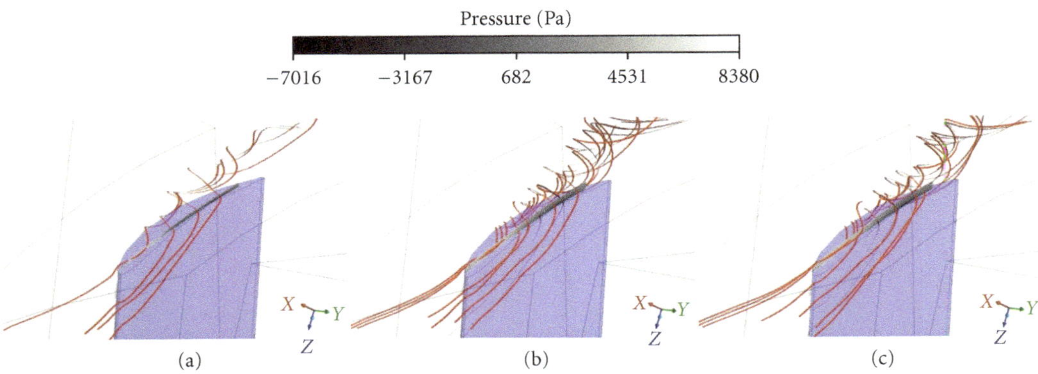

FIGURE 13: Streams of fluid crossing the gap between the top of the blade and the casing for the case of (a) 0.5 mm gap, (b) 1.0 mm gap and (c) 1.5 mm gap. Pressure distribution in the longitudinal plane is also shown.

the shape of the pressure contours at the blade tip cross-section surface, Figure 11(I), is similar to the pressure contours at the casing wall, Figure 11(II); however, the magnitude of the pressure varies slightly.

As a result of the low and high pressure regions around the blade, the flow escapes from the high pressure region to the low pressure region through the gap between the blade tip and the casing. This causes a blade tip vortex, which has been very well documented by previous researchers [11]. The development of this tip vortex can be seen in Figure 12(b), and is seen more clearly for each blade gap in Figure 13. It can be noted that the tip vortex becomes more developed for increasing tip gap. One aim of this study was to check that this vortex did not greatly affect the relationship between the pressure at the blade tip and the casing inner wall, at least for small tip clearances.

The angle of attack of the stator and rotor blades was selected to maintain smooth flow around the blade, with no separation, as shown in Figure 12(a). This smooth motion of the flow continued along the length of the rotor blade, excluding the top 5 mm of the blade tip, where separation of the flow was observed, as can be seen in Figure 12(b).

Now, as seen in Figure 11, the pressure appears to have a very similar form on the casing surface as it does at the blade tip. Whether or not the pressure profile around the rotor blade follows the blade motion as it vibrates is however not easily investigated by the use of Figure 11. This will now be investigated in more depth by looking at the pressure along three lines across the blade sector width during the blade's oscillation. These three lines were located at the blade tip, midway between the tip and at the casing surface; see Figure 4.

The pressure profile across the width of the rotor blade sector, over these lines, for three instances in time when the blade tip deflection is at its maximum, at its mean, and at its minimum deflection can be seen in Figures 14(a), 14(b), and 14(c), respectively. Figures 14(d), 14(e), and 14(f) then display the first harmonic, in space, of the pressure across the polylines for the corresponding displacements. Figures 14(d)–14(f) are simply the result of low pass filtering Figures 14(a)–14(c). Although the magnitude of the pressure distributions in Figures 14(a), 14(b), and 14(c) is not the same for the three pressure lines, the pressure distributions are generally of a similar shape and only differ at the top and bottom surface of the blade. Comparison of the dominant first harmonic of the pressure distribution shows an almost identical match between the pressures at the casing surface and that at the blade tip, shown in Figures 14(d), 14(e), and 14(f). It is definitely evident from observation of Figure 14 that the pressure distribution at the casing surface across the blade passage is very similar to that measured at a distance corresponding to the blade tip.

Figure 14 compared the pressure distributions across the rotor blade sector for different tip/casing gaps at a single instant during the blade oscillation. Figure 15 looks at the change in pressure distribution across the rotor blade sector at a single position, for example, blade tip during the blades range of motion. From observation of Figure 15 it can be seen that there is little phase shift between these first harmonics of pressure distribution; however,

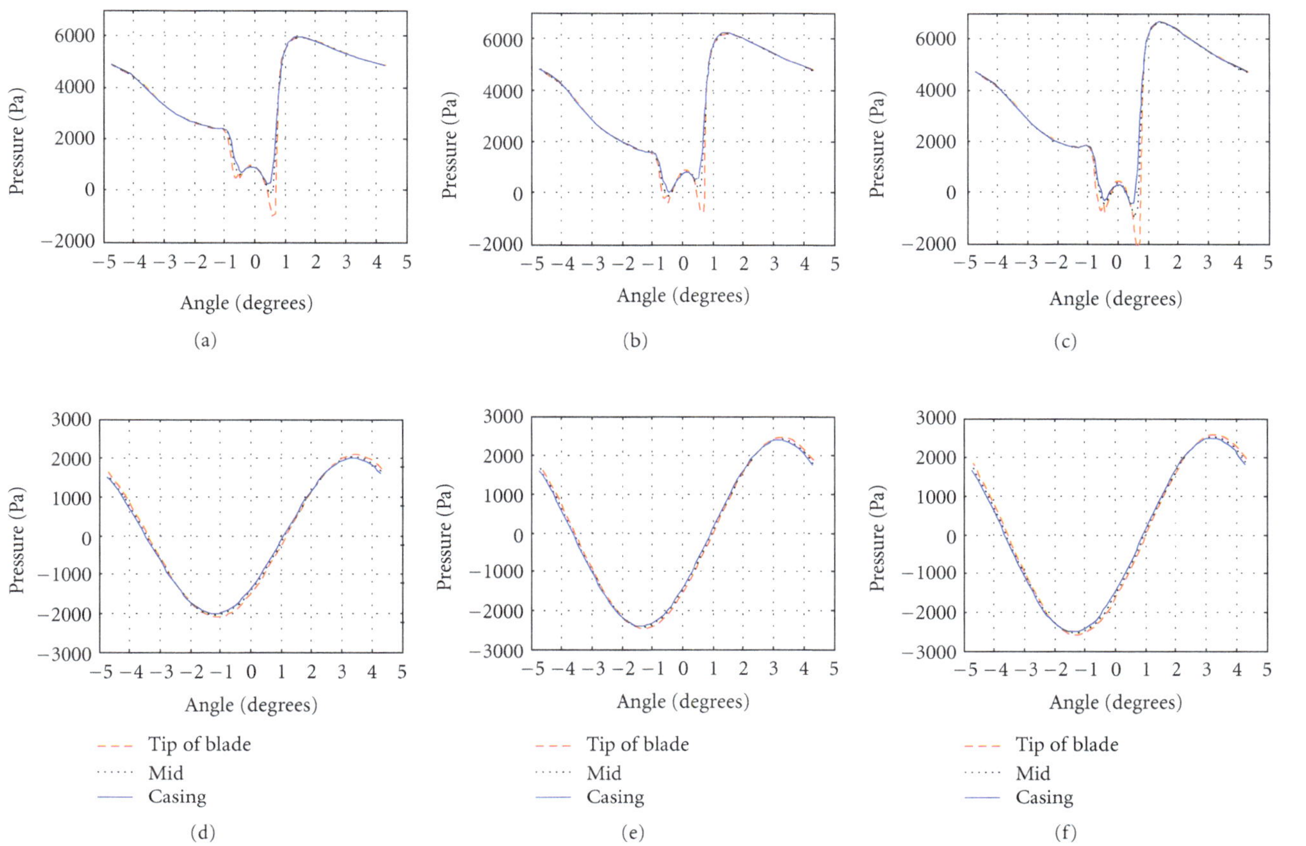

FIGURE 14: Pressure across the three polylines, shown in Figure 4(d), when the blade deflection is (a) at a maximum, (b) at a mean, and (c) at a minimum, for the case of 0.5 mm gap. (d), (e), and (f) are the first spatial harmonic of the pressure across the same polylines for the respective blade displacements. Note the angle given is relative to the centre of the aerofoil.

there is a relatively significant change in magnitude. Hence the pressure distribution is amplitude modulated by a pressure oscillation at a frequency corresponding to the blade passing period. This amplitude modulation of the pressure is something which should be considered in further analysis of the internal pressure signal of a gas turbine engine for possible use as a measure of the blade vibration.

The relationship of the pressure across the gap between the blade tip and casing surface is illustrated in Figure 16. See Figure 4 for the radial line for which the pressure distribution is plotted for the three blade deflections. The general shape of the pressure distribution along the line can be seen to be similar across the blade tip gap. Although the pressure distribution is similar in shape across the gap for the three blade deflection positions, the magnitude can be seen to differ. It is thought that this magnitude difference is a result of the different tip leakage (or flow across from one side of the blade to the other across the gap) resulting from the rotor and stator being at different angular offsets from each other at these times.

Figures 17 and 18 show the harmonic nature of the pressure at both the casing surface and the blade tip. In Figure 17 the pressure is observed over two stator passing periods at the casing surface as if a probe were inserted in the casing. The rotor blades will therefore be rotating past this point, while also oscillating back and forth about their equilibrium position. For a stationary probe located at the casing surface, the pressure around the blade will be modulated by both the vibrating motion and the motion of the rotation of the blade stage. This will cause higher harmonics of the stator passing frequency [1], which will be out of phase with each other. The pressure traces against time for the three different gaps between the blade tip and casing surface all seem to show this trend of higher harmonics of stator passing frequency. The first and second harmonics of the frequency corresponding to the stator passing period, 0.0031 s, are clearly visible.

Figure 18 shows the pressure at both the blade tip and the casing surface, as if a probe were placed in the fluid which followed the blade as it rotated. The pressure traces which follow the blade as it rotates also tend to vary harmonically with the stator passing period, dominated by the first harmonic and higher harmonics that are all generally in phase with each other. Once again the resulting pressures at the blade tip and casing surface show similar trends. The results for the 1.5 mm gap, however, do vary more significantly, which is believed to be due to the greater leakage of flow around the blade tip.

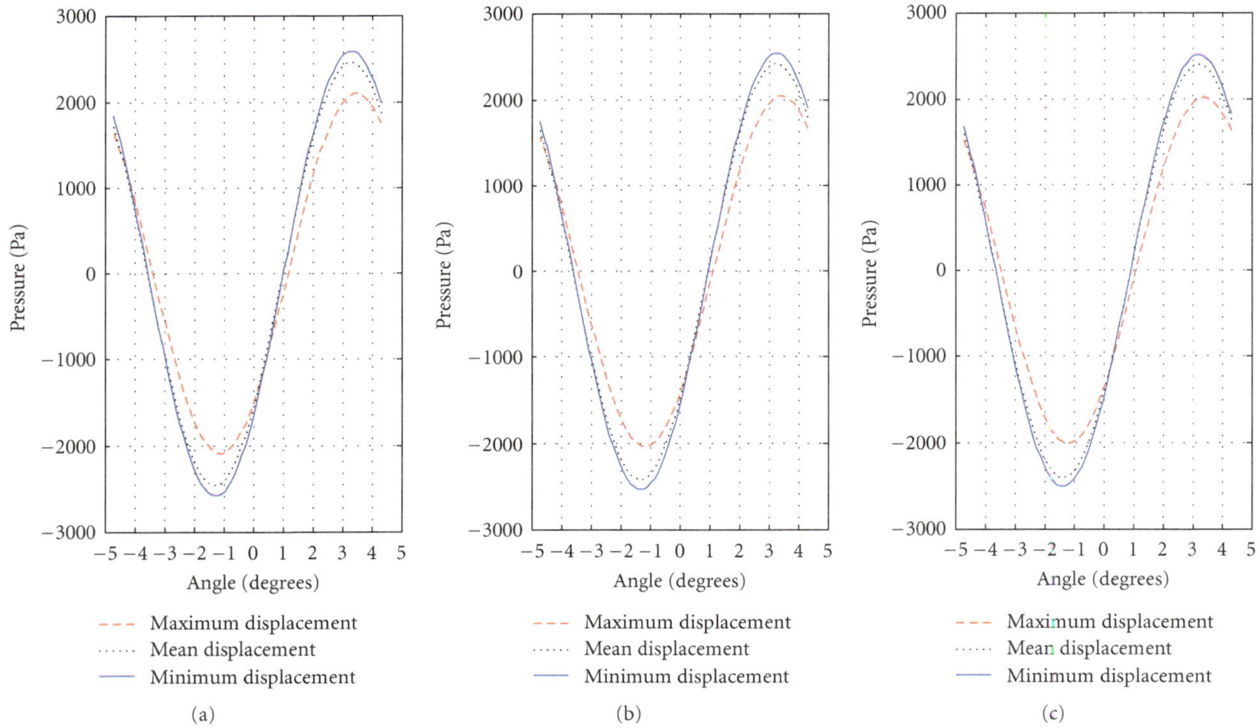

FIGURE 15: First spatial harmonic of the pressure across each of the three polylines, shown in Figure 4(d), when the blade deflection goes through its range of motion from maximum deflection, through mean deflection and to its minimum deflection position. (a) Polyline at tip of blade (b) polyline between surface and blade tip, and (c) polyline at casing surface.

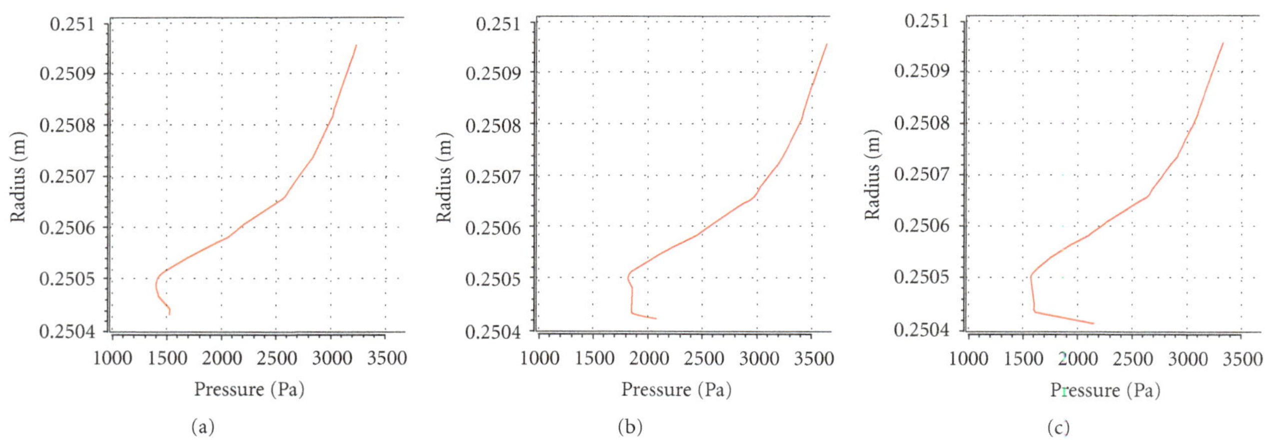

FIGURE 16: Pressure distributions along a line located at a radial direction from the blade tip to the casing, for the case of 0.5 mm gap, at (a) maximum, (b) mean value and (c) minimum blade displacement.

9. Conclusions

The work presented in this paper aimed to present the current state of the art of FSI numerical modelling of gas turbine engines and to develop a computationally efficient one-way FSI model which can be developed within the current commercial ANSYS software package. With this numerical model the pressure differences between the vibrating rotor blade tip and the casing surface were then investigated with two aims:

(i) to show if the pressure at the blade tip was significantly different from that at the casing surface;

(ii) to investigate if the pressure profile around the rotor blade followed its motion as it vibrates about its equilibrium position.

A three-dimensional CFD "tuned" one-way FSI model of a single stator and rotor stage, with vibrating rotor blades, was constructed. This model enforced the structural motion of the rotor blade. The rotor blade motion was calculated

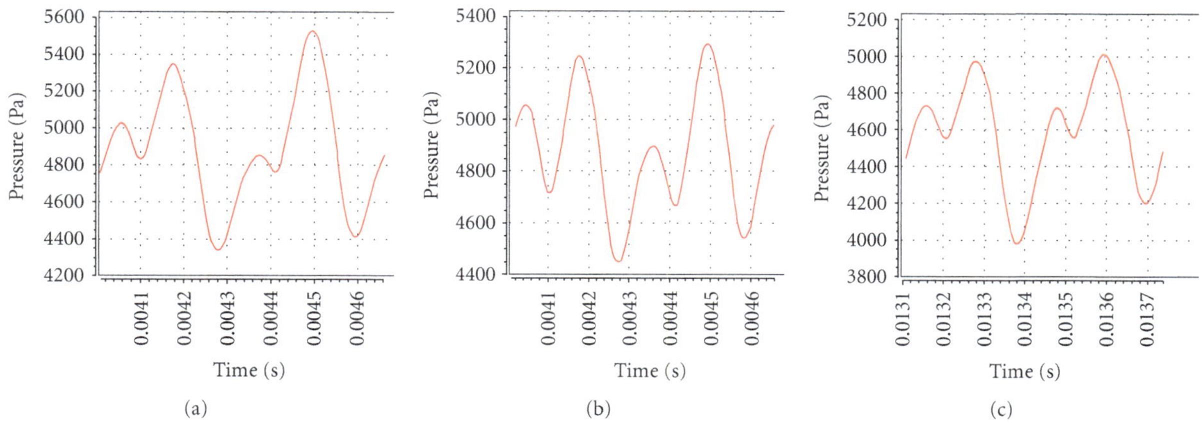

FIGURE 17: Pressure at a fixed point located at the casing wall for (a) 0.5 mm, (b) 1.0 mm, and (c) 1.5 mm gap.

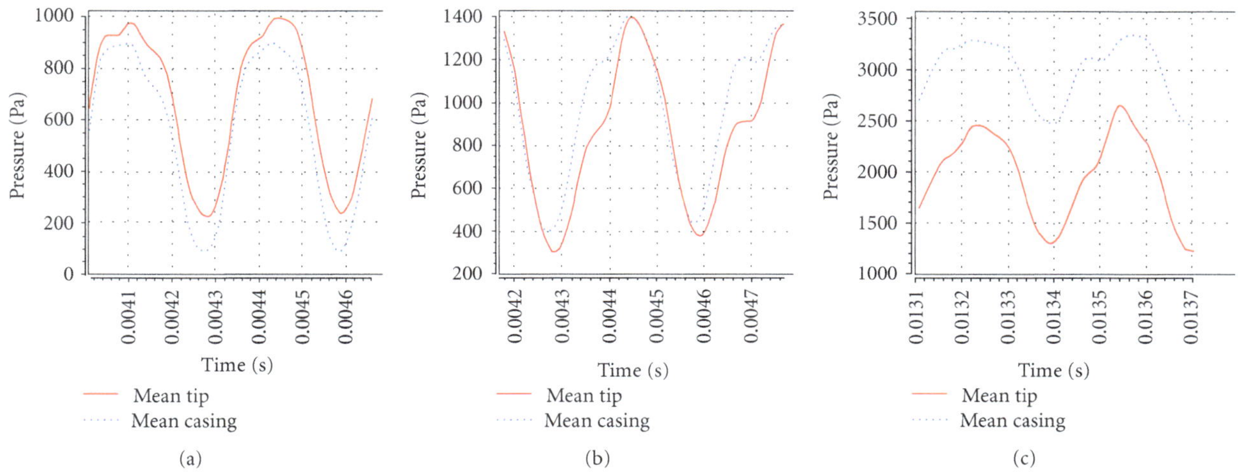

FIGURE 18: Pressure at the two points located at the mean value of the oscillated blade tip and at a corresponding point on the casing wall, in which both points are rotating with the same angular velocity as the rotor, for (a) 0.5 mm, (b) 1.0 mm, and (c) 1.5 mm gap.

from the pressure forces which would be present if there was no blade oscillation but still including the rotor/stator relative rotation. This allows the correct amplitude and phase of the structural motion to be used in the one-way CFD model. Setting up an FSI model in this manner gives the reduced computational expense of a one-way FSI model, but allows the correct deflection of the structure to be applied to the fluid domain boundaries.

Although the pressure at the casing surface did indeed differ from that at the rotor blade tip, the results presented in Figures 11, 14, 15, and 18 show generally that the pressure at the casing surface and rotor blade tip have similar trends and are representative of each other and are almost identical for the dominant first spatial harmonic of the pressure distribution. The difference which exists between these two locations is believed to be as a result of the flow which leaks over the blade tip and the blade tip vortices which develop. From the results presented it is also concluded that the pressure profile around the rotor blade tends to follow its motion. It is also shown that the internal pressure of the turbine is amplitude modulated by the blade passing period. However, due to both the integer number of stator to rotor blades and the relatively small dynamic blade deflection, further investigation into the importance of these two effects would need to be undertaken to make a more conclusive assessment.

References

[1] G. L. Forbes, *Non-contact gas turbine blade vibration monitoring using internal pressure and casing response measurements [Ph.D. dissertation]*, The University of New South Wales, 2010.

[2] G. L. Forbes and R. B. Randall, "Non-contact gas turbine blade vibration measurement from casing pressure and vibration signals," in *IFToMM-Rotordynamics*, Seoul, Korea, 2010.

[3] A. K. Slone, K. Pericleous, C. Bailey, and M. Cross, "Dynamic fluid-structure interaction using finite volume unstructured mesh procedures," *Computers and Structures*, vol. 80, no. 5-6, pp. 371–390, 2002.

[4] G. Morgenthal, "Fluid-structure interaction in bluff-body aerodynamics and long-span bridge design: phenomena and

methods," Tech. Rep. CUED/D-Struct/TR.187, Cambridge University, 2000.
[5] M. Heil, A. L. Hazel, and J. Boyle, "Solvers for large-displacement fluid-structure interaction problems: segregated versus monolithic approaches," *Computational Mechanics*, vol. 43, no. 1, pp. 91–101, 2008.
[6] P. Le Tallec and J. Mouro, "Fluid structure interaction with large structural displacements," *Computer Methods in Applied Mechanics and Engineering*, vol. 190, no. 24-25, pp. 3039–3067, 2001.
[7] H. Schumucker, F. Flemming, and S. Coulson, "Two-way coupled fluid structure interaction simulation of a propeller turbine," in *Proceedings of the 25th IAHR Symposium on Hydraulic Machinery and Systems*, IOP Publishing, 2010.
[8] T. Biesinger, C. Cornelius, and C. Rube, "Unsteady CFD methods in a commercial solver for turbomachinery applications," in *ASME Conference Proceedings*, vol. 7, pp. 2441–2452, 2010.
[9] G. L. Forbes, O. N. Alshroof, and R. B. Randall, "Gas Turbine blade vibration and casing wall pressure interaction," in *Proceedings of the Australasian Congress on Applied Mechanics (ACAM '10)*, Perth, Australia, 2010.
[10] O. N. Alshroof, G. L. Forbes, and R. B. Randall, "Relationship between the pressure at the casing wall and at the blade tip for a vibrating turbine blade," in *Proceedings of the 17th Australasian Fluid Mechanics Conference*, Auckland, New Zealand, 2010.
[11] D. L. Bell and L. He, "Three-dimensional unsteady flow for an oscillating turbine blade and the influence of tip leakage," *Journal of Turbomachinery*, vol. 122, no. 1, pp. 93–101, 2000.
[12] K. C. Hall and K. Ekici, "Computational methods for non-linear unsteady aerodynamics in turbomachinery," in *Proceedings of the 7th IFToMM-Conference on Rotor Dynamics*, Vienna, Austria, 2006.
[13] ANSYS, I., ANSYS CFX-Solver Modeling Guide Version 12.1, 2009.
[14] S. S. Rao, *Mechanical Vibrations*, Pearson Prentice Hall, Upper Saddle River, NJ, USA, 4th edition, 2004.
[15] ANSYS, I., Ansys 11, Structural Guide p. Section 4.3.4.

12

Experimental Investigation and Passive Flow Control of a Cavitating Centrifugal Pump

Spyridon D. Kyparissis and Dionissios P. Margaris

Fluid Mechanics Laboratory, Mechanical Engineering and Aeronautics Department, University of Patras, 26500 Patras, Greece

Correspondence should be addressed to Spyridon D. Kyparissis, kypariss@mech.upatras.gr

Academic Editor: N. Sitaram

Passive flow control techniques are used to improve the flow field and efficiency of centrifugal pumps and turbomachines, in general. An important phenomenon that mechanical engineers have to take into account is cavitation. It leads to the decrease of the pump performance and total head. In the present experimental study, a centrifugal pump is investigated in cavitating conditions. A passive flow control is realized using three different blade leading edge angles in order to reduce the cavitation development and enhance the pump performance. The experiments are carried out in a pump test rig specially designed and constructed, along with the impellers. The head drop and total efficiency curves are presented in order to examine the effect of the blade leading edge angle on the cavitation and pump performance. Finally, the vapour distribution along with the blades is illustrated for the tested blade leading edge angles.

1. Introduction

Cavitation is a very complex phenomenon including phase changes and viscous effects. When the local pressure within a flow with water as medium becomes equal to the vaporization pressure of the water, bubbles begin to form and are further transported to regions of higher pressure where they collapse. This bubble collapse can cause material damage and erosion.

Various researchers have considerably contributed in studying the cavitation in centrifugal pumps, applying passive flow control techniques and realizing a series of experiments. On the other hand, there is a limited number of works regarding the effect of the blade leading edge angle on the pump cavitation. Hirschi et al. [1] presented the results obtained with a 3D numerical method allowing the prediction of the cavitation behaviour of a centrifugal pump and compared this prediction to model tests. Hofmann et al. [2] studied experimentally the cavitation of two centrifugal pumps that can hold different runner geometries as well as different leading edge geometries within the same runner. Frobenius et al. [3] realized numerical simulations and experimental investigations of the cavitating flow through a centrifugal pump impeller of low specific speed. Coutier-Delgosha et al. [4] investigated a special test pump with two-dimensional curvature blade geometry in cavitating and noncavitating conditions using different experimental techniques.

Siljegovic et al. [5] concluded that passive flow control can effectively modulate flow conditions in a capillary driven microfluidic device. Ulas [6] designed and studied experimentally two cavitating venturis to deliver the desired mass flow rates for specific conditions. Escaler et al. [7] carried out an experimental investigation in order to evaluate the detection of cavitation in actual hydraulic turbines. Japikse et al. [8] realized a series of tests and improved the stability of compressors and turbopumps using passive flow control. Luo et al. [9] studied experimentally the effect of impeller inlet geometry on performance improvement for a centrifugal pump. Wu et al. [10] carried out an experimental study in order to analyze the cavitation of a centrifugal pump and its effect on hydrodynamic performance during transient operation. Kyparissis and Margaris [11–13] studied experimentally and computationally a centrifugal pump with double-arc synthetic blade design method in cavitating and noncavitating conditions, applying different rotational speeds.

The present experimental study and passive flow control are concerned with the effect of the blade leading edge angle on the cavitation of a centrifugal pump applying the double-arc synthetic blade design method [14–16], with rotational speed of 1200 rpm and flow rate of 35 m^3/h. The experiments are carried out in a pump test rig specially designed and constructed, along with the impellers. The test section allows optical observation of the flow field and cavitation with the aid of the stroboscopic light source. The head drop and total efficiency curves are presented in order to investigate the effect of the blade leading edge angle on the cavitation and pump performance. In addition, the vapour distribution along with the blades is depicted for the tested blade leading edge angles.

2. Experimental Rig and Procedure

The experimental study is performed using the pump test rig shown in Figure 1. The supplying tank (1) of the pumping installation is filled with water from the water supply. The pressure at the examined pump inlet is measured by the absolute pressure transducer (2). Furthermore, the study of cavitation is realized applying a wide range of pressures at the pump inlet that can be set by changing the pressure in the supply tank with the aid of a vacuum pump (3). The fluid temperature is measured by a temperature sensor (4) and is kept almost constant and equal to 25°C. The fluid level in the supply tank is read from the level meter (5) and the fluid-free surface pressure is measured by a vacuum-pressure gauge (6). The pump flow rate is measured using the electromagnetic flowmeter (7). Moreover, the flow rate regulation is obtained by throttling a butterfly valve (8) installed at the discharge pipe. The differential pressure between the pump inlet and outlet is measured by the differential pressure transducer (9). The pump is driven by a three-phase AC electric motor (10) of 2.2 kW and maximum rotational speed of 1410 rpm. The rotational speed can be continuously varied by an inverter (11). The pump suction cover (12) and a part of the suction pipe (13) are made from plexiglas. Transparent parts of the pump test rig allow observation of fluid flow inside the impeller and photography of the development of cavitation in flow passage of the impeller. The measurement data are acquired by the data acquisition system and LabVIEW platform (14). The main dimensions of the pump test rig and the regions where the pressure transducers are installed are shown in Figure 2.

For the realization of the cavitation test, the vacuum pump is set in operation and the air is removed from the tank, until a specific value of pressure is recorded by the vacuum-pressure gauge for every measurement. The pump rotational speed is kept constant and the butterfly valve is kept in a given position in order to have constant flow rate. For every measurement, the total head is computed according to the following formula:

$$H = \frac{\Delta p}{\rho g} + \Delta z + \frac{u_d^2 - u_s^2}{2g}. \quad (1)$$

The differential pressure Δp is measured by the differential pressure transducer (9). Moreover, the water density

Figure 1: The complete experimental setup along with the instrumentation system.

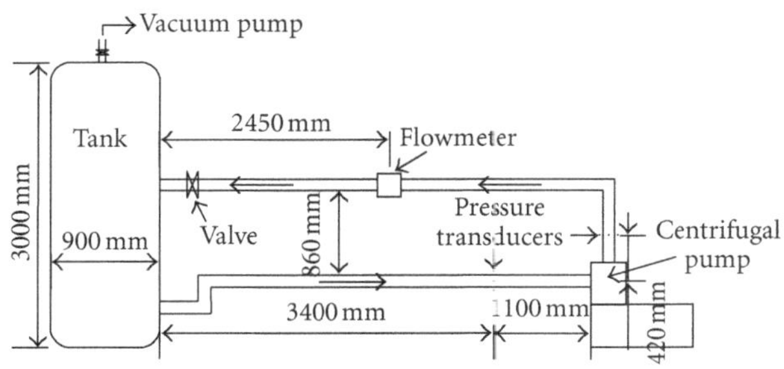

Figure 2: The main dimensions of the pump test rig.

for the temperature of 25°C is found $\rho = 997\,\text{kg/m}^3$. The discharge velocity u_d and the suction velocity u_s are calculated according to the corresponding flow rate Q measured by the electromagnetic flowmeter (7). The gravitational acceleration is determined as $g = 9.807\,\text{m/s}^2$ and the elevation difference between the pump inlet and the region where the discharge pressure is measured is calculated $\Delta z = 0.42$ m.

In addition, the net positive suction head available is computed according to the following equation:

$$\text{NPSHa} = \frac{p_s}{\rho g} + \frac{u_s^2}{2g} - \frac{p_v}{\rho g}. \quad (2)$$

The suction pressure p_s is measured by the absolute pressure transducer (2) and the vaporization pressure of the water for the temperature of 25°C is found $p_v = 3290.5$ Pa.

Finally, the total efficiency η is computed according to the following expression:

$$\eta = \frac{\rho g Q H}{\sqrt{3} V I}. \quad (3)$$

The voltage V and the amperage I are measured by the inverter (11).

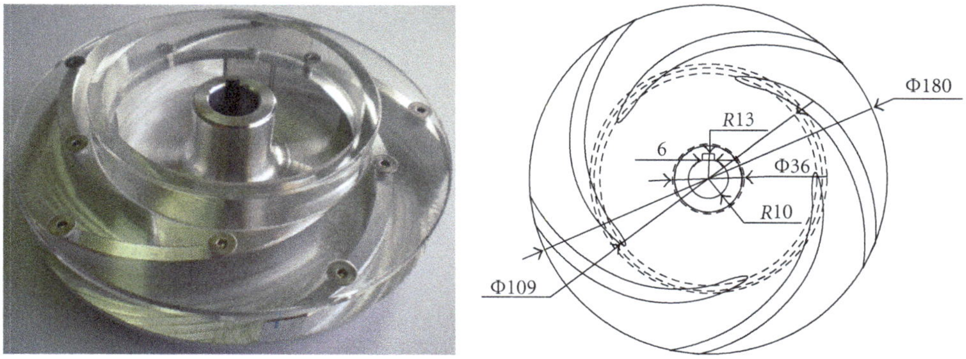

Figure 3: The examined impeller geometry and the scaled drawing (in mm) with blade leading edge angle of 9 deg.

Moreover, during every measurement, the cavitation is illustrated using the stroboscopic light. The NPSH value that the first bubble appears, according to the visualization of cavitation, is the net positive suction head required NPSHr.

The measuring equipment consists of an absolute and a differential pressure transducer, a temperature sensor, and an electromagnetic flowmeter. In order to measure the static pressure at the suction side of the pump, an absolute pressure transducer from ABB 2600T series, model 264NS with a base accuracy ±0.075% is applied. Moreover, the pressure difference between the discharge and suction region of the pump is measured using a differential pressure transducer from ABB 2600T series, model 264DS with a base accuracy ±0.075%. In addition, temperature is measured with the temperature sensor of the ABB SensyTemp TSP111 series. Furthermore, the measurement of the flow rate is realized using an electromagnetic flowmeter ABB FXE4000 COPA-XE/MAG-XE, model DE41F with a base accuracy ±0.5%.

The rotational speed of the three phase electric motor can be continuously varied by the inverter Fuji Electric FVR-E9S series.

The study of cavitation is realized using the vacuum pump series LABOPORT N 816.3 KN.45.18. Series N 816.3 diaphragm pumps are double-head, dry-running devices used in a wide range of laboratory applications. They transfer and pump down without contamination.

The measurement data from the pressure transducers, flowmeter, and temperature sensor are acquired by the analog input module cFP-AI-110. The National Instruments cFP-AI-110 is an 8-channel single-ended input module for direct measurement of millivolt, low voltage, or milliampere current signals from a variety of sensors and transmitters. It delivers filtered low-noise analog inputs with 16-bit resolution and 5 S/s sampling rate. Finally, this analog input module is connected to LabVIEW platform (14).

3. Blade Design Analysis

3.1. Impeller Geometry. Three impeller geometries have been constructed using aluminium alloy 7075-T6, which is composed of zinc as the primary alloying element. It is strong, with strength comparable to many steels and has good fatigue strength and average machinability, but has less resistance to corrosion than many other aluminium alloys. 7075-T6 is a heat temper grade of aluminium alloy 7075. It has an ultimate tensile strength of 510–538 MPa and yield strength of at least 434–476 MPa. In addition, the tested impellers have a removable transparent cover disk made of plexiglas in order to observe the fluid flow and the cavitation, as shown in Figures 3, 4, and 5.

Table 1: Main characteristics of the tested impellers.

Impeller characteristics	Values
Suction pipe diameter D_s, mm	100
Diameter of the impeller at the suction side D_1, mm	108
Diameter of the impeller at the pressure side D_2, mm	180
Impeller width at the suction side b_1, mm	15.6
Impeller width at the pressure side b_2, mm	15.6
Blade leading edge angle β_1, deg	9, 15, 21
Blade trailing edge angle β_2, deg	20
Maximum blade thickness S, mm	7.5
Number of blades z	5

The impellers have been designed according to the dimensions of the Caprari MEC—A2/80A volute, which is installed in the pump test rig. The main characteristics of the three tested impellers are presented in Table 1.

3.2. Double-Arc Synthetic Method: DASM. In the present work, the double-arc synthetic method (DASM) [14–16] is applied for the impellers blade design of the centrifugal pump. It is a simple design method that combines two Pfleiderer's double-arc methods [17]. There are published computational studies [14–16] that compare DASM with the Pfleiderer's [17] methods and it is observed that a centrifugal pump has better efficiency applying DASM.

As shown in Figure 6, the auxiliary circle C_a is drawn, according to the following formula:

$$d_1 = D_1 \sin \beta_1, \tag{4}$$

where D_1 is the diameter of the impeller at the suction side and β_1 is the blade leading edge angle.

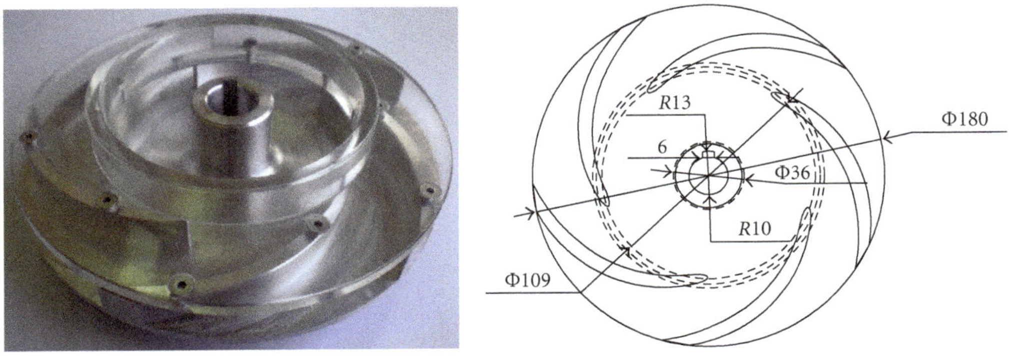

(a) (b)

FIGURE 4: The examined impeller geometry and the scaled drawing (in mm) with blade leading edge angle of 15 deg.

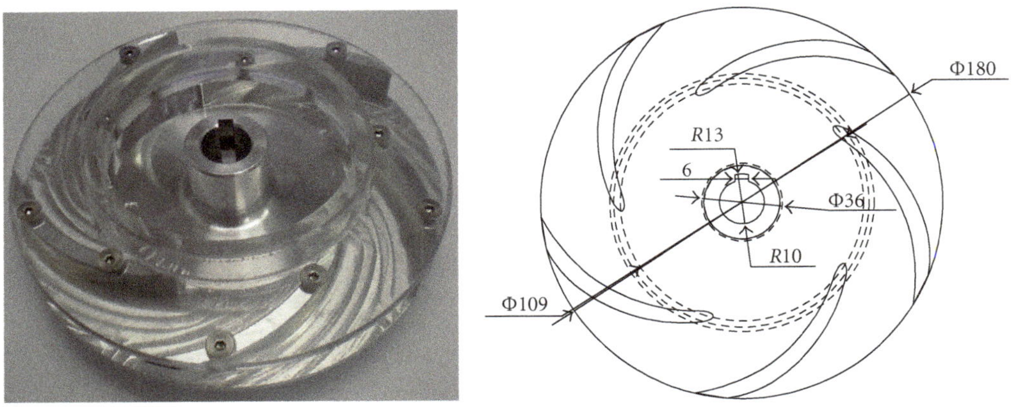

(a) (b)

FIGURE 5: The examined impeller geometry and the scaled drawing (in mm) with blade leading edge angle of 21 deg.

The periphery of the suction side is divided to five parts, just as the number of blades z. The tangents of the auxiliary circle from the points E and C are intersected to the point A. Afterwards, point A is shifted to the point L, which is the center of curvature for the first arc EY and the radius R_1 is the distance of the points E and L, as formulated in the following expression:

$$R_1 = 1.2(\mathrm{AE}). \quad (5)$$

The point P is at the extension of the line that connects the points A and C at distance equal to the distance of the points A and E. Applying DASM the point P is shifted to Y, according to the following expression:

$$\mathrm{EY} = 0.75\,\mathrm{EP}. \quad (6)$$

The angle between the lines OY and LY is defined as β and the distance between the points O and Y is defined as r, as shown in Figure 6. The radius R_2 of the second arc YH is the distance of the points K and Y and is obtained from the following expression:

$$R_2 = \frac{1}{2}\frac{(D_2/2)^2 - r^2}{(D_2/2)\cos\beta_2 - r\cos\beta}, \quad (7)$$

where D_2 is the diameter of the impeller at the pressure side and β_2 is the blade trailing edge angle.

4. Experimental Results

In the present study, the experimental investigation and passive flow control are realized in order to examine the pump performance and cavitation development for three different blade leading edge angles. Studying the noncavitating performance, the total head and the total efficiency as functions of the flow rate are illustrated in Figures 7 and 8, for the three examined blade leading edge angles and rotational speed of 1200 rpm. As shown in Figure 8, the flow rate to the best efficiency point is approximately 22 m^3/h for 9 deg, 34 m^3/h for 15 deg, and 43 m^3/h for 21 deg.

The total head drops for the rotational speed of 1200 rpm, flow rate of 35 m^3/h, and three examined blade leading edge angles of 9, 15, and 21 deg are depicted in Figure 9. The ordinate is the total head of the pump, while the abscissa is the net positive suction head available. There are four capital letters in parentheses corresponding to the vapour distributions in Figures 12, 13, 14, and 15. The filled points represent the experimental cases, where cavitation has been developed. It is noticed that as the NPSHa decreases the total head decreases slowly for all the tested blade leading edge angles. As the water in the suction pipe approaches the impeller eye, it has velocity and acceleration. In addition, it has to change its direction to enter the impeller, because

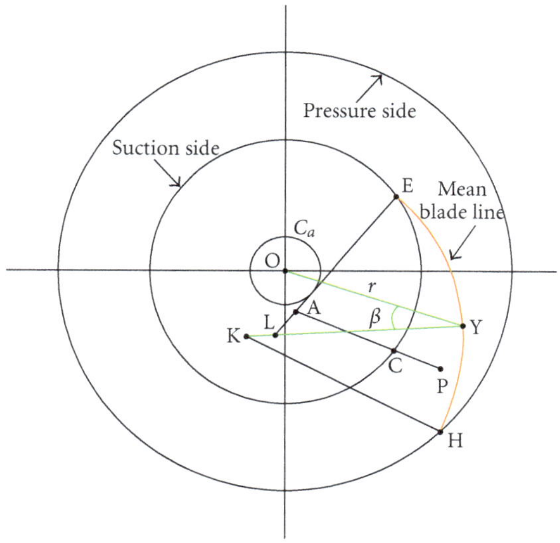

FIGURE 6: Representation of the mean blade line using DASM.

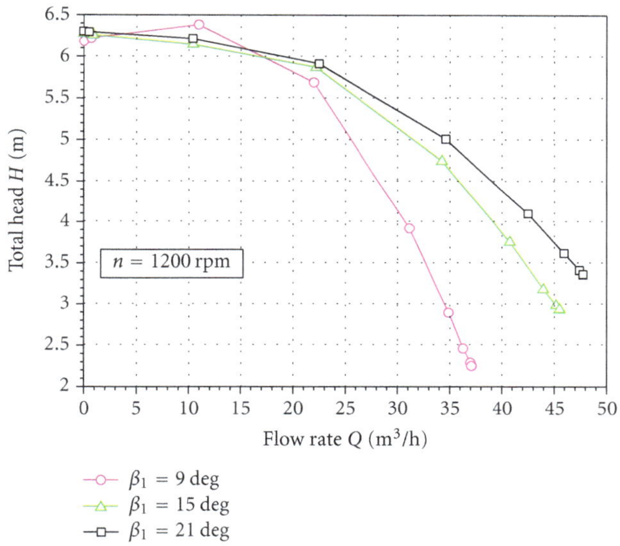

FIGURE 7: The total head as a function of the flow rate, studying the noncavitating performance.

of the blade leading edge angle. Losses in terms of total head occur due to each of the above reasons and because of friction. For the blade leading edge angle of 15 deg, no cavitation is noticed, for the examined pressures at the pump inlet. Nevertheless, there is head drop as the NPSHa decreases, because there are cavitation bubbles, very small and cannot be observed using the stroboscopic light. These bubbles block the flow passage between the blades and cause the head drop. Examining the blade leading edge angle of 9 deg, cavitation starts to be developed for the net positive suction head of 5.13 m, which is the corresponding NPSHr for 9 deg. In addition, testing the 21 deg, the cavitation begins for NPSH = 2.29 m, which is the corresponding NPSHr for 21 deg. Realizing a series of experiments the maximum total head drops for the three above mentioned studied

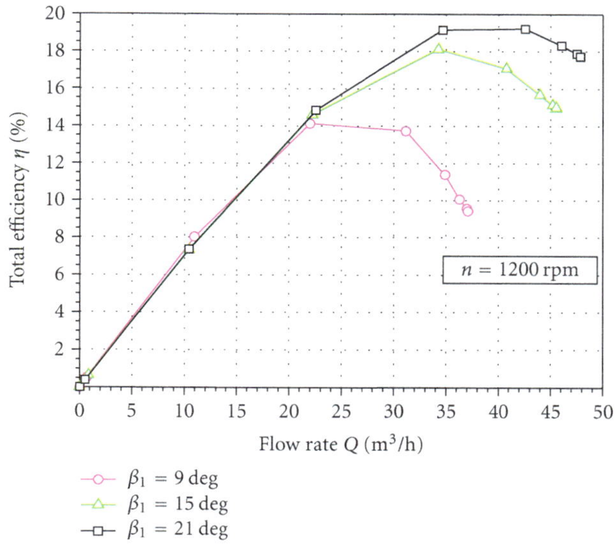

FIGURE 8: The total efficiency as a function of the flow rate, studying the noncavitating performance.

blade leading edge angles are 13.32%, 4.46%, and 6.80%, respectively. For the blade leading edge angles 9 and 21 deg, when cavitation starts to be developed the corresponding head drops are 5.47% and 5.49%, respectively. Thus, it is observed that for both the cases that cavitation begins, the total head is reduced by 5.5% instead of 3% that the Hydraulic Institute defines. The NPSHr can be from 2 to 20 times the 3% that the Hydraulic Institute defines, depending on pump design [18]. The cavity length when the cavitation bubbles appear is measured 4 and 7 mm for 9 and 21 deg, respectively. Moreover, it is noticed that as the blade leading edge angle increases the total head increases as well. Finally, as the net positive suction head available decreases the pump flow rate is kept almost constant, in spite of the cavitation development.

Furthermore, in Figure 10 the change of the total head with the blade leading edge angle is depicted, for three different values of the net positive suction head available. Figure 10 contributes to the recording of the total head values in respect of the change of the blade leading edge angle. It is observed that as the blade leading edge angle increases, the total head increases as well. For the blade leading edge angles between 15 and 21 deg, the total head is almost the same for the net positive suction head available values of 6 and 9 m. Finally, for the lowest examined value of the net positive suction head available, the total head has the lowest values for all the examined blade leading edge angles, because as the net positive suction head decreases the total head decreases as well, as shown in Figure 9.

In Figure 11 the change of the total efficiency with the net positive suction head available is presented, for the three examined blade leading edge angles, rotational speed of 1200 rpm, and flow rate of 35 m^3/h. It is observed that the change of the total efficiency with the NPSHa is similar to the change of the total head with the NPSHa, as shown in

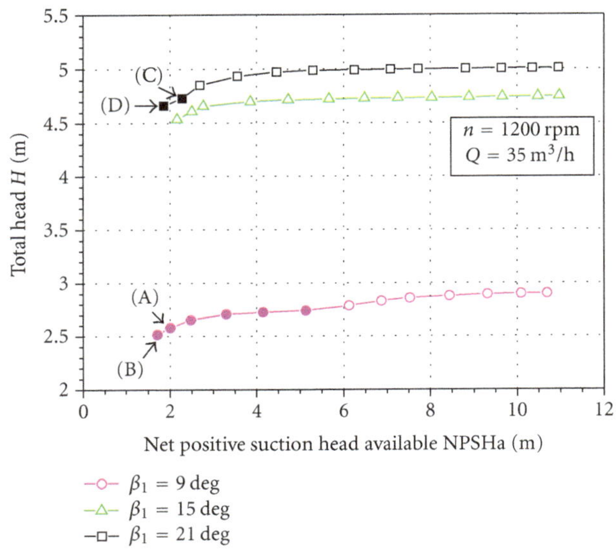

FIGURE 9: Total head drop curves, where cavitation has been developed to the filled points.

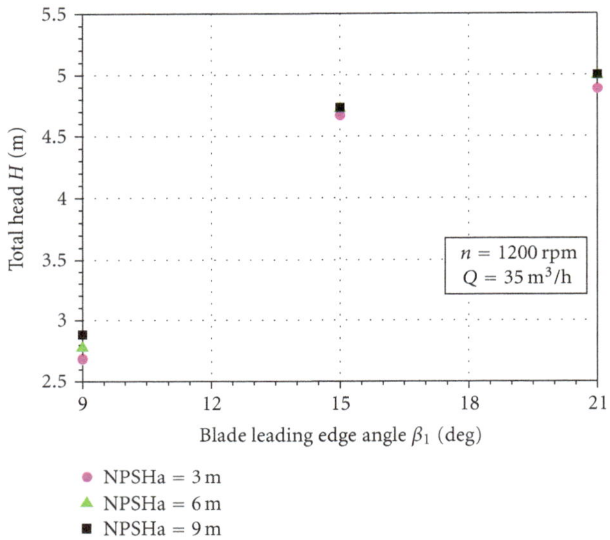

FIGURE 10: The change of the total head with the blade leading edge angle for three different values of the net positive suction head available.

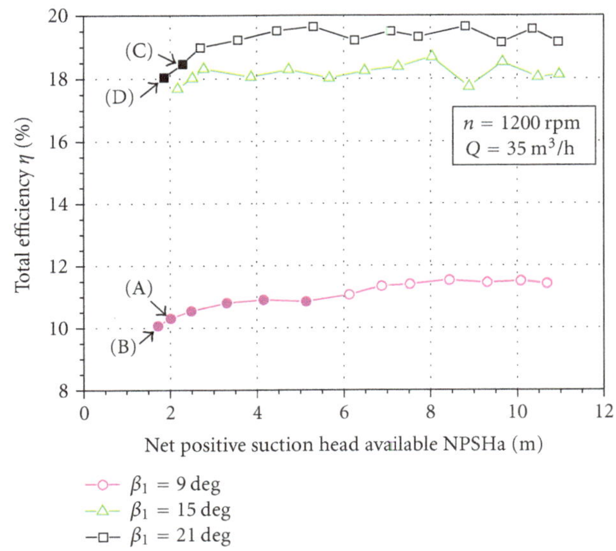

FIGURE 11: The change of the total efficiency with the net positive suction head available for rotational speed of 1200 rpm and flow rate of 35 m^3/h.

Figure 9. As the blade leading edge angle increases the total efficiency increases as well. In addition, for the lowest values of the NPSHa, much faster decrease of the total efficiency is observed, for all the tested blade leading edge angles. Examining the blade leading edge angles 15 and 21 deg, fluctuations are observed to the total efficiency, because of the unstable measurements of both the voltage and the amperage.

The uncertainty on the measurements has been evaluated from the resolution of the data acquisition system, realizing multiple experiments with deviation from the presented data, approximately ±0.10%. The relative uncertainty for the total head is 0.07% for all the examined blade leading edge angles. For the net positive suction head available, the relative uncertainty varies from 0.08% to 0.10% for 9 deg and 21 deg and 0.08% to 0.09% for 15 deg. Finally, for the total efficiency, the relative uncertainty varies from 0.54% to 0.55% for 9 deg and is 0.53% for 15 deg and 21 deg.

Furthermore, in the present experimental study the vapour distribution between the blades at the leading edge is illustrated using the stroboscopic light. The vapour is caused by the strong curvature of the streamlines along the front shroud, where low values of the net positive suction head prevail. Two of the cavitation cases of 9 deg are depicted in Figures 12 and 13, where a scaling bar is added to each figure, representing a length of 10 mm and a letter is added in parentheses corresponding to the letter in Figures 9 and 11. Thus Figure 12 corresponds to the point A and Figure 13 corresponds to the point B. Testing the blade leading edge angle of 9 deg, the cavitation is observed at the leading edge in the region of the pressure side. Moreover, it is noticed that as the net positive suction head decreases, lower pressures between the blades prevail and the cavity length increases. As the cavity length increases the total head and total efficiency decrease, as shown in Figures 9 and 11, due to flow blockage from cavitation vapour in the blades.

In Figures 14 and 15 the cavitation is developed at the leading edge in the region of the suction side for the blade leading edge angle of 21 deg. Figure 14 corresponds to the point C and Figure 15 corresponds to the point D. The region where the cavitation is developed is different from the corresponding of 9 deg. As the blade leading edge angle decreases, the passage between the blades decreases, lower pressures prevail in this region and the cavitation is developed at the pressure side. On the other hand, for blade leading edge angle of 21 deg, the lowest pressures prevail at the suction side of the blades and cavitation is developed in this region. Examining the blade leading edge angle of

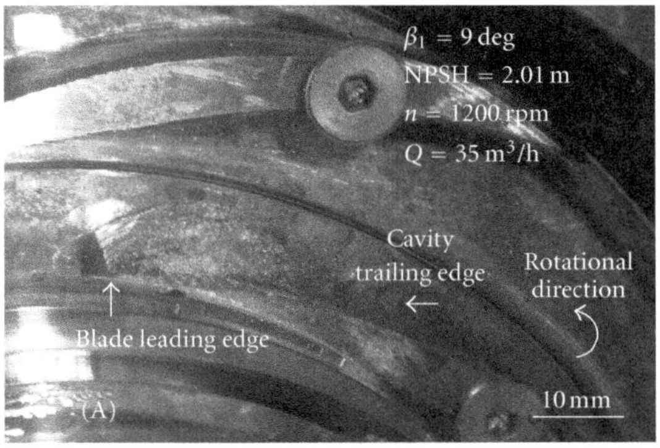

Figure 12: Illustration of cavitation for the blade leading edge angle of 9 deg and NPSH equal to 2.01 m.

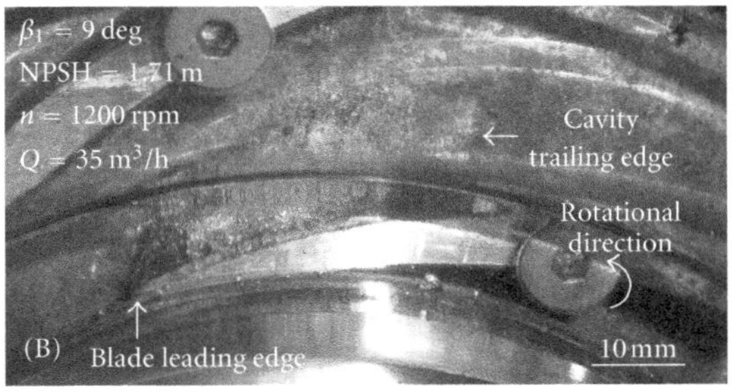

Figure 13: Illustration of cavitation for the blade leading edge angle of 9 deg and NPSH equal to 1.71 m.

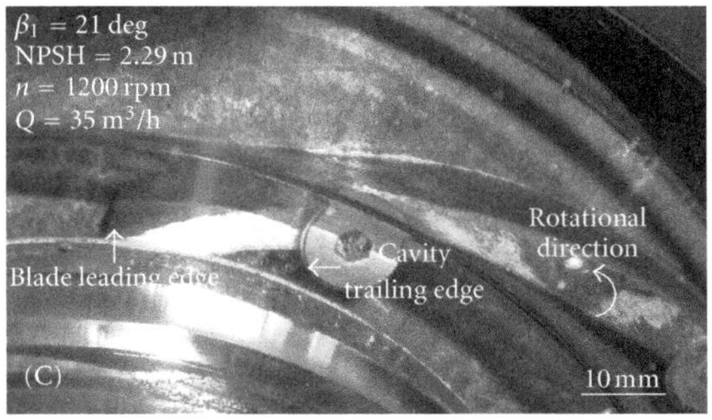

Figure 14: Illustration of cavitation for the blade leading edge angle of 21 deg and NPSH equal to 2.29 m.

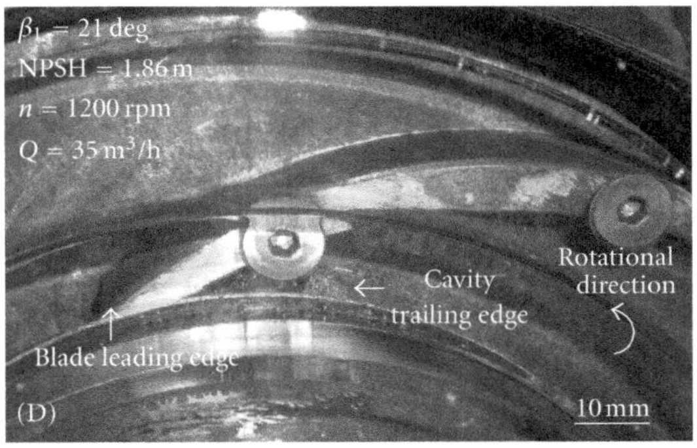

Figure 15: Illustration of cavitation for the blade leading edge angle of 21 deg and NPSH equal to 1.86 m.

15 deg, no cavitation is observed, because the testing flow rate of 35 m^3/h is almost equal to the corresponding of the best efficiency point for 15 deg. For 9 deg, the testing flow rate of 35 m^3/h is greater than its corresponding best efficiency point, thus the cavitation is developed at the pressure side of the blades. On the other hand, for 21 deg, the testing flow rate 35 m^3/h is less than its corresponding best efficiency point and the cavitation is developed at the suction side of the blades. Moreover, the cavity length for 21 deg is smaller than the corresponding of 9 deg, because lower net positive suction head values prevail for blade leading edge angle of 9 deg, as shown in Figures 9 and 11.

5. Conclusions

The present experimental study and passive flow control are concerned with the effect of the blade leading edge angle on the cavitation and performance of a centrifugal pump. A pump test rig has been developed in order to realize a passive flow control with different blade leading edge angles, allowing optical observation of the cavitation development with the aid of a stroboscopic light source.

The rotational speed of 1200 rpm and flow rate of 35 m^3/h are applied. Studying the effect of the blade leading edge angle on the cavitation, it is observed that as the blade leading edge angle increases, both the total head and the total efficiency increase for the examined operating conditions. Moreover, it is noticed that as the net positive suction head decreases the total head decreases slowly for all the examined blade leading edge angles.

Examining the blade leading edge angle of 9 deg, the cavitation is developed at the leading edge in the region of the pressure side. On the other hand, for 21 deg, the cavitation is developed at the leading edge in the region of the suction side. Finally, for the blade leading edge angle of 15 deg, there is no cavitation development and the danger of the pump damage and erosion is eliminated. Thus, the costs of maintenance of the centrifugal pumps are limited, applying the blade leading edge angle of 15 deg.

Further research work is planned to be accomplished experimentally and computationally, studying the effect of different blade design methods on the cavitation of a centrifugal pump and applying passive flow control with different impeller geometrical characteristics and operating conditions.

References

[1] R. Hirschi, P. Dupont, F. Avellan, J. N. Favre, J. F. Guelich, and E. Parkinson, "Centrifugal pump performance drop due to leading edge cavitation: numerical predictions compared with model tests," *Journal of Fluids Engineering*, vol. 120, no. 4, pp. 705–711, 1998.

[2] M. Hofmann, B. Stoffel, J. Friedrichs, and G. Kosyna, "Similarities and geometrical effects on rotating cavitation in two scaled centrifugal pumps," in *Proceedings of the 4rth International Symposium on Cavitation (CAV '01)*, pp. 20–23, California Institute of Technology, Pasadena, Calif, USA, June 2001.

[3] M. Frobenius, R. Schilling, J. Friedrichs, and G. Kosyna, "Numerical and experimental investigations of the cavitating flow in a centrifugal pump impeller," in *Proceedings of the ASME Joint U.S.-European Fluids Engineering Conference (FEDSM '02)*, pp. 361–368, Montreal, Canada, July 2002.

[4] O. Coutier-Delgosha, R. Fortes-Patella, J. L. Reboud, M. Hofmann, and B. Stoffel, "Experimental and numerical studies in a centrifugal pump with two-dimensional curved blades in cavitating condition," *Journal of Fluids Engineering*, vol. 125, no. 6, pp. 970–978, 2003.

[5] V. Siljegovic, N. Milicevic, and P. Griss, "Passive, programmable flow control in capillary force driven microfluidic networks," in *Proceedings of the 13th International Conference on Solid-State Sensors and Actuators and Microsystems (TRANSDUCERS '05)*, pp. 1565–1568, Seoul, Korea, June 2005.

[6] A. Ulas, "Passive flow control in liquid-propellant rocket engines with cavitating venturi," *Flow Measurement and Instrumentation*, vol. 17, no. 2, pp. 93–97, 2006.

[7] X. Escaler, E. Egusquiza, M. Farhat, F. Avellan, and M. Coussirat, "Detection of cavitation in hydraulic turbines," *Mechanical Systems and Signal Processing*, vol. 20, no. 4, pp. 983–1007, 2006.

[8] D. Japikse, K. N. Oliphant, and D. O. Baun, "Stability Enhancement of Compressors and Turbopumps by Passive Flow Control," Thermal and Fluids Analysis Works (TFAWS), College Park, MD, USA, 7—11 August 2006.

[9] X. Luo, Y. Zhang, J. Peng, H. Xu, and W. Yu, "Impeller inlet geometry effect on performance improvement for centrifugal pumps," *Journal of Mechanical Science and Technology*, vol. 22, no. 10, pp. 1971–1976, 2008.

[10] D. Wu, L. Wang, Z. Hao, Z. Li, and Z. Bao, "Experimental study on hydrodynamic performance of a cavitating centrifugal pump during transient operation," *Journal of Mechanical Science and Technology*, vol. 24, no. 2, pp. 575–582, 2010.

[11] S. D. Kyparissis and D. P. Margaris, "Experimental study in a centrifugal pump with double-arc synthetic blade design method in cavitating and non-cavitating conditions," in *Proceedings of the 4th International Conference on Experiments / Process / System Modeling / Simulation / Optimization, (IC-EpsMsO '11)*, Athens, Greece, July 2011.

[12] S. D. Kyparissis and D. P. Margaris, "Experimental investigation of cavitation in a centrifugal pump with double-arc synthetic blade design method," *International Review of Mechanical Engineering*, vol. 5, no. 5, pp. 884–892, 2011.

[13] S. D. Kyparissis and D. P. Margaris, "Experimental and computational study on the effects of the cavitation to the total head of a centrifugal pump," in *Proceedings of the 5th International Conference from Scientific Computing to Computational Engineering (IC-SCCE '12)*, Athens, Greece, July 2012.

[14] S. D. Kyparissis, E. C. Douvi, E. E. Panagiotopoulos, D. P. Margaris, and A. E. Filios, "CFD flowfield analysis and hydrodynamic double-arc blade design effects for optimum centrifugal pump performance," *International Review of Mechanical Engineering*, vol. 3, no. 3, pp. 284–292, 2009.

[15] S. D. Kyparissis, E. C. Douvi, E. E. Panagiotopoulos, D. P. Margaris, and A. E. Filios, "Parametric study performance of a centrifugal pump based on simple and double-arc blade design methods," in *Proceedings of the 3rd International Conference on Experiments / Process / System Modeling / Simulation / Optimization (IC-EpsMsO '09)*, pp. 8–11, Athens, Greece, July 2009.

[16] S. D. Kyparissis, D. P. Margaris, E. E. Panagiotopoulos, and A. E. Filios, "CFD analysis on the effects of the blade design method to the centrifugal pump performance," *International Review of Mechanical Engineering*, vol. 3, no. 5, pp. 553–561, 2009.

[17] C. Pfleiderer, *Die Kreiselpumpen Für Flüssigkeiten Und Gase*, Springer, Berlin, Germany, 1961.

[18] P. Girdhar and O. Moniz, *Practical Centrifugal Pumps: Design, Operation and Maintenance*, Elsevier, Oxford, UK, 2005.

13

Numerical Modeling of a Marine Propeller Undergoing Surge and Heave Motion

Spyros A. Kinnas, Ye Tian, and Abhinav Sharma

Ocean Engineering Group, Department of Civil, Architectural and Environmental Engineering, University of Texas at Austin, Austin, TX 78712, USA

Correspondence should be addressed to Spyros A. Kinnas, kinnas@mail.utexas.edu

Academic Editor: Moustafa Abdel-Maksoud

A boundary element method (BEM) and a vortex-lattice method (VLM) are extended in order to predict the unsteady performance of propellers subject to rigid body motions. The methods are applied in the case of prescribed surge and heave motions, and the results are compared with those from other methods.

1. Introduction

Due to the unsteady motions of a ship, the operating propeller is usually subjected to a time-dependent rigid body motion which can drastically change the hydrodynamic performance of the propulsion system and reduce the maneuverability of the ship. Incident waves tend to move the hull in 6 DOF over a range of frequencies which in turn oscillate the propeller in all possible degrees of freedom. In general, a propeller operates in time-varying wake field and thus generates vibratory forces causing severe wear and tear of the propeller system. For all such cases, it becomes important to consider the effect of unsteady contributions to the wake field caused by waves and wave-induced motions. The combined effect of these two contributions was investigated in the past by Sluijs [1] and by Jessup and Boswell [2]. Later, efforts were made by Parsons and Vorus [3] to assess added mass and damping coefficients for a vibrating propeller. Sasajima [4] adapted a quasi-steady approach to estimate propeller bearing forces while Breslin and Andersen [5] in favored fully unsteady approach as opposed to quasisteady approach for calculating time-varying forces on propellers.

In the recent past, Lee and Kinnas [6] developed unsteady wake alignment model to simulate propeller in nonaxisymmetric flows. Politis [7] had modeled the unsteady motion of propeller using a free wake model. Very recently, He [8] developed a full unsteady wake alignment model for hydrofoils and marine propellers which was incorporated by Sharma et al. [9, 10] in modeling the unsteady motion of hydrofoils and marine propellers.

In this paper, a computational approach to analyze the hydrodynamic performance of a marine propeller undergoing unsteady motion is presented. Firstly, an unsteady boundary element method (BEM) model for a general rigid body motion problem is proposed. The kinematic and dynamic boundary conditions of the BEM model are also incorporated into a Vortex Lattice Method (VLM) code. The BEM model is first applied to a pitching hydrofoil followed by validation of numerical results of the current model against those obtained from a commercial RANS solver. The problem of flow around a marine propeller performing surge and heave motion in unbounded fluid is then solved using both VLM and BEM. A fully unsteady wake alignment algorithm is implemented into the VLM code in order to satisfy the force-free condition on the propeller wake surface. Finally, a comparative study of unsteady propeller forces obtained from BEM/VLM, RANS, and other available numerical results is carried out and results are presented. The ultimate goal of the present research work is to develop an integrated 6 DOF propeller-hull motion model by including complete propeller-hull interaction in time. A comprehensive analysis of the flow around such system under most severe motions can be performed by the proposed model.

2. Unsteady BEM Model for an Obstacle with Rigid Body Motion

As shown in Figure 1, an obstacle is subject to rigid body motion under an inertial frame of reference. Consider a fluid particle which is at location $\mathbf{x_1}$ when $t = t_1$. After a small interval, when $t = t_2 = t_1 + \Delta t$, the particle is moved to a new location $\mathbf{x_2}$. At the same time, the point on the body which was at $\mathbf{x_1}$, when $t = t_1$ is moved to a new location $\mathbf{x_1'}$, due to rigid body motion. Two velocities, $\mathbf{U_b}$ and $\mathbf{q}$, are associated with the movement of the point on the body and the fluid particle, where $\mathbf{U_b}$ is the velocity due to the rigid body motion, which translated the point on the body from $\mathbf{x_1}$ to $\mathbf{x_1'}$, and $\mathbf{q}$ is the total velocity which moved the fluid particle from $\mathbf{x_1}$ to $\mathbf{x_2}$.

The velocity flow field can be decomposed into two components: inflow velocity and perturbation velocity, written as

$$\mathbf{q} = \mathbf{U}_\infty + \mathbf{u}, \quad (1)$$

where $\mathbf{q}$ is the total velocity, $\mathbf{U}_\infty$ is the inflow velocity, and $\mathbf{u}$ is the perturbation velocity due to the presence of obstacles.

2.1. Governing Equations. The methodology applied in this study is based on the inviscid potential flow theory:

$$\mathbf{u} = \nabla\phi, \quad (2)$$

where ϕ is the perturbation potential. The perturbation velocity field, therefore, is governed by the Laplace Equation:

$$\nabla^2\phi = 0. \quad (3)$$

If the inflow is also irrotational, we can also define the total potential Φ as

$$\mathbf{q} = \nabla\phi + U_\infty = \nabla\Phi. \quad (4)$$

Appling Green's second identity, the Laplace equation (3) can be written as following boundary integrated form:

$$2\pi\phi_p = \iint_{S_H}\left[\phi_p\frac{\partial G(p,p')}{\partial\mathbf{n}_{p'}} - \frac{\partial\phi_p}{\partial\mathbf{n}_{p'}}G(p,p')\right]dS + \iint_{S_W}\phi_w\left(y_{p'}\right)\frac{\partial G(p,p')}{\partial\mathbf{n}_{p'}}dS, \quad (5)$$

where S_H represents the surface of hydrofoil or propeller blades and S_W represents the surface of trailing edge wake. $G(p,p')$ is the Green's function, which is defined as $1/R(p;p')$ in 3D, and $2\ln R(p;p')$ in 2D.

2.2. Boundary Conditions

2.2.1. Kinematic Boundary Condition. Without viscosity, the solid surface has to be treated as slip wall, which implies that the fluid particles on the surface of the body can only be sliding on the surface. From Figure 1, we can easily obtain the expression of the kinematic boundary condition on the solid surface:

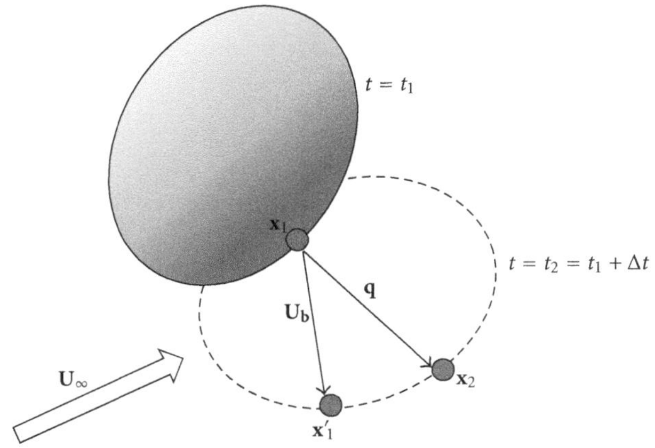

FIGURE 1: Schematic plot of an obstacle with rigid body motion.

$$\mathbf{n}\cdot(\mathbf{q} - \mathbf{U_b}) = \mathbf{0}, \quad (6)$$

which can be rewritten as

$$\frac{\partial\phi}{\partial n} = -(\mathbf{U}_\infty - \mathbf{U_b})\cdot\mathbf{n}. \quad (7)$$

It is convenient to define

$$\mathbf{U}_{\text{in}} = \mathbf{U}_\infty - \mathbf{U_b}. \quad (8)$$

Therefore, the kinematic boundary condition can be written as

$$\frac{\partial\phi}{\partial n} = -\mathbf{U}_{\text{in}}\cdot\mathbf{n}. \quad (9)$$

2.2.2. Dynamic Boundary Condition. The momentum equation for potential flow can be integrated into the form of unsteady Bernoulli's Equation:

$$\frac{P_\infty}{\rho} + \frac{1}{2}\mathbf{U}_\infty\cdot\mathbf{U}_\infty = \frac{P}{\rho} + \frac{\mathbf{q}\cdot\mathbf{q}}{2} + \frac{\partial\phi}{\partial t}, \quad (10)$$

where P_∞ is the reference pressure infinitely far upstream, and P is the pressure in the field. It is more convenient to treat the unsteady term in (10) with the material derivative by using following identity:

$$\frac{\mathbf{D}\phi}{\mathbf{D}t} = \frac{\partial\phi}{\partial t} + \mathbf{q}\cdot\nabla\phi. \quad (11)$$

Equation (10) becomes

$$\frac{P_\infty - P}{\rho} + \frac{1}{2}\mathbf{U}_\infty\cdot\mathbf{U}_\infty = \frac{\mathbf{q}\cdot\mathbf{q}}{2} + \frac{\mathbf{D}\phi}{\mathbf{D}t} - \mathbf{q}\cdot\nabla\phi. \quad (12)$$

Following the definition in Figure 1, as $\Delta t \to 0$, we have

$$\begin{aligned}\frac{\mathbf{D}\phi}{\mathbf{D}t} &= \frac{\phi(\mathbf{x}_2, t_2) - \phi(\mathbf{x}_1, t_1)}{\Delta t} \\ &= \frac{\phi(\mathbf{x}_2, t_2) - \phi(\mathbf{x}_1', t_2)}{\Delta t} + \frac{\phi(\mathbf{x}_1', t_2) - \phi(\mathbf{x}_1, t_1)}{\Delta t} \\ &= \nabla\phi \cdot \frac{x_2 - x_1'}{\Delta t} + \frac{\delta\phi}{\delta t} \\ &= \nabla\phi \cdot (\mathbf{q} - \mathbf{U_b}) + \frac{\delta\phi}{\delta t},\end{aligned} \tag{13}$$

where $\delta\phi/\delta t$ denotes the rate of change of the perturbation potential associated with a solid particle on the surface of the body.

Substituting (13) into (12), we obtain the dynamic boundary condition on the solid surface:

$$\frac{P_\infty - P}{\rho} + \frac{1}{2}\mathbf{U}_\infty \cdot \mathbf{U}_\infty = \frac{\mathbf{q}\cdot\mathbf{q}}{2} - \nabla\phi \cdot \mathbf{U_b} + \frac{\delta\phi}{\delta t}. \tag{14}$$

Adding $0.5\mathbf{U_b} \cdot \mathbf{U_b} - \mathbf{U}_\infty \cdot \mathbf{U_b}$ to both side of the equation above will lead to

$$\begin{aligned}\frac{P_\infty - P}{\rho} + \frac{1}{2}\mathbf{U}_{\text{in}} \cdot \mathbf{U}_{\text{in}} &= \frac{\mathbf{q}\cdot\mathbf{q}}{2} + \frac{\mathbf{U_b}\cdot\mathbf{U_b}}{2} \\ &\quad - (\mathbf{U}_\infty + \nabla\phi) \cdot \mathbf{U_b} + \frac{\delta\phi}{\delta t},\end{aligned} \tag{15}$$

which is

$$\frac{P_\infty - P}{\rho} = \frac{1}{2}(\mathbf{U}_{\text{in}} + \mathbf{u})^2 - \frac{1}{2}\mathbf{U}_{\text{in}}^2 + \frac{\delta\phi}{\delta t}. \tag{16}$$

Equation (16) is a general expression of the unsteady Bernoulli's equation for an obstacle with any rigid body motion. For a 2D hydrofoil with pitch motion, $\mathbf{U_b}$ is $\boldsymbol{\Omega} \times \mathbf{r}$, where $\boldsymbol{\Omega}$ angular velocity of pitching, and $\mathbf{r}$ is the vector from the center of pitching to the points on the surface of the hydrofoil, as shown in Figure 2.

As shown in Figure 3, in the case of a propeller under surge and heave motion, $\mathbf{U_b}$ and $\mathbf{U}_{\text{in}}$ are defined as

$$\begin{aligned}\mathbf{U}_b &= \boldsymbol{\omega}_{\text{prop}} \times \mathbf{r} + \mathbf{U}_{b,\text{surge}} + \mathbf{U}_{b,\text{heave}}, \\ \mathbf{U}_{\text{in}} &= \mathbf{U}_\infty - \boldsymbol{\omega}_{\text{prop}} \times \mathbf{r} - \mathbf{U}_{b,\text{surge}} - \mathbf{U}_{b,\text{heave}},\end{aligned} \tag{17}$$

where $\boldsymbol{\omega}_{\text{prop}}$ is the angular velocity of the rotation of the propeller; $\mathbf{r}$ is the radial vector from the axis of the propeller to the points on the blade surface; $\mathbf{U}_{b,\text{surge}}$ and $\mathbf{U}_{b,\text{heave}}$ are the surge and heave velocity of the propeller, respectively. The surge and heave motion of a propeller are incorporated into a BEM code, PROPCAV. It has to be pointed out that the formulation presented above is based on the assumption of inviscid and unseparated flow. If the amplitude of motion of the moving hydrofoils or propellers becomes too large, the boundary layer may separate even at the leading edge of the hydrofoil or the blade of propeller and then form leading edge vortex, which is beyond the scope of application of current models.

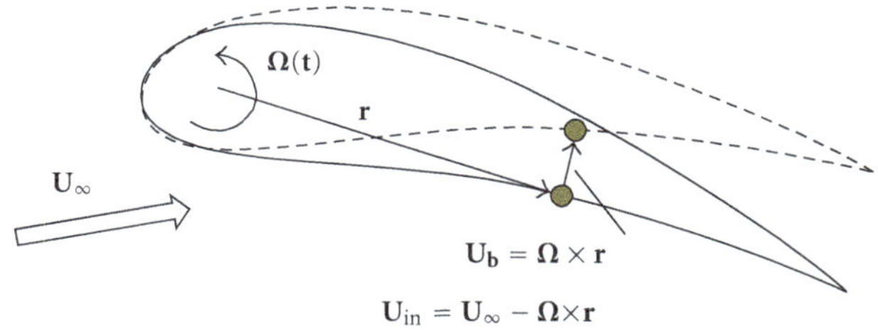

FIGURE 2: A hydrofoil with pitch motion.

3. Unsteady VLM Model for Propeller with Rigid Body Motion

The kinematic and dynamic boundary conditions described for the BEM model can be easily applied to vortex lattice method (VLM). In this study, a VLM code, MPUF3A, is modified to incorporate the correct kinematic and dynamic boundary conditions for a propeller with either surge or heave motions. He [8] developed a fully unsteady wake alignment model based on MPUF3A. The effects of the unsteady wake model on propellers undergoing heave motions are also investigated. The details on VLM and the unsteady wake alignment can be found in He [8].

4. Application to 2D Pitching Hydrofoils

The BEM model presented in this study is first applied to a pitching hydrofoil, which has an NACA66 thickness form with 5% thickness and 0% camber over chord length. The angular frequency, k of the pitching motion is π, and the rotation-axis origin is set as the leading edge of the hydrofoil. The angle between the hydrofoil and the inflow, or, the instantaneous angle of attack, α, is defined as follows:

$$\alpha = \alpha_0 \sin kt, \tag{18}$$

where α_0 denotes the amplitude of the angle between the hydrofoil and the inflow. RANS and also inviscid finite volume method (FVM) simulations are performed with commercial software FLUENT for the same configuration. The inviscid FVM simulation is carried out in order to assess viscous effects. The standard k-ε turbulence model with standard wall function is used to solve RANS equations. Two-dimensional unstructured grids were generated with structured prism boundary layer cells attached on the hydrofoil. In total, 407K number of cells is distributed over the entire domain for pitch motion while 390K cells are provided for heave motion. Obtained range of y^+ for both the motions is between 40 and 120. Dynamic mesh is used to simulate the pitch motion. Computational (CPU) time of 17 hours is taken for the RANS simulation on 4 Intel Xeon 2.54 GHz CPUs whereas 2D BEM model takes only 3 minutes on a single CPU for simulating the complete flow.

Figures 4 and 5 compare the unsteady lift and drag coefficients from 2D BEM model and inviscid FVM/RANS where $C_D = D/(0.5\rho U^2 c)$. Results from both of these methods show close level of agreement. However, a slight shift between

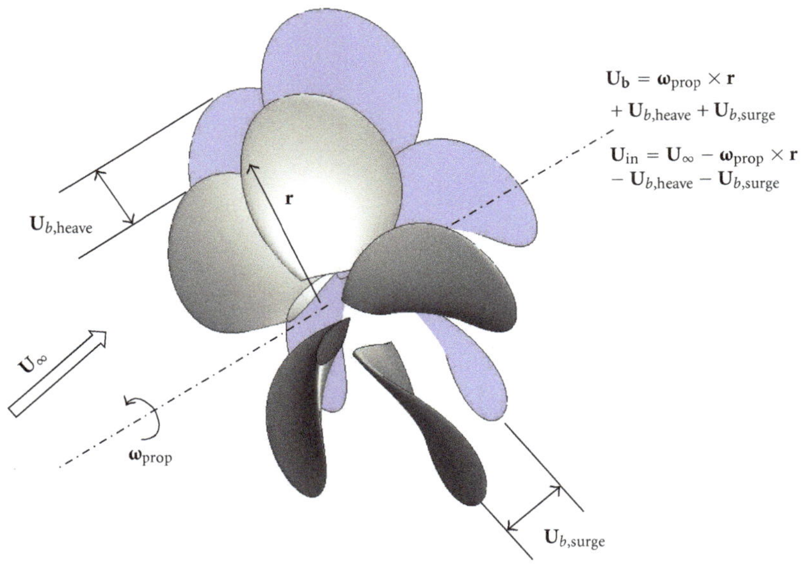

FIGURE 3: A propeller with surge and heave motion.

inviscid and viscous drag coefficient can be noticed. This shift is attributed to the viscous effects which are not included in the inviscid FVM simulation and 2D BEM model. The shift of the drag coefficient appears to be a constant value, which indicates that it can be modeled with a time-independent constant skin friction coefficient. In Figures 6 and 7, pressure distribution on the foil surface is compared at different flow time. The results suggest that except at the trailing edge of the hydrofoil, both 2D BEM model and RANS agree well with each other. In the unsteady BEM model, an iterative pressure Kutta (IPK) condition, proposed by Kinnas and Hsin [11], has been implemented to obtain zero loading at the trailing edge of the hydrofoil.

Figure 8 shows the comparison between unsteady wake profiles obtained from 2D BEM model and RANS. Both the trajectories and rollups are predicted accurately by 2D BEM model. As shown in Figure 9, range of y^+ is between 40 and 100, which is acceptable for standard wall function.

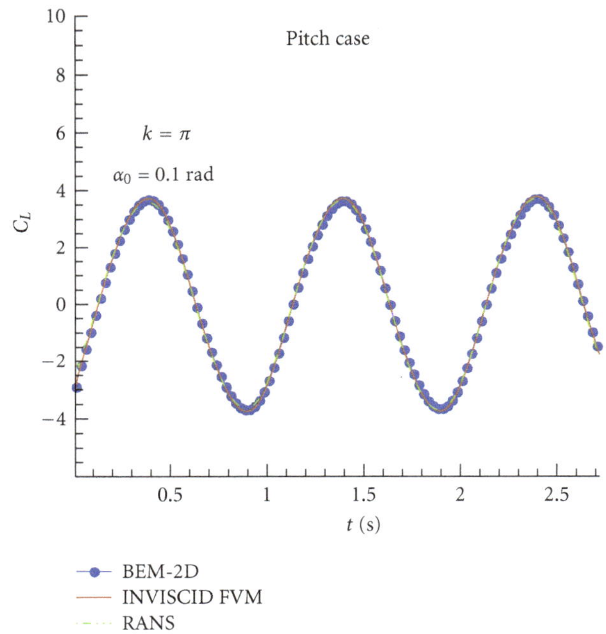

FIGURE 4: Lift coefficient versus time.

5. Application to Propellers

Marine propellers are subjected to spatially nonuniform unsteady inflow, which causes periodic loads on the propeller blades and shaft system. The unsteadiness in the inflow is primarily responsible for blade pressure fluctuations and cavity inception whose effect on the propulsion system could be significant. In addition to this, the transmission of unsteady forces through shafting into the vehicle can cause severe wear and tear of the system. Therefore, it becomes important to analyze the effects of unsteadiness in the inflow on the propulsive performance of a propeller for an efficient design of the hull propeller system.

5.1. Propeller Undergoing Pure Surge Motion. A model propeller DTNSRDC P4381 undergoing surge motion respectively is simulated. The geometry of this propeller is available to the author through the technical report by Boswell [12]. Fully wetted unsteady runs are performed at design advance coefficients ($J_s = 0.889$).

In this study, the surge motion is defined as harmonic oscillation:

$$\mathbf{U}_{b,\text{surge}} = \mathbf{U}_{\text{surge}} \sin\left(\omega_{\text{surge}} t\right), \tag{19}$$

where $\mathbf{U}_{\text{surge}}$ denotes the amplitude of the surge velocity, and ω_{surge} is the angular frequency of the surge motion. Table 1 shows the parameters of the surge motion in this case.

TABLE 1: Surge motion parameters.

J_S	$\mathbf{U}_{\text{surge}}/\mathbf{U}_\infty$	$\omega_{\text{surge}}/\omega_{\text{prop}}$	Time step size (VLM and BEM)	Time step size (RANS)
0.889	0.20	1.0	Fixed ($\Delta t/T_{\text{surge}} = 0.016$)	Adaptive ($\Delta t/T_{\text{surge}} = 5.62e-04$)

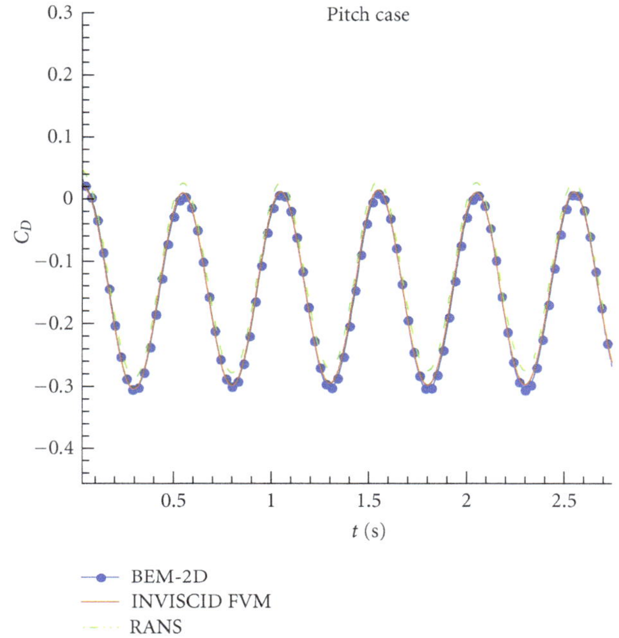

FIGURE 5: Drag coefficient versus time.

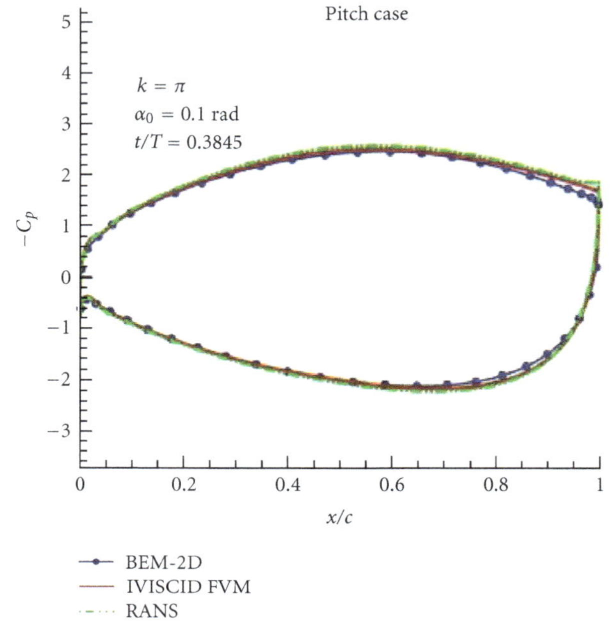

FIGURE 6: Pressure distribution on the pitching hydrofoil at $t/T = 0.3845$.

In VLM, the propeller blade is discretized using 20 panels in chordwise and 18 panels in spanwise direction while in BEM, discretization is done using 60 panels in chordwise and 20 panels in spanwise direction. Grids are distributed uniformly along the spanwise direction while in the chordwise direction full cosine distribution is provided. Thickness/loading coupling has been implemented in VLM to incorporate the thickness.

RANS simulation for the same propeller undergoing the same surge motion is also carried out. Modeling of surge motion of propeller in RANS solver is performed by leveraging moving mesh feature of FLUENT and using UDF to give surge motion to the complete domain. Total numbers of 1.75 million structured hexahedral cells are provided in a periodic domain. 192 hours are taken by RANS to simulate 9 seconds of flow on 24 Intel Xeon 2.54 GHz CPUs, on the other hand, VLM takes only 12 minutes and BEM takes 15 minutes of CPU time for the complete simulation.

A detailed convergence study of both the BEM and VLM codes can be found in Sharma et al. [10].

Figures 10 and 11 show the comparison between time variation of the thrust and torque coefficients among different numerical methods. BEM and VLM compare well with each other but show some differences with RANS. It is also noticed that the time average KT over one cycle of surge oscillation is growing for RANS, while it is constant for BEM

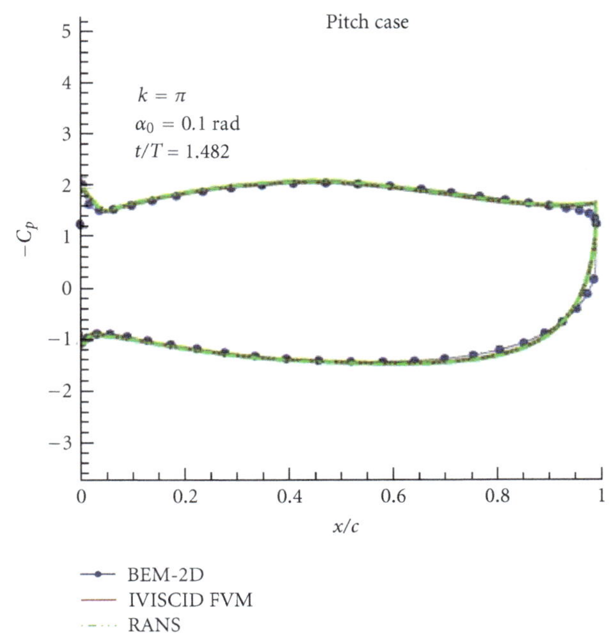

FIGURE 7: Pressure distribution on the pitching hydrofoil at $t/T = 1.482$.

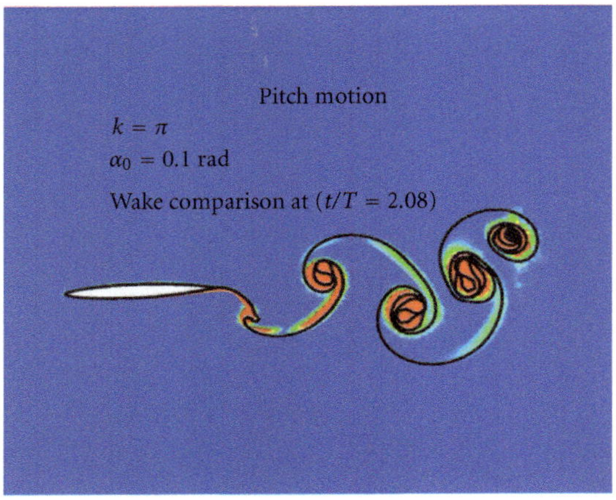

FIGURE 8: Comparison of wake pattern for pitch motion where $k = \pi$ and pitch amplitude = 0.1 rad. BEM is shown by solid curve, and RANS by colormap.

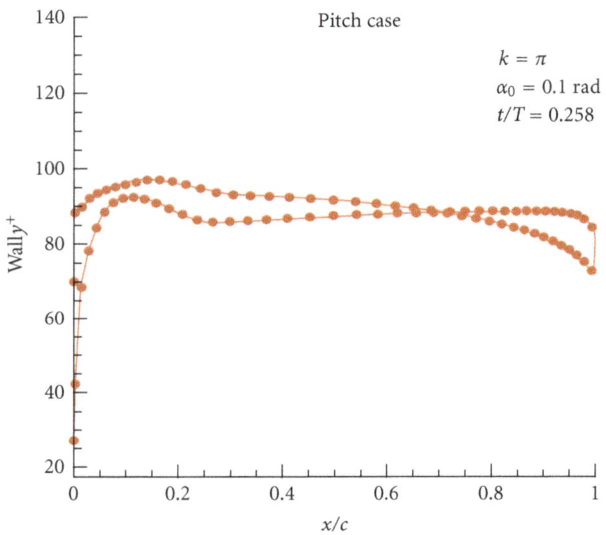

FIGURE 9: y^+ on the hydrofoil surface.

and VLM. Similarly, mean KQ for RANS is growing, but it remains constant for BEM or VLM.

5.2. Propeller Undergoing Pure Heave Motion. A model propeller DTMB P4718 undergoing heave motion is simulated. The performance of this propeller underdoing heave motion was investigated by Politis [7] independently. Fully wetted unsteady runs by VLM are performed at design advance coefficients ($J_s = 0.751$). In this study, the heave motion is in the form of harmonic oscillation:

$$\mathbf{U}_{b,\text{heave}} = \mathbf{U}_{\text{heave}} \sin(\omega_{\text{heave}} t), \tag{20}$$

where $\mathbf{U}_{\text{heave}}$ denotes the amplitude of the heave velocity, and ω_{heave} is the angular frequency of the heave motion. Table 2 shows the parameters of the heave motion in this case.

TABLE 2: Heave motion parameters.

J_S	$\mathbf{U}_{\text{heave}}/\mathbf{U}_\infty$	$\omega_{\text{heave}}/\omega_{\text{prop}}$	Time step size (VLM/BEM)
0.751	1.045	0.5	Fixed ($\Delta t/T_{\text{heave}} = 0.013$)

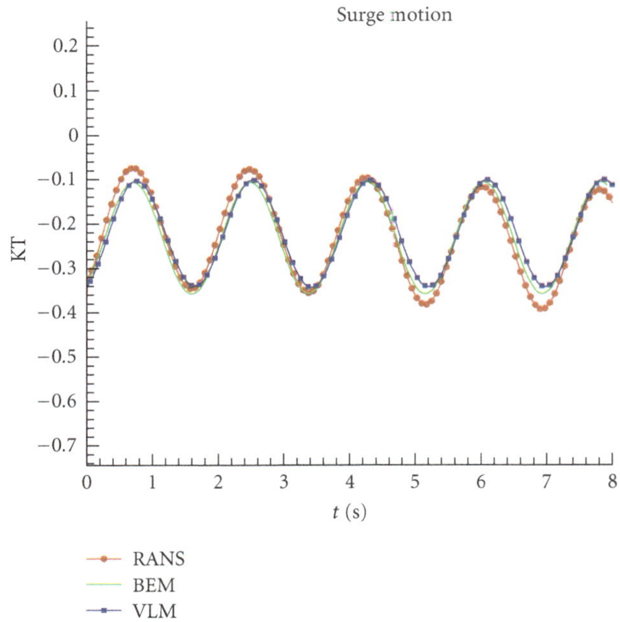

FIGURE 10: Unsteady thrust coefficient.

In this case, the spatial discretization on the propeller blade is the same as that in the case of surge motion. The results from VLM are compared with those obtained from Politis [7], as shown in Figure 12. The unsteady single-blade thrust coefficients from the present methods appear to be fairly close to each other and also close to the Politis' results. In the case of heave motion, the VLM code with fully unsteady wake alignment (FWA) is also tested. Although in many situations, such as inclined shaft propeller and propeller at low advance ratio, the fully aligned wake model could significantly improve the results, in this case, it does not make a big difference. The reason could be that the propeller is still working near the design condition, in which the pitch angle of the wake is relatively small.

6. Conclusions and Future Work

A time-dependent 2D BEM model has been developed and validated against RANS solver to analyze the nature of unsteady forces, pressure distributions, and wake profiles for an oscillating hydrofoil. Simulations of heaving and pitching motion are performed for a set of parameters to avoid inception of leading edge vortex (LEV) which can drastically change pressure distribution on the surface of hydrofoil. In the future, the LEV model developed by Tian and Kinnas [13] will be incorporated to analyze flapping hydrofoil for a wider range of parameters.

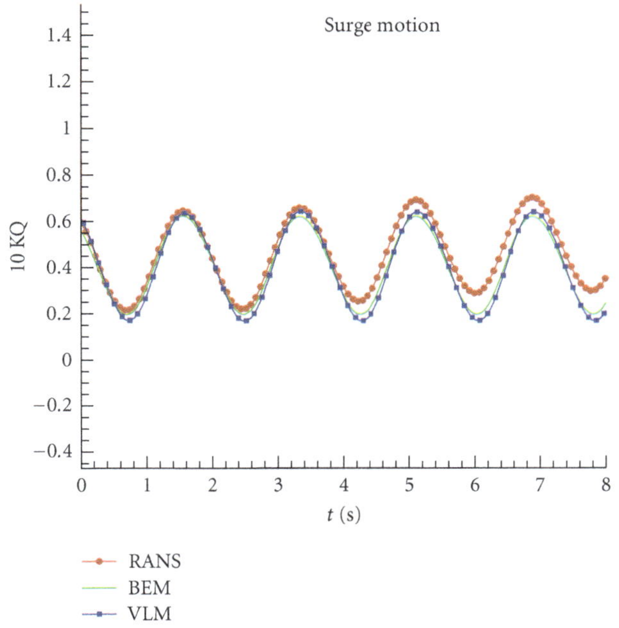

FIGURE 11: Unsteady torque coefficient.

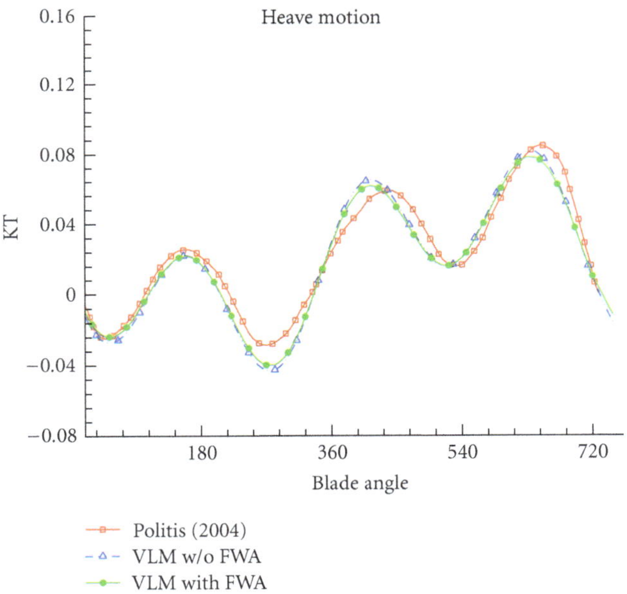

FIGURE 12: Comparison of KT from MPUF-3A and Politis [7]. Unlike other KT plots, direction of thrust is considered positive, that is, along the direction of ship's forward motion.

The unsteady hydrodynamic analysis of propeller under surge and heave motion has been performed using vortex lattice method (VLM) and boundary element method (BEM). A fully unsteady wake alignment algorithm is implemented in the VLM code. On comparing the results from the present method to those from Politis [7], fair-to-moderate level of agreement has been found between the two. Surge motion of a propeller has successfully been simulated using RANS solver; however, on comparing time-varying KT and KQ with those obtained from BEM or VLM, differences were found for some cases. Discrepancies observed while comparing the results from the VLM and the BEM codes need to be investigated further.

The numerical methods proposed in this study are not intended as a substitution of RANS simulations. The effects of turbulence and vortical separated flows are difficult to be handled by the methods presented in this paper. In these cases, RANS or LES could give results with more physical representations of the problem, at the same time, with significantly increased computational cost. The uncertainties in the numerical solutions have to be addressed as well. The study by Sharma [9] showed that current methods converged to the same results as refining the spatial and temporal resolution. Direct correlation of the results from current methods with any experimental data would be greatly helpful to validate the current methods. In the current study, only two simple harmonic motions are considered. In the future, it is worthwhile to analyze the propeller undergoing surge and heave motions with more harmonic components and understand the corresponding propeller performance in the frequency domain. The present method will also be extended to analyze pitch, yaw, and sway motion of the propeller.

Having a strong background in simulating unsteady motion of hydrofoils and marine propellers, a more comprehensive approach is required in dealing with the minute details of flow around the oscillating hull propeller system. The long-term objective of the current research work is to integrate the present model with hull motion model followed by coupling with Reynolds Averaged Navier Stokes (RANSs) method to develop a complete 6 DOF hull-propeller system motion model. Coupling with RANS by Kinnas et al. [14] can be improved to provide accurate propeller-hull interaction in time. In this approach, the propeller forces are represented by spatially distributed and time-dependent body forces in the RANS code to account for the effects of propeller-induced flows. In the cases that wake alignment is important, unsteady wake alignment could be implemented using the velocity field from RANS. The treatment of the different time step sizes of the hull motion and the propeller subsystem has to be investigated further. Development of the required model will analyze the complete flow around hull-propeller system and predict forces, moments, and pressure distribution with utmost accuracy.

Acknowledgments

Support for this research was provided by the U.S. Office of Naval Research (Contract nos. N00014-07-1-0616 and N00-014-10-1-0931) and Phases V and VI of the *Consortium on Cavitation Performance of High Speed Propulsors* with the following current members: American Bureau of Shipping, Daewoo Shipbuilding and Marine Engineering Co. Ltd., Kawasaki Heavy Industry Ltd., Rolls-Royce Marine AB, Rolls-Royce Marine AS, Samsung Heavy Industries Co. Ltd., SSPA AB, Sweden, VA Tech Escher Wyss GmbH, Wärtsilä Propulsion Netherlands B.V., Wärtsilä Propulsion Norway

AS, Wärtsilä Lips Defense S.A.S., and Wärtsilä CME Zhenjiang Propeller Co. Ltd.

References

[1] M. F. van Sluijs, "Performance and propeller load fluctuations of a ship in waves," TNO 163S, Netherlands Ship Research Centre, 1972.

[2] S. D. Jessup and R. J. Boswell, "The effect of hull pitching motions and waves on periodic propeller blade loads," in *Proceedings of the 14th Symposium on Naval Hydrodynamics*, Ann Arbor, Mich, USA, 1982.

[3] M. G. Parsons and W. S. Vorus, "Added mass and damping estimates for vibrating propellers," in *Proceedings of the Propellers Symposium Transactions SNAME*, Virginia Beach, Va, USA, 1981.

[4] T. Sasajima, "Usefulness of quasi-steady approach for estimation of propeller bearing forces," in *Proceedings of the Propellers Symposium Transactions SNAME*, 1978.

[5] J. P. Breslin and P. Andersen, *Hydrodynamics of Ship Propellers*, Cambridge University Press, Cambridge, Mass, USA, 1st edition, 1994.

[6] H. Lee and S. A. Kinnas, "Unsteady wake alignment for propellers in nonaxisymmetric flows," *Journal of Ship Research*, vol. 49, no. 3, pp. 176–190, 2005.

[7] G. K. Politis, "Simulation of unsteady motion of a propeller in a fluid including free wake modeling," *Engineering Analysis with Boundary Elements*, vol. 28, no. 6, pp. 633–653, 2004.

[8] L. He, *Numerical simulation of unsteady rotor/stator interaction and application to propeller/rudder combination [Ph.D. thesis]*, Department of Civil Engineering, UT Austin, Austin, Tex, USA, 2010.

[9] A. Sharma, "Numerical modeling of a hydrofoil or a marine propeller undergoing unsteady motion via a panel method and RANS," Tech. Rep. 11-02, Department of Civil Engineering, UT Austin, Austin, Tex, USA, 2011.

[10] A. Sharma, L. He, and S. A. Kinnas, "Numerical modeling of a hydrofoil or a marine propeller undergoing unsteady motion," in *Proceedings of the 2nd International Symposium on Marine Propulsors*, Hamburg, Germany, 2011.

[11] S. A. Kinnas and C. Y. Hsin, "Boundary element method for the analysis of the unsteady flow around extreme propeller geometries," *AIAA Journal*, vol. 30, no. 3, pp. 688–696, 1992.

[12] R. J. Boswell, *Design, Cavitation Performance, and Open-Water Performance of a Series of Research—Skewed Propellers*, Naval Ship Research and Development Center, Department of the Navy, 1971.

[13] Y. Tian and S. A. Kinnas, "Modeling of leading edge vortex and its effects on propeller performance," in *Proceedings of the 2nd International Symposium on Marine Propulsors*, Hamburg, Germany, 2011.

[14] S. A. Kinnas, S. H. Chang, Y. H. Yu, and L. He, "A hybrid viscous/potential flow method for the prediction of the performance of podded and ducted propellers," in *Proceedings of the Propeller/Shafting Symposium*, Williamsburg, Va, USA, 2009.

Control of Surge in Centrifugal Compressor by Using a Nozzle Injection System: Universality in Optimal Position of Injection Nozzle

Toshiyuki Hirano,[1] Takanori Uchida,[2] and Hoshio Tsujita[3]

[1] Production Systems Engineering Course, Tokyo Metropolitan College of Industrial Technology, 1-10-40 Higashiohi, Shinagawa-ku, Tokyo 140-0011, Japan
[2] Graduate School of Engineering, Hosei University, 3-7-2 Kajinocho, Koganei-shi, Tokyo 184-8584, Japan
[3] Department of Mechanical Engineering, Faculty of Science and Engineering, Hosei University, Tokyo, Japan

Correspondence should be addressed to Toshiyuki Hirano, hirano@s.metro-cit.ac.jp

Academic Editor: N. Sitaram

The passive control method for surge and rotating stall in centrifugal compressors by using a nozzle injection system was proposed to extend the stable operating range to the low flow rate. A part of the flow at the scroll outlet of a compressor was recirculated to an injection nozzle installed on the inner wall of the suction pipe of the compressor through the bypass pipe and injected to the impeller inlet. Two types of compressors were tested at the rotational speeds of 50,000 rpm and 60,000 rpm with the parameter of the circumferential position of the injection nozzle. The present experimental results revealed that the optimum circumferential position, which most effectively reduced the flow rate for the surge inception, existed at the opposite side of the tongue of the scroll against the rotational axis and did not depend on the compressor system and the rotational speeds.

1. Introduction

In recent years, the global approach to environmental protection has progressed and the measures against the problems such as the exhaust gas of cars and the fossil fuels depletion are urgent tasks. A turbocharger, as a part of the technology dealing with environmental problems, is expected to clean the exhaust gas from automobile engines by improving combustion efficiency, and to contribute to the reduction of fuel consumption through downsizing of engine leading to the weight reduction. Therefore, the application of a turbocharger to various types of vehicle is developing rapidly. Moreover, since the centrifugal compressor, which is a main component of turbocharger, possesses higher pressure ratio in a single stage and consequently contributes to the reduction in size and weight, it is widely used for many kinds of industrial machines as well as turbochargers. Therefore, the extension of the stable operation range and the improvement of the supercharging characteristics of a centrifugal compressor are required dramatically.

The operation of a centrifugal compressor of turbocharger at a lower flow rate close to the maximum pressure ratio induces instability phenomena such as rotating stall and surge. Especially, the surge may generate enough intense vibration and noise to destroy the whole pipeline system including a compressor. Moreover, it strongly influences the performance characteristics of the compressor. As a result, the stable operation range is inevitably restricted. Therefore, several investigations [1–6] have been carried out to control the inception of instability phenomena for the purpose of the extension of stable operation range of a centrifugal compressor to the lower flow rate.

In the previous studies on the control of surge in a centrifugal compressor, the injection nozzle system was employed, in which a part of the flow in the discharge duct was recirculated to the inlet of the impeller, and controlled

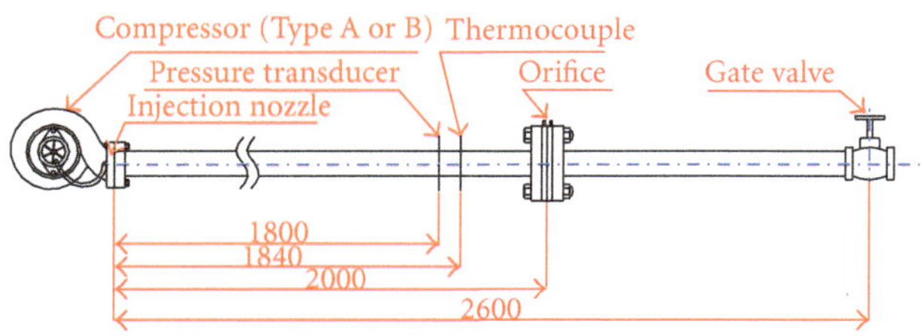

FIGURE 1: Experimental apparatus (unit: mm).

(a) Type A

(b) Type B

FIGURE 2: Tested impeller (unit: mm).

the inceptions of rotating stall and surge [7]. Moreover, optimization of the parameters affecting on the performance of the injection nozzle system such as a position of injection has been performed [8]. Furthermore, the dependency of the optimum injection position on the rotational speed of the impeller has been investigated [9]. However, since these studies have been performed for the same turbocharger system, the universality in the characteristics of the injection nozzle has not been clarified yet.

In this study, the optimum circumferential position of injection nozzle, which most effectively reduced the flow rate for the surge inception, was investigated for two types of compressors at rotational speeds of 50,000 rpm and 60,000 rpm to examine the universality of the optimum circumferential position of injection nozzle. In addition, the influences of the injection on the fluctuating property of the flow field before and after the surge inception were investigated by examining the frequency of static pressure fluctuation on the wall surface.

2. Experimental Apparatus

The experimental apparatus used in this study is shown in Figure 1. The configurations and the specifications of centrifugal impellers for two types of centrifugal compressors Type A and Type B are shown in Figure 2 and Table 1, respectively. The outlet blade angle is measured from a radial direction. The compressed air supplied from the screw compressor was used to drive the turbine impeller which revolved the compressor impeller through the corotating axis. A part of the air at the exit of the compressor scroll, which was compressed by the centrifugal impeller and the vaneless diffuser, was recirculated into the injection nozzle installed at the impeller inlet through the bypass pipe. Then, the remaining air was discharged through the delivery duct. The inner diameter of the injection nozzle used in this study is 4 mm. The injection nozzle is installed on the inner wall of the suction pipe of compressor, which is movable in the circumferential direction as shown in Figure 3.

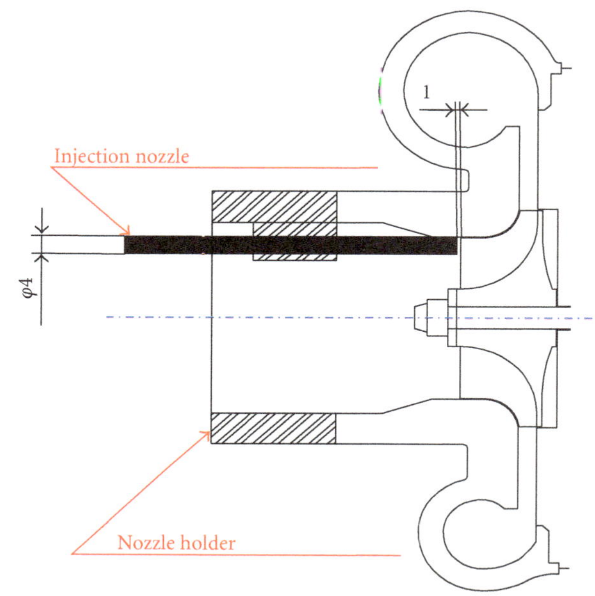

FIGURE 3: Injection nozzle (unit: mm).

3. Experimental Method

The origin of a circumferential position of the injection nozzle indicated by TT(0) is located at the corresponding position to the scroll tongue portion as shown in Figure 4. The circumferential position of the injection nozzle TT has the positive value in the rotational direction of the centrifugal

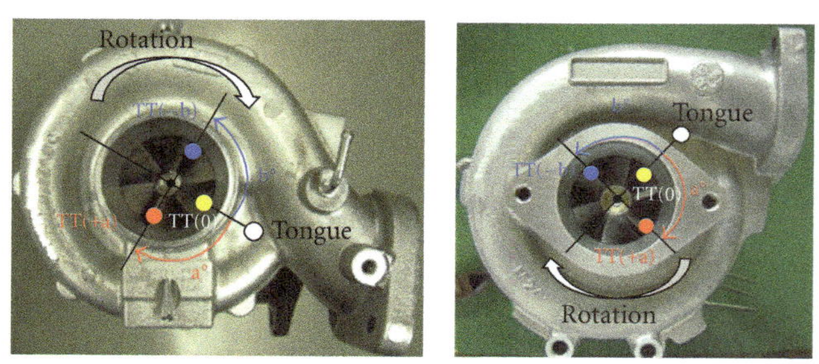

(a) Type A (b) Type B

FIGURE 4: Definition for position of injection nozzle.

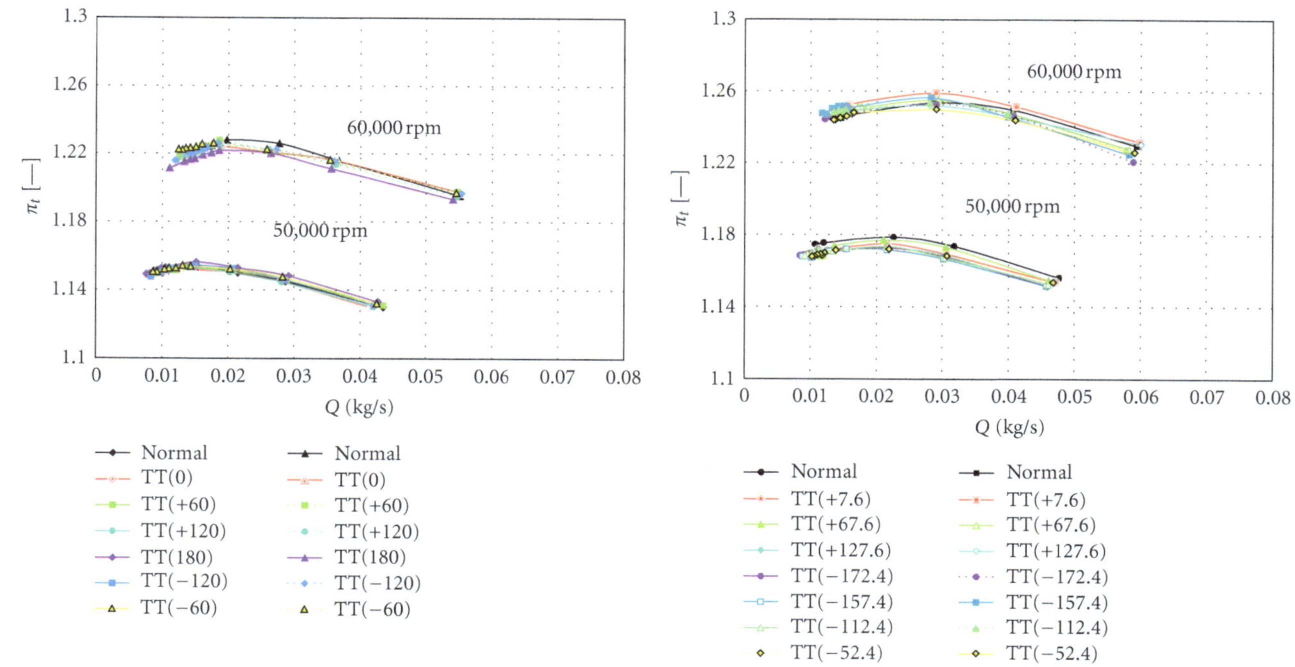

(a) Type A

(b) Type B

FIGURE 5: Performance characteristics.

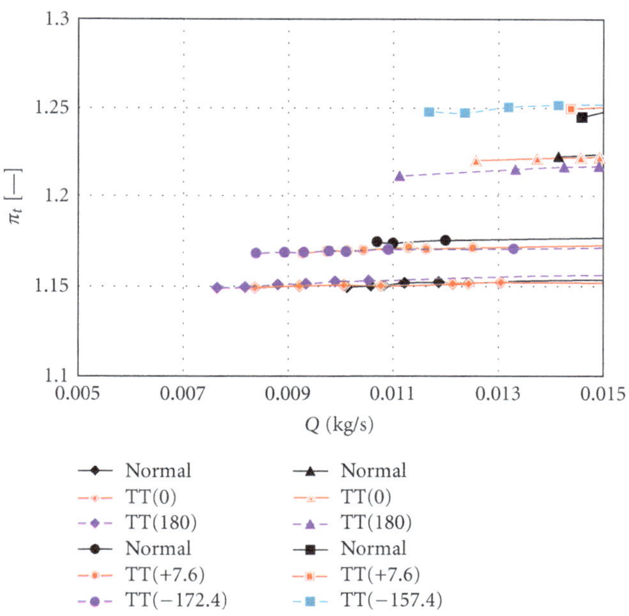

FIGURE 6: Performance characteristics at low flow rate.

impeller. In the following, the case with the injection nozzle is named "Injection," and that without the injection nozzle is named "Normal." The performance tests of compressors were performed at the corrected rotational speed of 50,000 rpm and 60,000 rpm. In each experiment, the flow rate was decreased gradually from the full opening condition by closing the gate valve installed at the downstream side of the delivery duct, until the surge inception. In this study, the preexperiments were performed to relate the degree of the gate valve opening to the flow rate. The limiting flow rate for the surge inception is defined as the flow rate at the degree of valve opening opened by its minimum revolution from that at which a sudden increase of the pressure fluctuation caused by the surge is observed. In this study, the surge inception was detected by the observation of the pressure fluctuation caused by the surge, which was suddenly increased by gradually closing the valve by its minimum revolution. In the experiments for Injection, the experiments were carried out at every 30 degrees of TT from TT(0). Moreover, the experiments for Type B were made at 15-degree intervals around the optimum TT which remarkably reduced the limiting flow rate for surge inception. The exit of the injection nozzle was located at 1 mm upstream of the inlet of the impeller for every experimental condition of Injection. The thermocouples were used to measure the temperature at both the inlet and outlet of the compressor. The static pressure at delivery duct and the pressure difference across the orifice were measured by the pressure transducer. The uncertainty for the measured pressure data is ±5 kPa. The corrected flow rate Q and the pressure ratio π_t were defined by the following equations:

$$Q = Q_0 \frac{P_{a0}}{P_a} \sqrt{\frac{T_1}{T_0}} \text{ [kg/s]}, \quad (1)$$

$$\pi_t = \frac{P_t}{P_a} \text{ [−]}, \quad (2)$$

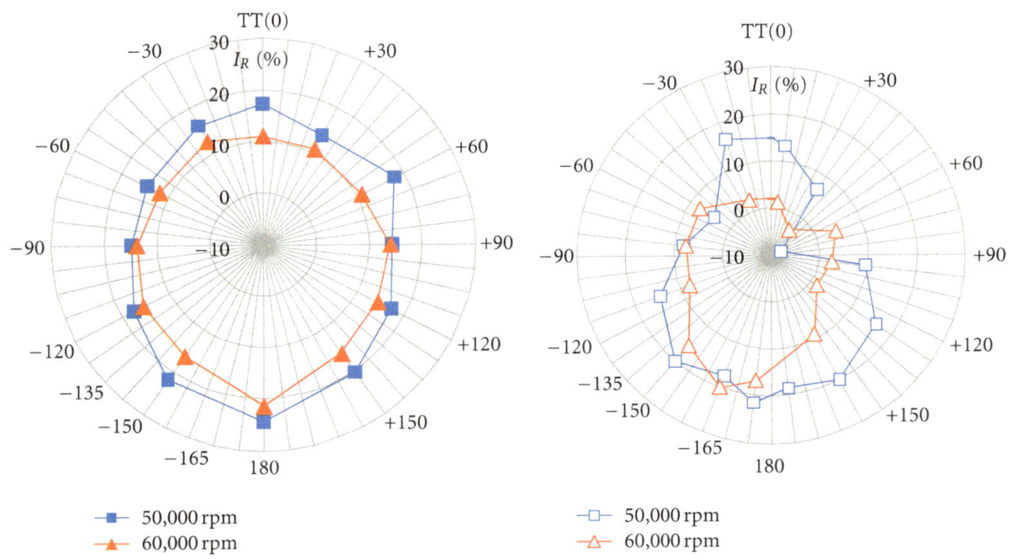

FIGURE 7: Improvement rate of surge margin.

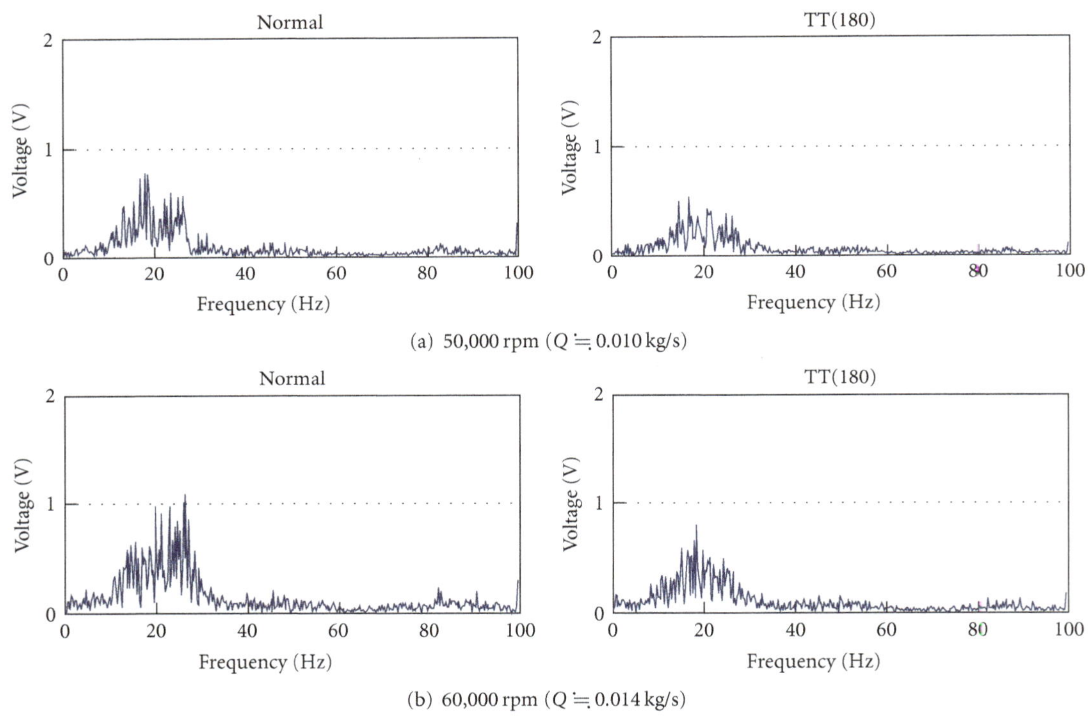

FIGURE 8: Spectrum of pressure fluctuation at delivery duct.

where Q_0 is the measured flow rate, P_{a0} is the standard atmospheric pressure, P_a is the measured atmospheric pressure, T_1 is the measured temperature, T_0 is the standard temperature, and P_t is the measured total pressure at the compressor outlet. In order to investigate the unstable phenomena, the wall static pressure fluctuation was measured at 1800 mm downstream of the compressor exit. The frequency characteristics of the wall static pressure fluctuation were analyzed by performing FFT. Flow sensor was installed at the bypass pipe to measure the flow rate ejected from the injection nozzle.

4. Results and Discussion

4.1. Performance Characteristics. Figure 5 shows the performance characteristics for Normal and Injection of Type

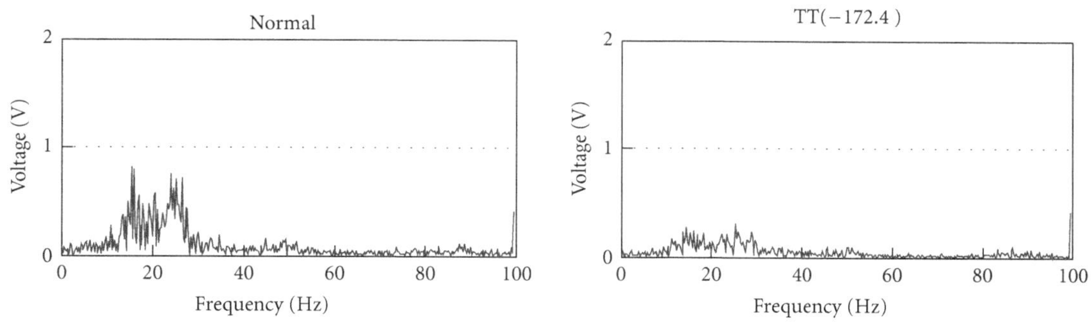

(a) 50,000 rpm ($Q \risingdotseq 0.011$ kg/s)

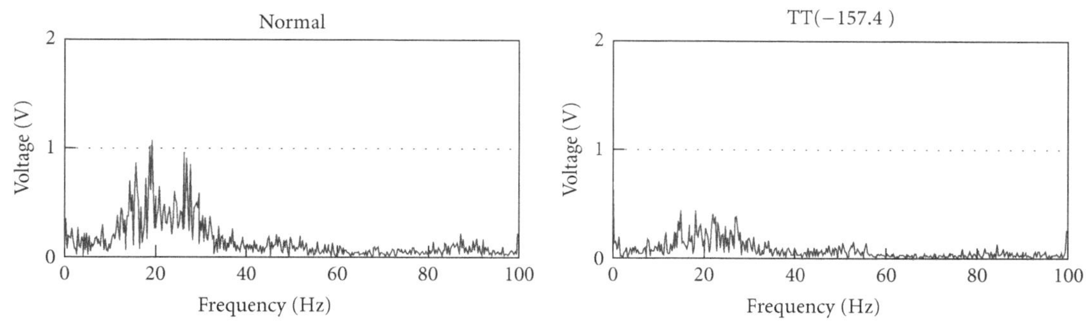

(b) 60,000 rpm ($Q \risingdotseq 0.011$ kg/s)

FIGURE 9: Spectrum of pressure fluctuation at delivery duct for Type B.

TABLE 1: Specifications of impeller.

Type of compressor	A	B
Inlet diameter (mm)	43.38	45
Outlet diameter (mm)	56	60
Number of blades	12	12
Outlet blade angle (deg.)	40	60
Inlet blade height (mm)	15.4	14
Outlet blade height (mm)	4.08	5.6
Blade thickness (mm)	0.4	0.675

A and Type B at rotational speeds of 50,000 rpm and 60,000 rpm. Figure 6 shows the performance characteristics at the low flow rate region for TT(0) and TT(180) of Injection, which remarkably reduced the limiting flow rate for surge inception as shown later, with that for Normal.

The performance curves for Injection almost coincide with those for Normal in both Type A and Type B for every rotational speed (Figure 5). Therefore, it is apparent that the influence of the injection system on the performance characteristics of the compressor is negligible. Moreover, the limiting flow rate for surge inception is reduced by using the nozzle injection system as shown in Figure 6. Therefore, the nozzle injection system is recognized to have ability to extend the stable operating range to the lower flow rate region. It is considered that the injection on the inner wall of the suction pipe of the impeller would reduce the reversed flow region distributed on the shroud side wall from the diffuser exit up to the impeller inlet which appeared before the surge inception and consequently extend the limiting flow rate for surge inception to the lower flow rate region.

4.2. Optimal Injection Position. In order to evaluate the effect of injection on the reduction of limiting flow rate for surge inception, the improvement rate of surge margin I_R was defined by the following equation:

$$I_R = \frac{Q_N - Q_I}{Q_N} \times 100\ [\%], \tag{3}$$

where Q_N and Q_I mean the limiting flow rate for surge inception in Normal and Injection, respectively. Figure 7 gives the relationship between the injection position TT and the improvement rate I_R.

The improvement rates I_R at 50,000 rpm are relatively higher than those at 60,000 rpm in both Type A and Type B. The I_R in Type A exhibits the improvement rate more than 10% for every TT, and the maximum I_R more than 20% is achieved especially for TT(180) at each rotational speed (Figure 7(a)). Similar tendency is also observed in Type B (Figure 7(b)), and the maximum improvement rate more than 20% is located around TT(180) at both 50,000 rpm and 60,000 rpm. These results suggest that the optimum circumferential position of the injection nozzle, which produces the maximum improvement ratio, is located around TT(180), which is the opposite side of the tongue of the scroll against the rotational axis, without the dependencies on the compressor type and the rotational speed. In general, the number of stall cell in the diffuser under the rotating stall condition varies from one to two according to the decrease

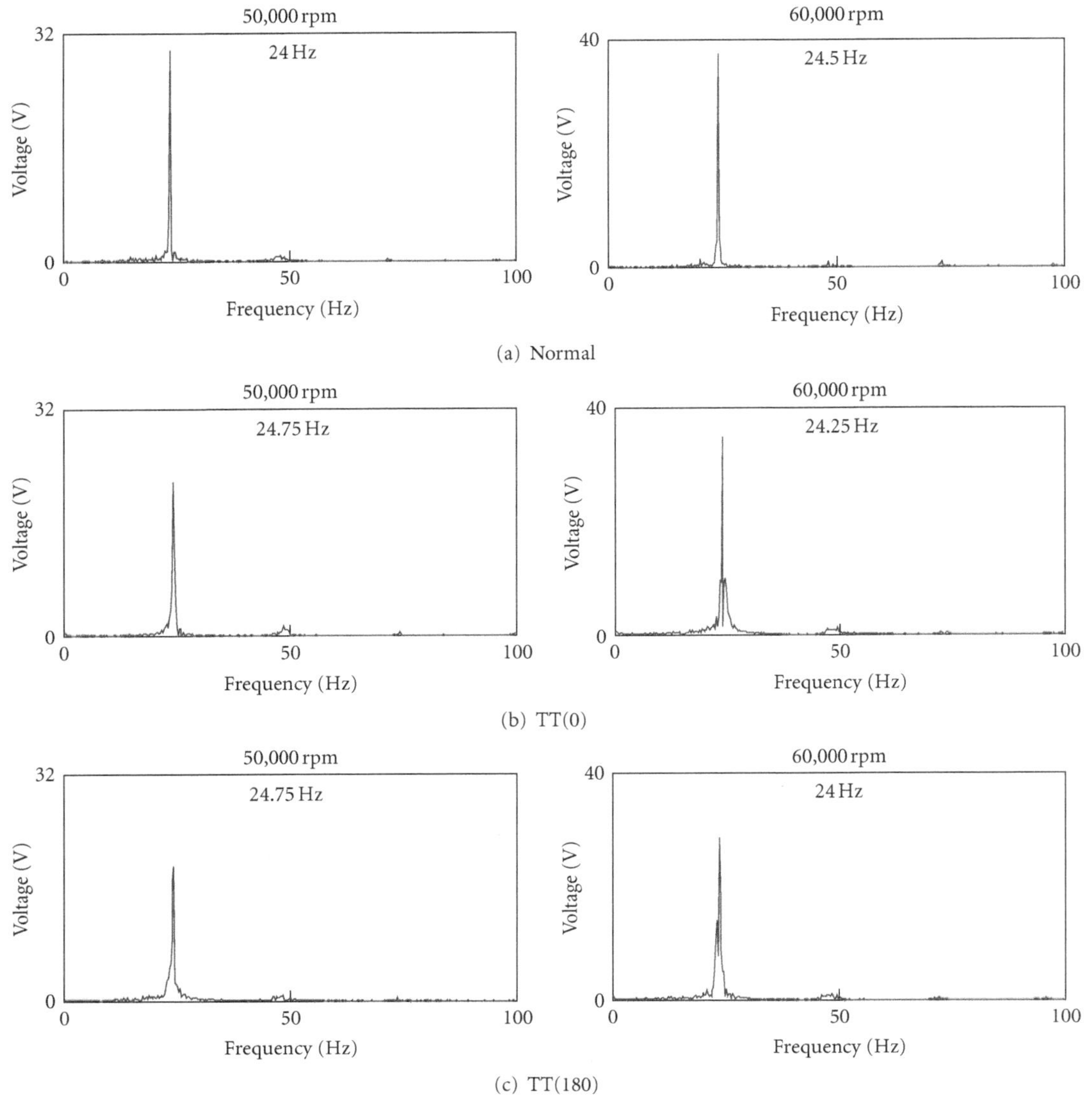

FIGURE 10: Spectrum of pressure fluctuation at delivery duct during surge for Type A.

of the flow rate. Moreover, in the case of the compressor with a scroll, the circumferential distribution of the static pressure on the diffuser endwall becomes nonaxisymmetric and is strongly distorted by the decrease of the flow rate. This distortion of the pressure distribution in the diffuser influences the flow field in the impeller, and consequently on the flow rate for surge inception. Therefore, it is considered that the optimum circumferential position of the injection nozzle has a strong relation with the behavior of the stall cell and the formation of circumferential distribution of the static pressure in the diffuser.

4.3. Frequency Characteristic

(a) Before Surge Inception. Figures 8 and 9 show the frequency characteristics of the static pressure fluctuation on the inner wall of the delivery duct at 1800 mm downstream of the compressor exit for Normal and the optimum TT of Injection in Type A and Type B, respectively. The flow rate in the results shown in Figures 8 and 9 is the nearest one to the limiting flow rate for surge inception for Normal. The peaks of the spectrum of pressure fluctuation for Normal at each rotational speed in Type A and Type B are observed around 15 to 30 Hz. The peaks for the optimum TT of Injection are lower than those in Normal. The surge frequencies in Type A and Type B are about 24.5 Hz and 23.5 Hz, respectively, irrespective of the rotational speed, which are different from the frequencies for the peaks observed in Figures 8 and 9. So, the peaks in Figures 8 and 9 are considered to be associated with the unstable phenomenon such as the rotating stall which appears before the surge inception. Therefore, it is considered that the optimum TT of Injection enhances the improvement rate of surge margin by suppressing the unstable phenomenon appearing before the surge inception.

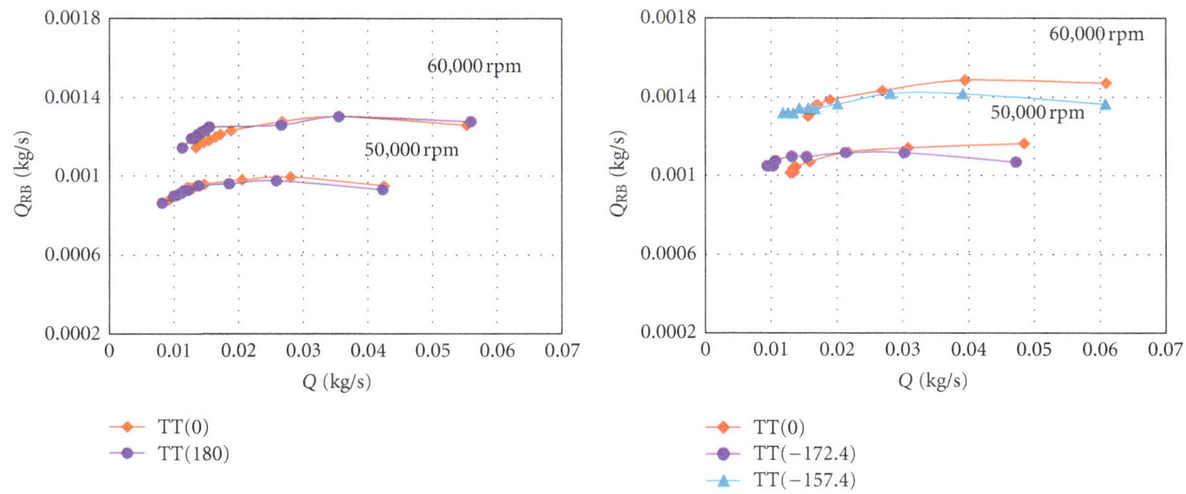

(a) Type A (b) Type B

FIGURE 11: Injection mass flow rate.

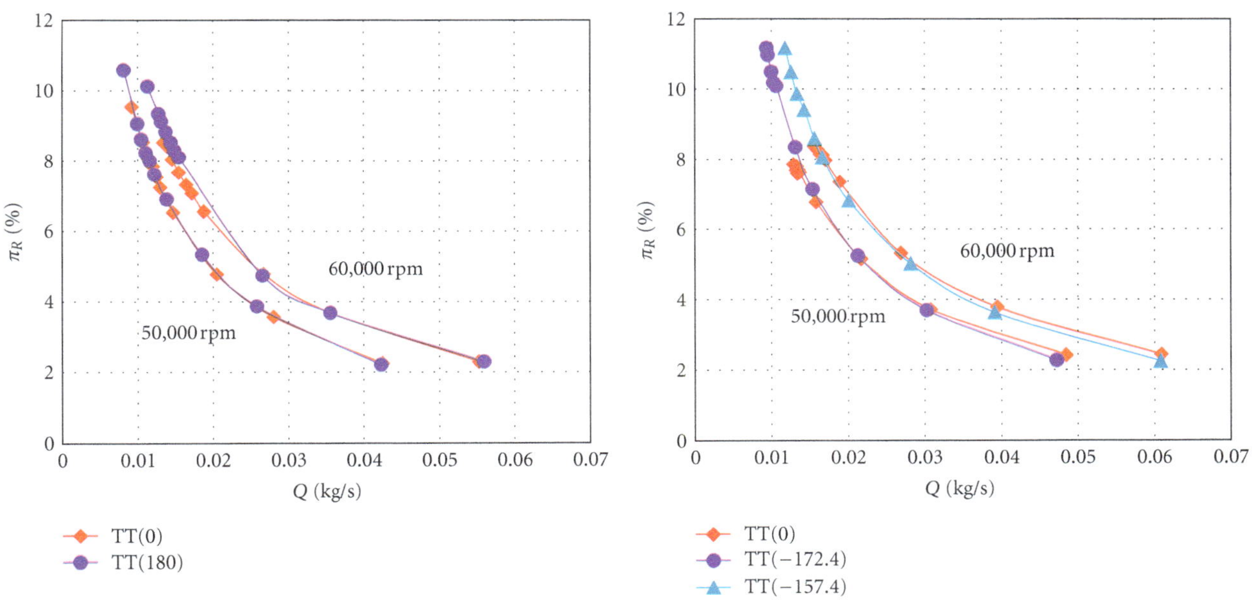

(a) Type A (b) Type B

FIGURE 12: Recirculation ratio.

(b) After Surge Inception. In order to examine the effects of the injection on the flow field under the surge condition, the frequency characteristics for Normal, TT(0), and the optimum TT(180) of Injection in Type A at $Q = 0.000$ kg/s after the surge inception are given in Figure 10. "$Q = 0.000$ kg/s" means the fully closed vane condition.

It is recognized that the peak of spectrum for the surge frequency of Injection becomes less than that of Normal after the surge inception. Moreover, the optimum TT(180) more reduces the peak of spectrum than TT(0). This means that Injection also reduces the strength of the surge, and the optimum circumferential position of injection for the surge reduction coincides with that for the reduction of the limiting flow rate for the surge inception.

4.4. Injection Flow Rate. Figure 11 shows the injection mass flow rate Q_{RB} for TT(0) and the optimum TT(180) of Injection at 50,000 rpm and 60,000 rpm in Type A and Type B. The injection flow rate in each condition is almost constant over the full operating range, but increased by the increase of the rotational speed. This is caused by the increase of pressure ratio due to the increase of the rotational speed. On the other hand, the injection flow rate is almost unaffected by the circumferential position of injection TT at each rotational speed.

In order to evaluate the ratio of the injection mass flow rate to the mass flow rate discharged from the delivery duct, the recirculation ratio π_R was defined by the following equation:

$$\pi_R = \frac{Q_{RB}}{Q_N} \times 100\ [\%]. \quad (4)$$

Figure 12 shows the recirculation ratio π_R for TT(0) and the optimum TT(180) of Injection at 50,000 rpm and 60,000 rpm in Type A and Type B. The influence of the circumferential position of injection TT on the recirculation ratio π_R is very small. The recirculation ratio is increased by the decrease of the flow rate in every experimental condition. Moreover, the comparison between the rotational speeds indicates that the recirculation ratio π_R is increased by the increase of the rotational speed corresponding to the tendency observed for the injection mass flow rate (Figure 11).

5. Conclusion

The following conclusions were obtained by the present study.

(1) With the nozzle injection system the stable operating range of the compressor improves and that is a direct effect on the performance characteristic.

(2) The optimum circumferential position of the injection nozzle, which produces the maximum improvement rate, is located around the opposite side of the tongue of the scroll against the rotational axis irrespective of the compressor type and the rotational speed.

(3) The optimum circumferential position of Injection enhances the improvement rate of surge margin by suppressing the unstable phenomenon appearing before the surge inception.

(4) The injection flow rate is almost constant over the full operating range, but increased by the increase of the rotational speed.

(5) The nozzle injection system has also the ability to reduce the strength of the surge.

(6) The optimum circumferential position of Injection for the surge reduction coincides with that for the reduction of the limiting flow rate for the surge inception.

References

[1] A. Suzuki, H. Tsujita, and S. Mizuki, "Passive control of surge for centrifugal compressor by using resonator," in *Proceedings of the 4th International Symposium on Experimental and Computational Aerothermodynamics of Internal Flows*, vol. 2, pp. 226–233, 1999.

[2] H. Tamaki, X. Zheng, and Y. Zhang, "Experimental investigation of high pressure ratio centrifugal compressor with axisymmetric and non-axisymmetric recirculation device," in *Proceedings of the ASME International Gas Turbine Institute (IGTI '12)*, 2012.

[3] S. Mizuki, H. Tsujita, and Y. Hishinuma, "Control of surge for centrifugal compression system by using a bouncing ball," ASME Paper 2000-GT-429, 2000.

[4] F. Willemas and B. De Jager, "One-sided control of surge in a centrifugal compressor system," ASME Paper 2000-GT-527, 2000.

[5] G. J. Skoch, "Experimental investigation of diffuser hub injection to improve centrifugal compressor stability," NASA/TM 2004-213182, 2004.

[6] H. Chen and V. Lei, "Casing treatment & inlet swirl of centrifugal compressors," in *Proceedings of the ASME International Gas Turbine Institute (IGTI '12)*, 2012.

[7] R. Gu and M. Yashiro, "Surge control for centrifugal compressor of turbocharger," *JSAE Paper*, vol. 36, no. 2, pp. 83–88, 2005.

[8] R. Gu, H. Ikeda, A. Yoshida, H. Tsujita, and S. Mizuki, "Effect of air injection on performance characteristics of small centrifugal compressor," in *Proceedings of the Turbomachinery Society of Japan*, May 2007.

[9] R. Gu, S. Mizuki, and H. Tsujita, "Surge control of centrifugal compressor by inducer tip injection," in *Proceedings of the International Gas Turbine Congress (IGTC '07)*, Tokyo, Japan, 2007.

15

Heat Transfer Augmentation Technologies for Internal Cooling of Turbine Components of Gas Turbine Engines

Phil Ligrani

Parks College of Engineering, Aviation, and Technology, Saint Louis University, 3450 Lindell Boulevard, McDonnell Douglas Hall Room 1033A, St. Louis, MO 63103, USA

Correspondence should be addressed to Phil Ligrani; p_ligrani@msn.com

Academic Editor: J.-C. Han

To provide an overview of the current state of the art of heat transfer augmentation schemes employed for internal cooling of turbine blades and components, results from an extensive literature review are presented with data from internal cooling channels, both with and without rotation. According to this survey, a very small number of existing investigations consider the use of combination devices for internal passage heat transfer augmentation. Examples are rib turbulators, pin fins, and dimples together, a combination of pin fins and dimples, and rib turbulators and pin fins in combination. The results of such studies are compared with data obtained prior to 2003 without rotation influences. Those data are comprised of heat transfer augmentation results for internal cooling channels, with rib turbulators, pin fins, dimpled surfaces, surfaces with protrusions, swirl chambers, or surface roughness. This comparison reveals that all of the new data, obtained since 2003, collect within the distribution of globally averaged data obtained from investigations conducted prior to 2003 (without rotation influences). The same conclusion in regard to data distributions is also reached in regard to globally averaged thermal performance parameters as they vary with friction factor ratio. These comparisons, made on the basis of such judgment criteria, lead to the conclusion that improvements in our ability to provide better spatially-averaged thermal protection have been minimal since 2003. When rotation is present, existing investigations provide little evidence of overall increases or decreases in overall thermal performance characteristics with rotation, at any value of rotation number, buoyancy parameter, density ratio, or Reynolds number. Comparisons between existing rotating channel experimental data and the results obtained prior to 2003, without rotation influences, also show that rotation has little effect on overall spatially-averaged thermal performance as a function of friction factor.

1. Introduction

The techniques employed to enhance convective heat transfer rates for internal cooling of turbine airfoils of gas turbine engines include rib turbulators, pin fins, dimpled surfaces, surfaces with arrays of protrusions, swirl chambers, and surface roughness [1]. According to Ligrani et al. [1], all of these devices act to increase secondary flows and turbulence levels to enhance mixing, in some cases, to form coherent fluid motions in the form of streamwise-oriented vortices. Such vortices and secondary flows not only act to increase secondary advection of heat away from surfaces, but also to increase three-dimensional turbulence production by increasing shear and creating gradients of velocity over significant flow volumes. These then give larger magnitudes of turbulence transport over larger portions of the flow fields. All of the devices mentioned also provide some heat transfer augmentation by increasing surface areas for convective heat transfer. The overall objective for such internal cooling technologies is optimal thermal protection with minimal use of coolant air with coolant mass flow rates as small as possible, and minimal pressure drop penalties within coolant passages. Appropriate levels of thermal protection are quantified by sufficiently low absolute temperature levels, as well as relatively uniform temperature distributions, as characterized by sufficiently small temperature gradients within solid components.

Figure 1 shows a typical internal cooling arrangement for a multipass turbine blade from Han et al. [2, 3]. With such an arrangement, the cooling air enters the airfoil through the

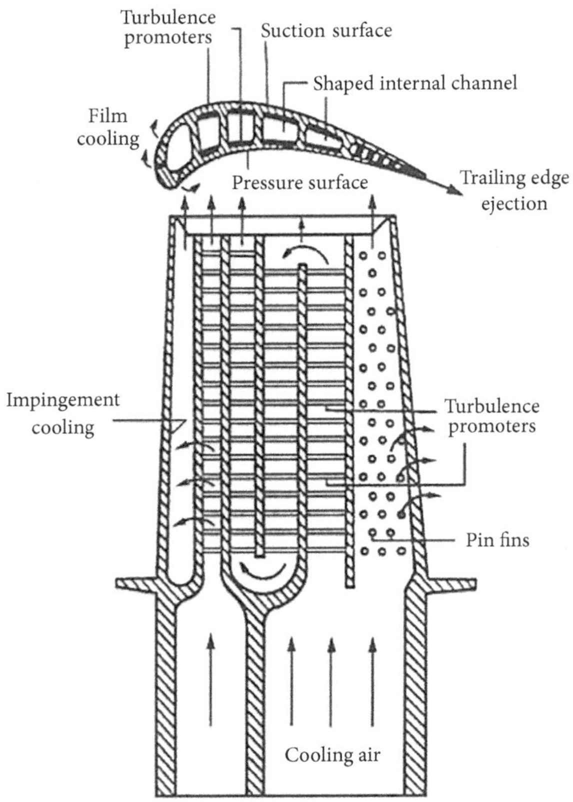

FIGURE 1: Typical internal cooling arrangement for a multipass turbine blade from Han et al. [2, 3].

blade root and then passes through one of several internal passages. As a result of the blade shape, the cooling passages at many locations are generally confined, and are therefore, difficult to cool. Such cooling passages also generally utilize complex cross-sectional shapes, because they must be compatible with the external contours of the turbine airfoils. Features that provide the highest heat transfer coefficients for a given flow rate are required in the leading and trailing edge cavities of airfoils that may or may not be film cooled. The leading edge geometry is characterized by a small coolant metal area to hot gas metal area ratio, while a thin trailing edge creates geometric constraints on passage sizes and accessibility for cooling air. For this reason, impingement cooling and swirl chambers are typically employed near the leading edge of the airfoil, either together or individually, and pin fins, and more complex rib turbulator shapes are typically found near the trailing edge regions of the airfoil. Rib turbulators, or turbulence promoters, are also located in passages near the mid-chord parts of the airfoil. Dimples and/or shaped roughness elements are also sometimes employed in all interior parts of airfoils, either by themselves or in conjunction with other devices. After passing over or through these devices, the cooling air then either exits through film cooling holes, or though exit passages which lead to another part of the airfoil.

Numerous and important constraint requirements are imposed upon the development of internal cooling configurations for gas turbine engines, such as the arrangement presented in Figure 1. This is a consequence of the complexity and sophistication of turbine component design. Satisfying these constraints requires simultaneous consideration of a variety of multidisciplinary technological issues. These include manufacturing, available volume space, airfoil shape, airfoil surface cooling requirements, aerodynamic losses, coolant availability, second law losses, and pressure drop penalties, in addition to heat transfer and thermal considerations. As a result, the current trend for advanced turbine component cooling design is a simultaneous development of external thermal protection schemes, like film cooling, with internal cooling technologies, including consideration of the adjacent conjugate conduction variations in adjacent solid components, as well as associated aerodynamic pressure loss penalty issues. One way that designers, manufacturers, and researchers attempt to meet these myriad requirements is development of internal cooling technologies which employ combinations of different devices together within one coolant passage. Examples include Murata et al. [4] who employ rib turbulators, pin fins, and dimples together, Rao et al. [5] who utilize a combination of pin fins, and dimples, and Siw et al. [6] who use rib turbulators and pin fins in combination within the same coolant passage.

One key objective of the present paper is discussion of new internal passage heat transfer augmentation technologies, such as these, as they are employed for internal cooling of turbine components of gas turbine engines. The overall paper is presented in three major parts. Presented first is a brief review of developments up to 2003, considering investigations which do not include the effects of rotation. An overview of more recent developments since 2003 (without rotation issues considered) is presented next. The third and final part of the paper then addresses the effects of rotation on heat transfer augmentation technologies for internal cooling.

Included within the discussions are comparisons of heat transfer augmentations, friction factor augmentations, and thermal performance parameters provided by different devices used for internal cooling of turbine airfoils, including rib turbulators, pin fins, dimpled surfaces, surfaces with arrays of protrusions, swirl chambers, and surface roughness. To characterize the heat transfer augmentations, Nu/Nu_o, the ratio of the Nusselt number of the channel with augmenters divided by the Nusselt number of a similar smooth channel (at the same experimental conditions) is employed. Likewise, the increase in pressure losses is represented by f/f_o, the ratio of the friction factor measured in a channel with the augmenters divided by the friction factor of a similar smooth channel at the same experimental conditions. Such Nusselt number and friction factor ratios provide means to compare different techniques, as well as the performance of a given technique at different experimental conditions.

The overall intent of the discussions and presentation is development of improved cooling configurations, which lead to overall improvements in overall gas turbine engine efficiency, reliability, and durability. In addition, the area is important because schemes for internal cooling of turbine airfoils can also be used in other devices, such as combustion chambers, air-conditioners, radiators, microscale heat exchangers, macroscale heat exchangers, devices for electronics cooling, and biomedical devices, to name only a few.

2. Developments up to 2003

Within this part of the paper, heat transfer augmentation technologies developed prior to 2003, generally without consideration of rotation effects, are briefly summarized. The discussion is presented in five parts: pin fin arrays, dimpled surfaces, rib turbulators, swirl chambers, and overall comparisons.

2.1. Pin Fin Arrays. When employed for internal cooling, pin fins or pedestals are generally arranged into arrays and extend between two opposite walls of an internal cooling passage. Brigham and VanFossen [7] and Ligrani and Mahmood [8] describe typical arrangements. According to Ligrani et al. [1], pin fins are generally used in the parts of turbine airfoils where higher levels of heat transfer augmentation are required and where high pressure drops are tolerated, and in many cases, even desired. Trailing edges of airfoils fall into this category. High coolant Reynolds numbers, associated with high coolant pressure levels, are desirable for heat transfer. However, by the time the coolant reaches the trailing edge, it has often acquired a significant amount of heat, and the process of extracting heat from trailing edge components is more challenging compared to other parts of the airfoil. At the same time, manufacturing limitations on the width of trailing edge slots prevent adequate restriction for the coolant, thus requiring that the flow is metered upstream. Pin fins or pedestals provide appropriate structural support near the trailing edge and are effective for allowing the pressure throughout most of the cooling circuit to remain high, while providing the necessary restriction near the trailing edge, to limit the coolant consumption to the desired level.

The large majority of existing pin fin investigations only focus on heat or mass transfer. Considered are the influences of pin geometry, pin materials, pin fin array configuration, channel geometry, different Reynolds numbers, and other parameters [12–15]. VanFossen [16] presents heat transfer coefficients measured on circular pin fins made of different materials, which are placed in rectangular cross-sectional channels. Brigham and VanFossen [7] report large overall heat transfer and mass transfer augmentations for circular pin fins arranged in in-line and staggered arrays. Simoneau and VanFossen [17] measure pin surface heat transfer coefficients and streamwise turbulence intensities as the test section gas temperature is changed. Later investigations by Lau et al. [18] and McMillin and Lau [19] consider streamwise pressure variations in a channel with circular pin fins. Chyu [20], Chyu and Goldstein [21], and Chyu et al. [22] report local and spatially averaged mass transfer coefficients for different pin fin arrays, which provide evidence of large overall mass transfer augmentations for certain in-line and staggered circular arrays. Grannis and Sparrow [23] numerically simulate the two-dimensional flow and pressure fields around diamond-shaped pin fins. Olson [24] presents normalized pressure drop and wall Nusselt number data as dependent upon Reynolds number and wall heat flux magnitude.

The effects of different pin fin shapes on heat transfer and flow in internal passages are considered by a number of investigators [8, 20, 22, 23, 25–30]. A variety of pin fin shapes are investigated, including circular [8, 20, 25–30], pins with end-wall fillets [20], diamond [23, 25–27], three-dimensional protruding elements [26], cubic [25, 26], and elliptical [28, 29]. Of these studies, the one described by Hwang and Lu [27] is especially unique because three pin fin configurations, each arranged in a staggered array, are considered in a trapezoidal duct, both with and without lateral flow ejection.

In regard to pin fin flow structure, Chyu and Natarajan [26] present mass transfer distributions and surface flow streakline patterns (obtained using an oil-graphite technique) around several different shapes of single protruding elements placed on an endwall, including a cylinder, a cube, a diamond, a pyramid, and a hemisphere. The vortices and mixing present just downstream of pin fins are further elucidated by results presented by Uzol and Camci [29]. According to Won et al. [30], the most important features downstream of an individual pin are two large vortex pairs, which are formed by the blockage effects of the pin fins. The secondary flows within and around these vortices are especially intense relative to other secondary flows which are present. Additional description of the different flow structures which develop near one endwall junction of one pin fin (as it is contained within a pin array) is provided by Ligrani et al. [1], including a typical surface Nusselt number ratio distribution, produced on the endwall by these flow structures [8].

2.2. Dimpled Surfaces. Dimples are arrays of indentations along surfaces. These are most commonly spherical in shape, although a variety of other shapes have also been employed, including triangular and tear-drop. Arrays of dimples are beneficial method for internal cooling because they produce multiple vortex pairs which augment local Nusselt number distributions as they advect downstream. They are notable for the low pressure drop penalties which they produce, which is because they do not protrude into the flow to produce significant amounts of form drag. With this benefit, dimples offer advantages for cooling later turbine stages where lower pressure cooling air is employed. They are also advantageous because the pressure drop which they produce through an airfoil passage is relatively low, which allows favorable pressure margins to be maintained in parts of the airfoil interior which are further downstream [1]. Another benefit is dimple manufacture removes material from internal cooling passage components, in contrast to rib turbulators and pin fins, which require addition of extra material and weight.

One of the first published studies of the effects of hemispherical dimples on flow structure and heat transfer is described by Murzin et al. [31]. These investigators describe the flow over and within shallow spherical depressions and conclude that this flow is mostly symmetric, and consists of a stable recirculatory flow inside of the depression. Gromov et al. [32] describe symmetric and nonsymmetric streamlines and flow patterns produced by hemispherical cavities with a variety of sizes. Cells of fluid motion are described in the form of tightening spirals, helical streamlines, and horseshoe shaped vortices. Kesarev and Kozlov [33] present distributions of local heat transfer coefficients inside a single hemispherical cavity, and indicate that the convective heat transfer from the cavity is higher, especially on the downstream portion, than that from the surface of a plane circle of the same diameter as the cavity diameter. Afanasyev et al. [34] describe the heat transfer enhancement mechanism for flows over walls indented with regular arrays of spherical pits with several different shapes. Enhancements of 30–40 percent, with pressure losses that are not increased appreciably relative to a smooth surface, are reported. Belen'kiy et al. [35] describe heat transfer intensification from a tube surface fitted with a staggered array of concave dimples on surfaces annular internal passages. Terekhov et al. [36] present experimental measurements of flow structure, pressure fields, and heat transfer in a channel with a single dimple on one surface. Different magnitudes and frequencies of flow oscillations are described, along with heat transfer and pressure loss dependence on dimple geometry. From flow visualizations, Zhak [37] describes a variety of vortex structures in different shaped rectangular cavities. Schukin et al. [38] present results which are directly applicable to gas turbine cooling, including average heat transfer coefficients from the measurements on a heated plate downstream of a single hemispherical cavity in a diffuser channel and in a convergent channel. Bearman and Harvey [39] investigate cross flows over cylindrical surfaces with an array of dimples. The geometry of the dimples corresponds to the optimum concavity depth found by Kimura and Tsutahara [40] for minimum drag on cylinders. At high Reynolds numbers, significantly lower drag coefficients are present with arrays of dimples compared to smooth surfaces. Kithcart and Klett [41] compare heat transfer and skin friction of flows over surfaces on one wall of a rectangular channel with hemispherical dimples, hemispherical protrusions, and rectangular protrusions. The authors conclude that all three geometries produce similar levels of heat transfer augmentation.

More recently, Chyu et al. [42] present data showing the effects of Reynolds number on local heat transfer coefficient distributions on surfaces imprinted with staggered arrays of two different shapes of concavities: spherical and tear-drop. Their measurements show distributions of heat transfer coefficients everywhere on surfaces which are significantly higher than values in channels with smooth surfaces. Lin et al. [43] present computational simulation results of the flow structure and resulting heat transfer distributions for the same surface geometries and flow conditions. Flow streamlines and temperature distributions are presented which provide insight into the flow structural characteristics produced by the dimples. Another study, described by Gortyshov et al. [44], employs spherically shaped dimples, called spherical intensifiers, placed at different relative positions on the opposite sides of a narrow channel. The investigators observe that an increase in relative dimple depth produces increases in surface heat transfer rates, as well as increases in streamwise pressure penalties. Moon et al. [45] give heat transfer and friction factor data, for ratios of channel height to dimple print diameter from 0.37 to 1.49, that illustrate the effects of channel height on a surface with a staggered pattern of dimples. According to these investigators, improvements in heat transfer intensification and pressure losses remain at approximately constant levels over the ranges of Reynolds number and channel height investigated. Chen et al. [46] employ hemispherical-shaped protrusions on the insides of tubes to augment surface heat transfer rates. Six different protrusion arrangements are employed, which increase surface heat transfer rates above those present in smooth tubes by 25 to 137 percent when compared at the same Reynolds number. Data from Moon and Lau [47] are based on measurements made only at a few discrete locations on their test surface, whereas the other studies obtained results from averages made from spatially-resolved results, measured over continuous sections of test surfaces.

Mahmood et al. [48] and Ligrani et al. [1] describe the mechanisms responsible for local and spatially averaged heat transfer augmentations on flat channel surfaces with an array of dimples along one wall. In the Mahmood et al. [48] study, the ratio of channel height to dimple print diameter is 0.5. Key flow features responsible for local heat transfer augmentations include (i) shedding of multiple vortex pairs from dimples, (ii) the strong secondary fluid motions within these vortex pairs and associated vortical fluid, (iii) shear layer reattachment within each dimple, and (iv) periodicity and unsteadiness associated with vortex pair shedding and the flows within individual dimples, which is produced as flow is ejected and then inrushes to each dimple. This unsteadiness causes the thermal boundary layer which forms downstream of individual dimples to be periodically reinitialized, and increases flow mixing over length-scales approximately equal to the dimple print diameter. The effects on thermal transport are especially pronounced near downstream rims of dimples as well as on flat surfaces downstream of and between dimples [1]. Mahmood and Ligrani [49] consider the effects of the ratio of inlet stagnation temperature to local surface temperature for ratios of channel height to dimple print diameter of 0.20, 0.25, 0.50, and 1.00. According to these authors, visualized smoke patterns show that vortex pairs, which are periodically shed from the dimples, become stronger as the ratio of channel height to print diameter decreases. However, regardless of the ratio of channel height to dimple print diameter, local Nusselt numbers increase substantially as the ratio of inlet stagnation temperature to local surface temperature decreases.

Burgess and Ligrani [51] present spatially resolved Nusselt numbers measured on the dimpled test surface placed on one wall of a channel with a height to dimple print diameter ratio of 1.0, and ratios of dimple depth to dimple print diameter of 0.1, 0.2, and 0.3. The opposite channel wall is smooth. Reynolds number based on channel height for the measurements is 20,000. According to these investigators, lower Nusselt number ratios are located over the upstream halves of the depressions. Local Nusselt number ratios are then higher in the downstream halves. The highest values are then located near the downstream rims of each dimple, both slightly within each depression, and on the flat surface just downstream of each dimple. In general, spatially averaged Nusselt number augmentations from dimples increase (at a particular Reynolds number), as the ratio of dimple depth to dimple print diameter increases from 0.1 to 0.3. However, in application to component internal cooling, a limit often exists regarding maximum allowable manufactured dimple depth, and thus, dimple depth relative to the thickness of associated turbine blade and vane material is an important consideration. Consistent with the results of Kesarev and Kozlov [33] and Schukin et al. [38], most local values in the concave cavities are higher than values measured a smooth channel at the same Reynolds number and temperature ratio. Additional information on the flow structure due to dimple depressions on a channel surface is provided by Ligrani et al. [52]. Here, detailed descriptions of the behavior of the primary and secondary vortex pairs, which are shed from the dimples, are described. Most importantly, the locations of the primary and secondary vortex pairs coincide closely with locations where the normalized longitudinal Reynolds normal stress is increased. Frequent, continuous, unsteady interactions are also described between the flows in adjacent dimples, where the frequency of the large-scale unsteadiness scales on time-averaged bulk velocity and dimple print diameter.

Mahmood et al. [53] and Ligrani et al. [54] consider heat transfer and flow structure in a channel with dimples and protrusions on opposite walls. Similar arrangements, but only with measurements of friction factors and some flow structural characteristics, are considered by Kithcart and Klett [41]. According to Ligrani et al. [54], instantaneous flow visualization images and surveys of time-averaged flow structure show that the protrusions result in added vortical, secondary flow structures, and flow mixing. As a result, local friction factors and local Nusselt numbers are augmented compared to a channel with no protrusions on the top wall. Mahmood et al. [53] indicate that important Nusselt number variations are observed as the array of protrusions is changed with respect to the locations of the dimples. With protrusions, form drag and channel friction are increased. As a result, thermal performance parameters are then generally slightly lower when protrusions and dimples are employed, compared to a channel with a smooth-dimple arrangement. Additional information on the flow structure resulting from an array of surface dimples is provided by Ligrani et al. [1]. According to Isaev et al. [55], the flow inside of a spherical indentation dimple is also quite complex, with nonsymmetric flow patterns sometimes present, even though the dimple is symmetric. According to these investigators, "large-scale vortex rings, inside which particles perform reciprocating motions are incorporated into the spiral flows outgoing from singularities at the surface of the well."

2.3. Rib Turbulators. Rib turbulators are often in the form of rectangular cross-sectional bars mounted along the surface, which are often angled with respect to the bulk flow direction. Because they protrude into the flow, they act to trip the flow, mix the flow, and also generate vortices and three-dimensional velocity gradients. Rib turbulators or trip strips are general purpose heat transfer augmentors in cooled airfoils. By varying the main geometric parameters, trip strip height, channel blockage, orientation, and spacing, it is possible to optimize the cooling scheme in such a way that the airfoil mid-body is not overcooled, while providing the necessary exit temperatures at the leading edge and trailing edge where film cooling may be required [1]. Due to casting limitations, ribs are rarely rectangular in cross-section when employed within turbine components. Instead, they often maintain more rounded profiles due to manufacturability limitations. A significant number of experimental and numerical studies address the effects of rib turbulators on heat transfer in internal channels [1, 56–100]. Considered are single pass and multipass channels, square and rectangular channels, and channels with and without rotation.

Several of the earliest experimental studies consider single-pass, stationary channels with no rotation [62, 68–71, 78, 92, 93]. Of these studies, Casarsa et al. [62] characterize the velocity and heat transfer fields in an internal cooling channel with 90° ribs which produce 30 percent blockage. Included are heat transfer enhancement magnitudes, time-averaged distributions of mean velocity components, and time-averaged distributions of normalized root-mean-square velocity fluctuations. Mahmood et al. [78] present spatially resolved Nusselt numbers and flow structure are for a stationary channel with an aspect ratio of 4 and angled rib turbulators inclined at 45° with perpendicular orientations on two opposite surfaces. The ratio of rib height to hydraulic diameter is 0.078, the rib pitch-to-height ratio is 10, and the blockage provided by the ribs is 25 percent of the channel cross-sectional area. According to these authors, spatially resolved local Nusselt numbers are highest on tops of the rib turbulators, with lower magnitudes on flat surfaces between the ribs, where regions of flow separation and shear layer reattachment have pronounced influences on local surface heat transfer behavior. Also important are intense, highly unsteady secondary flows and vortex pairs, which increase secondary advection and turbulent transport over the entire channel cross-section.

Other studies of stationary channels, prior to 2003, with no rotation, consider single-pass and multipass channels. Thurman and Poinsatte [97] measure heat transfer and bulk air temperature in a three-pass duct with orthogonal ribs and bleed holes both located on one wall. According to these investigators, changing the locations of the ribs relative to the holes produces large changes to surface heat transfer coefficient distributions. Cho et al. [63] employ continuous and discrete, parallel and cross arrays of ribs in a single-pass square duct. Discrete ribs with gaps in between are found to produce more uniform heat transfer coefficient distributions than continuous ribs. According to these authors, remarkable enhancements of local and spatially averaged surface heat transfer rates are possible with rib turbulators, in spite of lowered local Nusselt numbers at certain locations along ribbed surfaces.

Computational studies of flows and heat transfer in ducts with rib turbulators consider straight single-pass ducts [56, 72, 73, 86, 89, 99], two-pass ducts [57], two-pass ducts with U-shaped channels in between [60, 76, 84, 88, 90], 90°

orthogonal ribs [72, 73, 84, 86, 99], 45° angled ribs [56, 57, 60, 76, 88–90], 45° V-shaped ribs [73], and rotation [56, 57, 60, 72, 76, 84, 88, 90]. In one case [89], the angled ribs placed on two opposite walls of the channel are rounded. Jia et al. [73] and Watanabe and Takahashi [99] include experimentally measured friction factors, Nusselt numbers, and streamwise velocity fluctuations, which are compared to numerically predicted results. Rigby and Bunker [85] employ a full three-dimensional Navier-Stokes equation solver to investigate the same complex trailing edge passage of a high pressure turbine blade investigated experimentally by Bunker et al. [61].

In regard to flow structure, flow recirculation patterns around a single rib are positioned just downstream of each rib, and cover a considerable portion of the ribbed wall. Each of these recirculation zones is often considered to be a potential hot spot because it is associated with locally lower surface Nusselt numbers [1, 78]. Downstream of the flow recirculation zone, downstream of each rib, is a line across the surface where streamlines reattach. Weak secondary flows often form downstream of this reattachment line for 90° rib turbulators [63]. With angled ribs, additional secondary flows are induced by flow skewing [1, 62]. Additional data regarding the effects of the cross-rib arrangement on flow structure are presented by Mahmood et al. [78] for a 45° rib arrangement, by Jia et al. [73] for a square cross-section with ribs on one wall, and by Watanabe and Takahashi [99] for a channel with 90° rib turbulators. Data which provide quantitative detail on the recirculation zone which forms just above a rib, the recirculation zone which forms near the corner just upstream of a rib, the large recirculation zone which is present downstream of a rib, and the spanwise vortex, or second smaller separation zone, which is present in the vicinity of the downstream bottom corner of a rib are also given by Casarsa et al. [62]. Note that the sudden expansion downstream of the rib results in a large recirculation zone. A shear layer forms between this recirculating flow, and the high-speed mainstream bulk flow, located just above [62, 99]. According to Casarsa et al. [62], the shear layer reattachment line exhibits a particular shape near the lateral wall because side wall boundary layers are altered by the presence of the rib. When the boundary layer is deviated by the rib, new streamwise vorticity is generated inside of it. The flow then rolls up on itself. The result is local deformation of the reattachment line from a straight line [62].

2.4. Swirl Chambers. Swirl chambers are internal flow passages arranged with either spinning vanes, internal inserts, or inlets and outlets configured to produce large-scale swirling of the flow (relative to the chamber dimensions) generally about the principal chamber axis [105]. This large-scale swirling and the Görtler vortex pairs that are formed enhance surface heat transfer rates. Here, the heat transfer and fluid mechanics within swirl chambers are described as they are used to augment heat transfer rates for internal turbine blade cooling, an application introduced by Glezer et al. [106, 107]. Other investigations prior to 2003 employ a variety of different methods to induce large-scale swirling, generally, with an overall goal of augmentation of heat transfer rates [105–127]. Of these investigations, Hwang and Cheng [118] employ multiple swirling jets to augment heat transfer in a triangular duct. In some cases, the highest local heat transfer enhancements are produced on a bottom surface by wall jets, whereas in others, the highest local enhancements are produced on a target wall by impingement jets. Local and spatially averaged surface Nusselt numbers are given as they vary with Reynolds number and three different jet inlet angles [118].

The configuration employed by Hedlund and Ligrani [115], Hedlund et al. [116], and Ligrani et al. [123] models an internal passage used to cool the leading edge of a turbine blade. Multiple inlets are employed at fixed locations, to induce swirling, which makes this swirl chamber geometry different from the arrangements used in many other investigations. Hedlund et al. [116] and Hedlund and Ligrani [115] show that important variations of surface Nusselt numbers and time-averaged flow characteristics are present due to arrays of Goertler vortex pairs, especially near the inlets of the swirl chambers, where Nusselt numbers are highest. Nusselt numbers then decrease and become more spatially uniform along the interior surface of the chamber as the flows advect away from each inlet. Numerous convoluted Görtler vortex pairs of different size near the concave surface of the chamber are evidenced by smoke visualizations, as well as by time-averaged surveys of flow characteristics, which show regions of positive and negative time-averaged circumferential vorticity adjacent to each other near the concave surface. Important Nusselt number variations due to the Görtler vortex pairs are evident at nearby surface locations, where higher and lower local Nusselt numbers correspond to Görtler vortex pair downwash and upwash regions, respectively.

2.5. Overall Comparisons. Figures 2 and 3 present comparisons of globally averaged Nusselt number ratios $\overline{\overline{\mathrm{Nu}}}/\mathrm{Nu}_o$ and friction factor ratios f/f_o for rib turbulators, pin fins, swirl chambers, dimple-smooth arrangements, dimple-protrusion arrangements, dimple-dimple arrangements, surface roughness, and smooth-walled channels. The f/f_o abscissa scale in Figure 3 ranges from 0 to 20 (compared to 0 to 80 in Figure 2) to provide a detailed view of variations over this friction factor ratio range [1].

According to Ligrani et al. [1], of the techniques considered, swirl chambers give some of the highest levels of heat transfer augmentation when compared at the same value of friction factor ratio. In some cases, several of the rib turbulator configurations produce comparable Nusselt number ratios. However, pressure losses and friction factor ratios associated with swirl chambers and rib turbulators are also relatively high. Many of the rib turbulator configurations produce levels of heat transfer augmentation that are higher than those produced by the dimple-protrusion-smooth configurations. However, generally, the increases in pressure losses due to the rib turbulators are also larger than those associated with the dimple-protrusion-smooth configurations. f/f_o magnitudes are as high as 71 for channels with rib turbulators, and as high as 25 for swirl chambers,

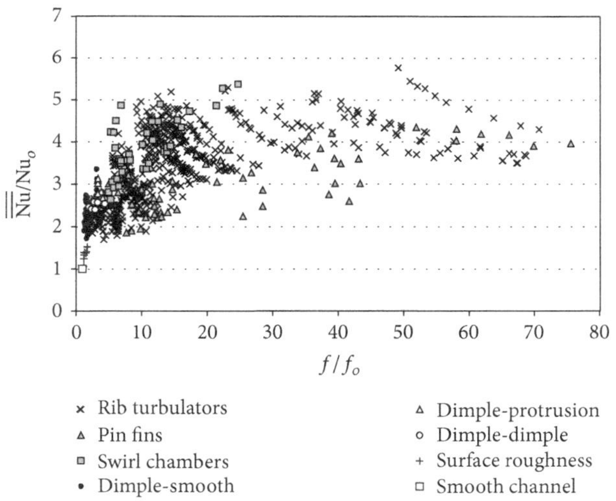

FIGURE 2: Comparison of relative performance of rib turbulators, pin fins, swirl chambers, dimple-smooth arrangements, dimple-protrusion arrangements, dimple-dimple arrangements, surface roughness, and smooth-walled channels from investigations prior to 2003. Globally averaged Nusselt number ratios $\overline{\overline{\text{Nu}}}/\text{Nu}_o$ are presented as dependent upon friction factor ratios f/f_o [1].

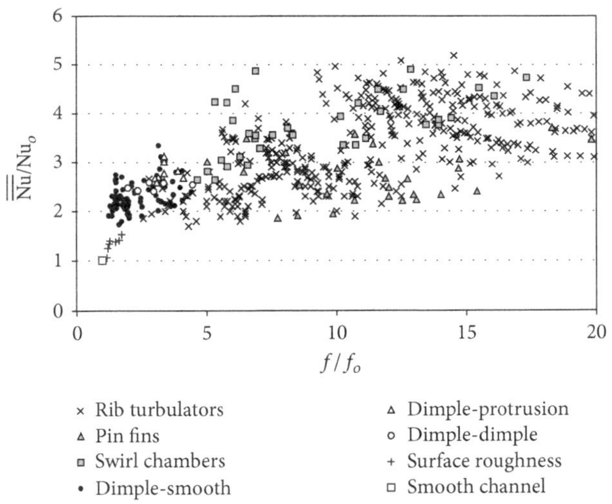

FIGURE 3: Comparison of relative performance of rib turbulators, pin fins, swirl chambers, dimple-smooth arrangements, dimple-protrusion arrangements, dimple-dimple arrangements, surface roughness, and smooth-walled channels from investigations prior to 2003. Globally averaged Nusselt number ratios $\overline{\overline{\text{Nu}}}/\text{Nu}_o$ are presented as dependent upon friction factor ratios f/f_o [1].

whereas magnitudes for dimple-protrusion-smooth configurations reach only up to 4.5. Nusselt number ratios in channels with pin fins lie near the bottom of or below the values for rib turbulators when compared at the same f/f_o value. Pin fin friction factor ratios are then as high as 76 [1].

Also included in Figures 2 and 3 are Nusselt number ratios and friction factor ratios for a smooth channel, and for a channel with three-dimensional, irregular surface roughness [129, 130]. Such surface roughness augmenters work as microturbulators and microvortex generators. These act to increase mixing and levels of turbulence transport locally near surfaces. With this mechanism, $\overline{\overline{\text{Nu}}}/\text{Nu}_o$ range from 1.06 to 1.53, and f/f_o range from 1.18 to 1.75 in Figures 2 and 3 [1].

The data in these two figures are presented again in Figures 4 and 5. In the first of these figures, globally averaged thermal performance parameters $\overline{\overline{\text{Nu}}}/\text{Nu}_o/(f/f_o)^{1/3}$ are given as dependent upon the friction factor ratio f/f_o. The form of this performance parameter is suggested by Gee and Webb [131] to provide a heat transfer augmentation quantity ($\overline{\overline{\text{Nu}}}/\text{Nu}_o$) and a friction factor augmentation quantity ($(f/f_o)^{1/3}$), where each is given for the same ratio of mass flux in an internal passage with augmentation devices to mass flux in an internal passage with smooth surfaces. According to Figure 4, the highest $\overline{\overline{\text{Nu}}}/\text{Nu}_o/(f/f_o)^{1/3}$ thermal performance parameters are produced by swirl chambers and dimple-smooth arrangements, and the lowest are produced by channels with arrays of pin fins. Results for rib turbulators, dimple-protrusion arrangements, dimple-dimple arrangements, surface roughness, and smooth-walled channels then generally lie between these distributions [1].

The $\overline{\overline{\text{Nu}}}/\text{Nu}_o/(f/f_o)$ parameter, presented in Figure 5, is referred to as the Reynolds analogy performance parameter. The highest values of this parameter are produced by dimple-smooth arrangements, and the lowest are produced by pin fins. Results for rib turbulators, swirl chambers, dimple-protrusion arrangements, dimple-dimple arrangements, surface roughness, and smooth-walled channels then generally lie between these distributions [1].

3. Developments Since 2003

Within this part of the paper, heat transfer augmentation technologies developed since 2003, generally without consideration of rotation effects, are discussed. The material is presented in five parts: pin fin arrays, dimpled surfaces and surfaces with protrusion arrays, rib turbulators, combination devices, and overall comparisons.

3.1. Pin Fin Arrays. Recent investigations of arrays of pin fins within internal ducts consider the influences of a variety geometric and flow parameters. These include pin fin shape [9], pin fin height [132], streamwise and spanwise pin-fin spacings [133, 134], streamwise pin fin spacing and Reynolds number [135], and combinations of pin fins with dimples [5, 136]. Also addressed in recent studies are turning flow effects on arrangements utilized for trailing edge cooling [137], high solidity pin fin arrays with incremental replenishment [138], effects of particle deposition and erosion characteristics on heat transfer in pin fin arrays [139], and effects of pin fin array detached spaces [140].

Examples from investigations of shaped pin fin arrays are shown in Figures 6 and 7 from Siw et al. [9]. Considered are triangular, semicircular, and circular pin fins. The corresponding three sets of dimensional heat transfer coefficient results are given in Figure 7. These data sets are obtained at the

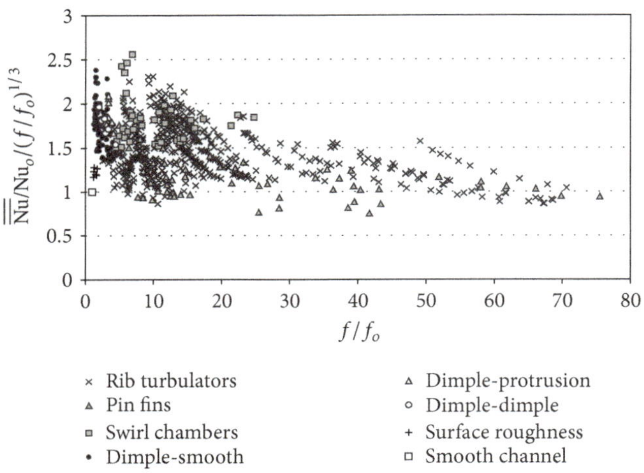

FIGURE 4: Comparison of relative performance of rib turbulators, pin fins, swirl chambers, dimple-smooth arrangements, dimple-protrusion arrangements, dimple-dimple arrangements, surface roughness, and smooth-walled channels from investigations prior to 2003. Globally averaged thermal performance parameters $\overline{\overline{\mathrm{Nu}}}/\mathrm{Nu}_o/(f/f_o)^{1/3}$ are presented as dependent upon friction factor ratios f/f_o [1].

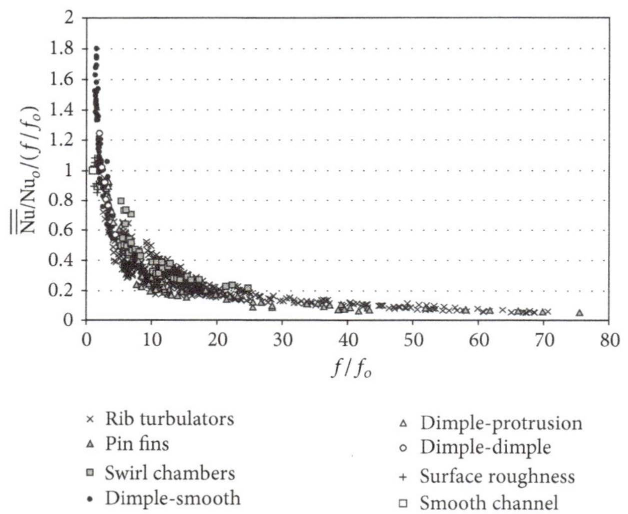

FIGURE 5: Comparison of relative performance of rib turbulators, pin fins, swirl chambers, dimple-smooth arrangements, dimple-protrusion arrangements, dimple-dimple arrangements, surface roughness, and smooth-walled channels from investigations prior to 2003. Globally averaged thermal performance parameters $\overline{\overline{\mathrm{Nu}}}/\mathrm{Nu}_o/(f/f_o)$ are presented as dependent upon friction factor ratios f/f_o [1].

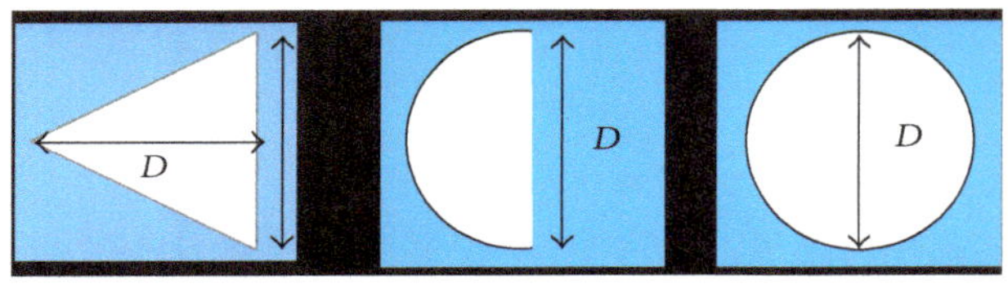

FIGURE 6: Pin fins element configurations employed in the investigation by Siw et al. [9].

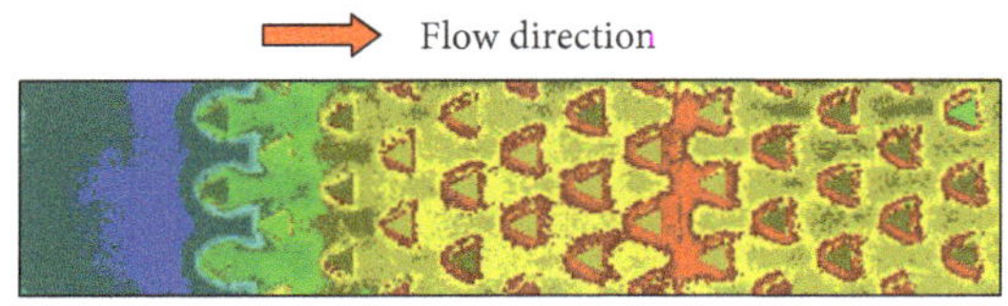

(a) Triangular pin fin array

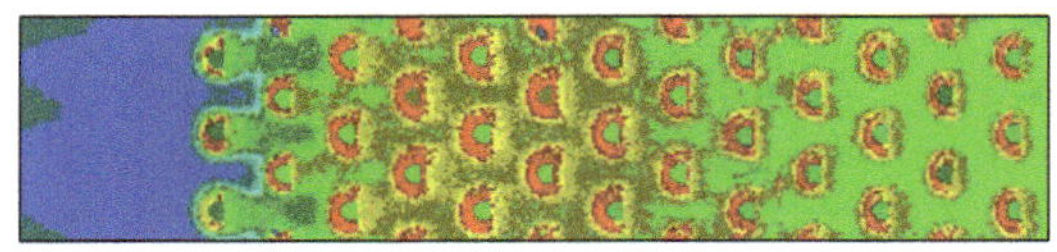

(b) Semicircular pin fin array

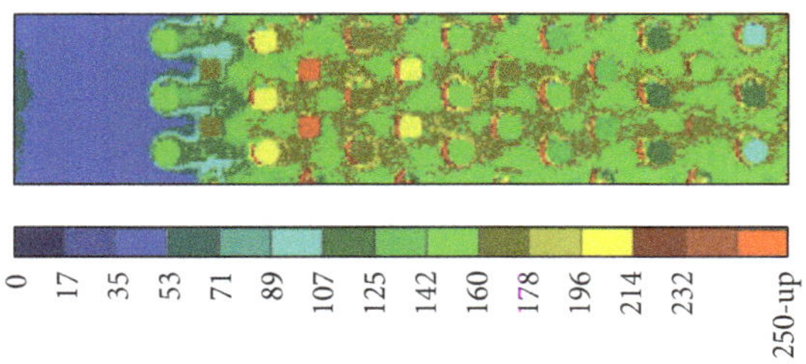

(c) Circular pin fin array

FIGURE 7: Dimensional heat transfer coefficient distributions in $\mathrm{W/m^2K}$ units for the configurations shown in Figure 6 from Siw et al. [9].

same Reynolds number of 25,000 and with the same interpin spacing, where Reynolds number is determined using inlet duct hydraulic diameter, and bulk spatially averaged air inlet velocity. Local qualitative heat transfer coefficient variations for the circular and semicircular arrays are similar, with the highest local maxima present near the third and fourth streamwise rows, just upstream of regions where values decrease somewhat to approximate fully developed values [9]. With such fully developed conditions, local periodic values repeat themselves with streamwise development, with local periodic variations which depend upon streamwise pin fin spacing. Local values are augmented in the first few rows of both the circular and semicircular arrays because of the development and presence of horseshoe vortices around each pin fin.

As the flow develops farther downstream, local heat transfer coefficients on endwall regions around the triangular pin fins are higher than values at similar locations for the circular and semicircular arrangements. Overall, the highest overall heat transfer coefficients, which are spatially averaged over all pin fin and endwall surfaces, is produced by the triangular pin fin arrangement. According to Siw et al. [9], this is a result of the sharp edges on the triangular elements which generate wakes with increased shear, more flow mixing, and increased turbulence transport. The triangular array also gives more uniform heat transfer coefficient distributions with spatial location on the endwall [9].

3.2. Dimpled Surfaces and Surfaces with Protrusion Arrays. Recent investigations of arrays of dimples and protrusions consider flow structure from dimpled surfaces [141], V-shaped dimples [142], and effects of arrays of dimples or protrusions on heat transfer augmentation on the turn-region of a U-shaped channel used for cooling of a blade tip region [10]. Examples from this latter investigation, from Sundén et al. [10], are presented in Figures 8 and 9. The U-shaped channel arrangements, with either dimples or protrusions, are shown in Figure 8. Note that a smooth-tip two-pass channel is used as a baseline result for performance comparison. Numerically predicted dimensionless surface temperature distributions for the two arrangements are given in Figure 9, with a dimple result presented on the left and a protrusion result presented on the right. Here, inlet Reynolds number is 200,000, where Reynolds number is determined using inlet duct hydraulic diameter, and bulk spatially averaged air inlet velocity. In addition, higher numerical values of the quantity presented in Figure 9 indicate higher absolute surface temperatures. With both arrangements, local surface temperatures on the left-hand sides of each image are lower because of the impingement of cold supply fluid. Higher temperatures are then present near the side corners, farther downstream, because of local recirculating flows. In general, a larger region with lower temperatures is produced by the protrusion array arrangement, rather than by the dimpled surface arrangement. According to the investigators [10], both arrangements give increased turbulent mixing, relative to the baseline channel, with heat transfer enhancement ratios as high as 2.0. Both the dimple and protrusion arrangements are thus believed to provide effective means to increase heat transfer enhancements, with relatively small additional pressure losses.

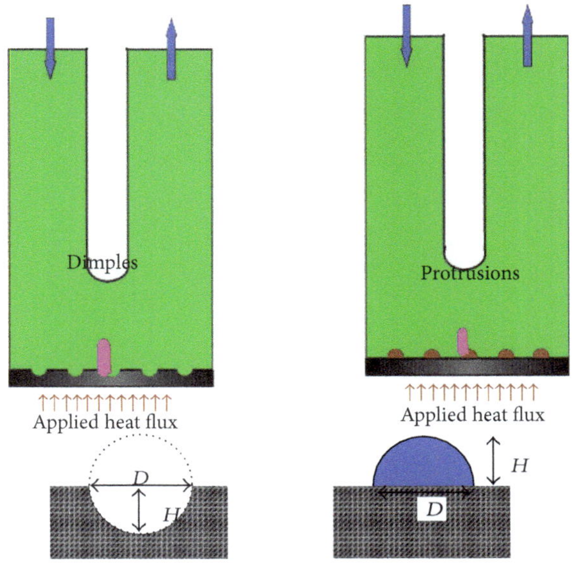

FIGURE 8: Dimple and protrusion configurations investigated in a two-pass channel for tip-wall cooling by Sundén et al. [10].

3.3. Rib Turbulators. Recent investigations of rib turbulators address different rib configurations [143–154], as well as a variety of other geometric parameters, such as continuous and truncated ribs [155, 156], semiattached rib configurations [157], and effects of a turning vane arrangement within a ribbed internal cooling channel [11, 149, 158]. A variety of internal flow arrangements are addressed, including extraction flow effects and bleed holes [159, 160], different crossflow schemes [161], conjugate heat transfer with crossing jets [162], and impingement and effusion cooling with ribbed surfaces [163, 164]. Other recent investigations focus on the effects of rib arrays in combination with steam cooling [165, 166] and mist/air cooling [167].

Examples from the investigation of Chen et al. [11] are given in Figures 10 and 11. Within this investigation, the effects of turning vane configuration are considered by employing the three U-shaped internal cooling passage configurations which are shown in Figure 10: (a) test section 1, without a turning vane in the passage turn, (b) test section 2, with turning vane in the passage turn, and (c) test section 3, with turning vane and ribs in the passage turn. Figure 11 presents the experimentally measured Nusselt number ratio distribution on the ribbed wall for configuration (c) for $Re = 30,000$, where Reynolds number is based upon inlet duct hydraulic diameter, and bulk spatially averaged air inlet velocity. Enhancements from the ribs in the turn region are apparent within Figure 11. However, according to Chen et al. [11], distributions for configuration (c) before and after the turn are similar to distributions for configuration (b). This leads to the conclusion that the ribs within the turn have little influence on Nusselt number ratio distributions before and after the turn. When all three configurations are considered, the overall spatially averaged Nusselt number ratio distribution is highest with configuration (a). The lowest overall pressure loss coefficient is then produced by configuration (b).

3.4. Combination Devices. Important advances in the development of internal cooling technologies for turbine components result from simultaneous use of different augmentation technologies together. A number of recent investigations consider combinations of two different types of devices [5, 6, 136, 168, 169]. For example, Amano et al. [168] investigate turbine blade cooling channels with dimples and rib-turbulators, Qui et al. [169] and Siw et al. [6] consider cooling channels with pin fins and rib turbulators, and Rao et al. [5, 136] employ combinations of pin fins and dimples. Use of three different devices simultaneously is discussed by Lan et al. [170] and by Murata et al. [4], who investigate internal passage cooling with combinations of rib turbulators, dimples, and protrusions. Another recent investigation focuses on internal and external cooling methods in combination, by addressing the interactions between internal ribs and external coolant film [171]. The combination and interaction of internal ribs with impingement and/or effusion cooling is described by Yang et al. [164] and Wang et al. [172].

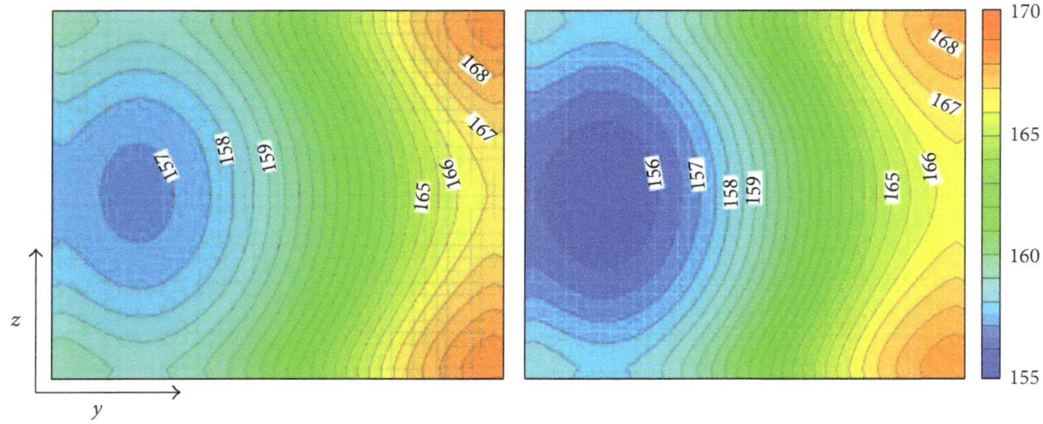

FIGURE 9: Surface temperature distributions for the dimple (left) and protrusion (right) configurations shown in Figure 8, as investigated by Sundén et al. [10].

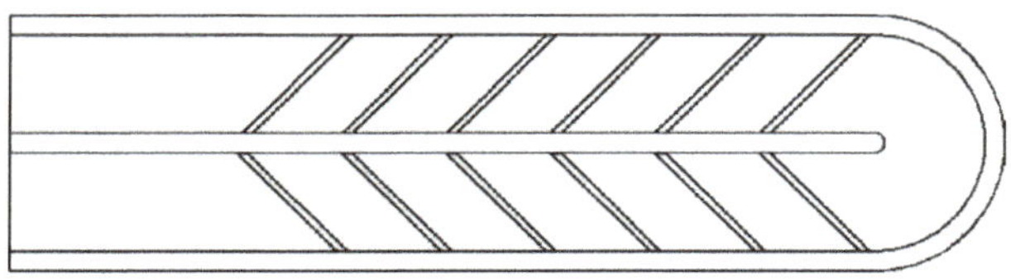

(a) Test section 1: without turning vane

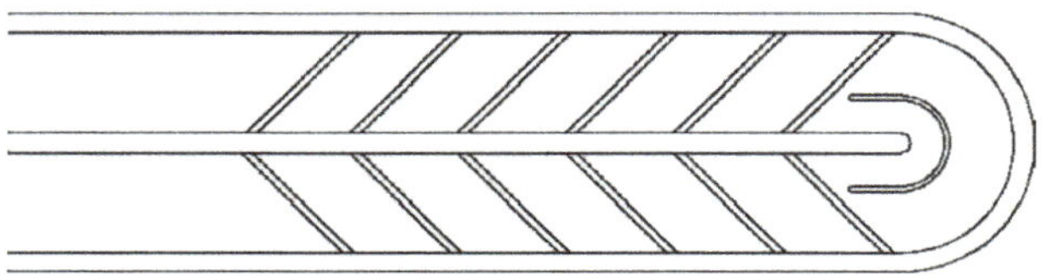

(b) Test section 2: with turning vane in turn

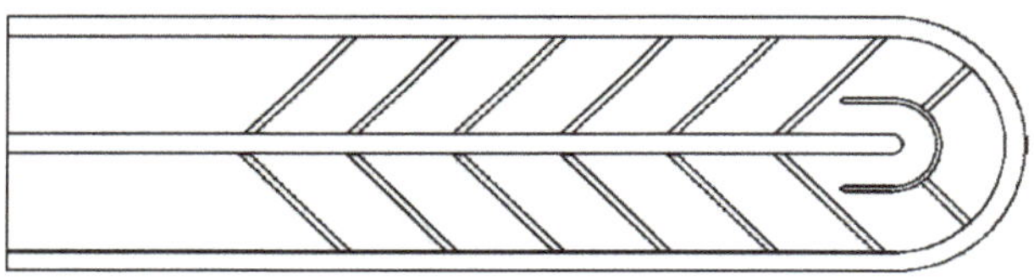

(c) Test section 3: with turning vane and ribs in turn

FIGURE 10: Vane turning test section arrangements, investigated by Chen et al. [11].

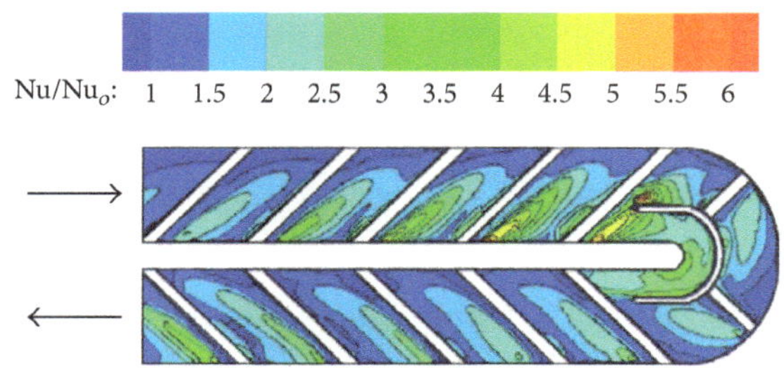

FIGURE 11: Normalized Nusselt number distribution on the ribbed wall for test section 3, experimental data for Re = 30,000, from Chen et al. [11].

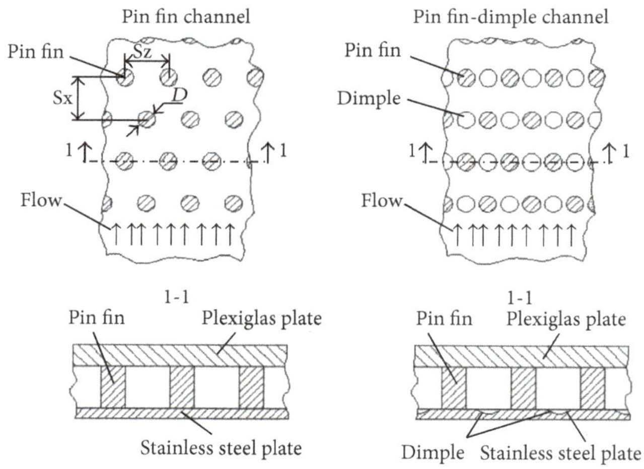

FIGURE 12: Pin fin and pin fin-dimple channel configurations considered within the channel flow investigation by Rao et al. [5].

Utilizing pin fins and dimpled surfaces together is illustrated by the results which are given in Figures 12 and 13 from Rao et al. [5]. The pin fin and pin fin-dimple channel configurations which are investigated are shown in Figure 12. According to the investigators, the spacing and arrangement of the pin fin array is an optimal array arrangement for turbine airfoil cooling. The resulting endwall Nusselt number ratio distribution, presented in Figure 13, is obtained with a constant heat flux surface boundary condition for the pin fin-dimple channel with Re = 18,600, where Reynolds number is determined using inlet duct hydraulic diameter, and bulk spatially averaged air inlet velocity. Important Nusselt number ratio increases are evident near the upstream edges of the pin fins and near the downstream edges of the dimples. The former variations are mostly due to the horseshoe vortices which form at the pin fin/endwall junction [1, 8], whereas the latter are a result of the different vortex structures which are periodically shed from the individual dimples [1, 48, 52]. The results from this investigation show that overall endwall heat transfer enhancements are larger for the pin fin-dimple channel, relative to the channel with

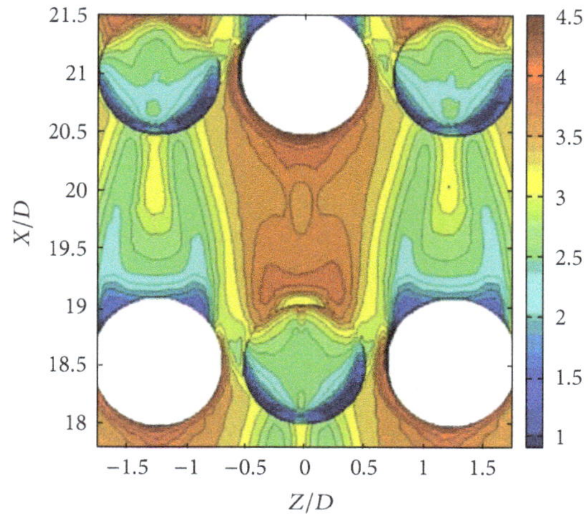

FIGURE 13: Local Nusselt number ratios on the endwall of the pin fin-dimple channel for Re = 18,600 from the investigation by Rao et al. [5].

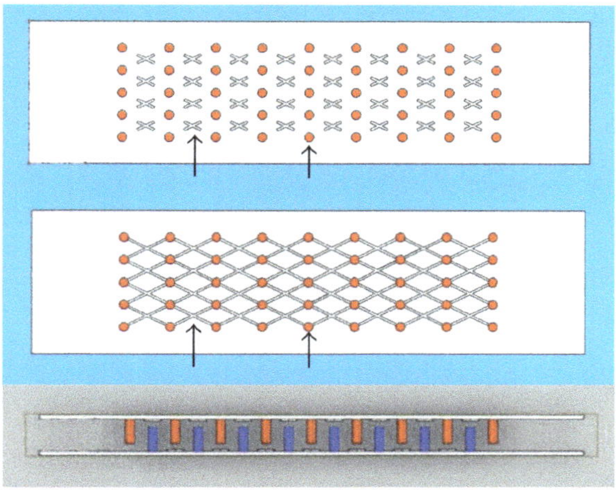

FIGURE 14: Broken rib, full rib array, and side view of staggered pin fin configuration from Siw et al. [6].

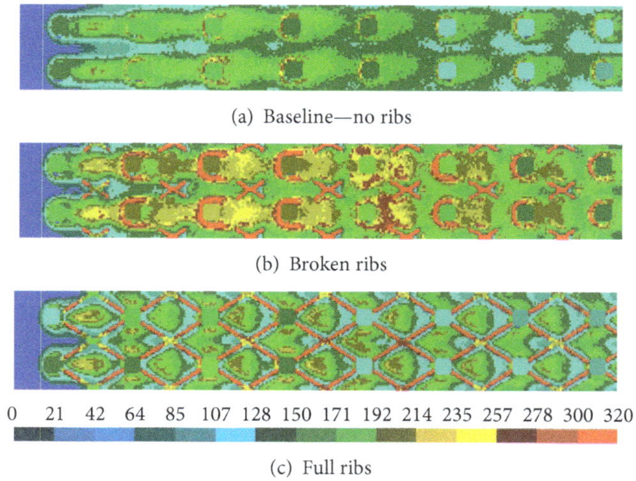

FIGURE 15: Local dimensional heat transfer coefficient distributions (W/m^2K) for configurations shown in Figure 14 for Re = 25,000 from Siw et al. [6].

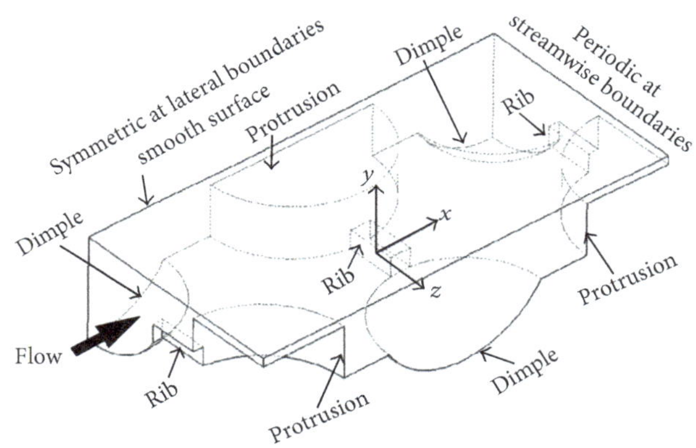

FIGURE 16: Rib, dimple, protrusion arrangement, illustrating domain for numerical analysis, employed in the investigation by Murata et al. [4].

pin fins only. The investigators [5] attribute this to increased turbulent mixing near the wall caused by strong vortical flow shedding from the dimples.

Examples which illustrate the combined use of pin fins and ribs are presented in Figures 14 and 15, from Siw et al. [6]. The broken rib array, full rib array, and side view of the staggered pin fin configuration, which are utilized, are shown in Figure 14. The resulting dimensional heat transfer coefficient distributions for a Reynolds number of 25,000 are then given in Figure 15 (where Reynolds number is determined using inlet duct hydraulic diameter, and bulk spatially averaged air inlet velocity). Here, the ratio of detached spacing between the pin-tip and the endwall, and pin diameter is 2. The ratio of pin height and pin diameter is also 2. Inspection of Figure 15 reveals that the largest heat transfer coefficient values are generally present with the broken rib arrangement, most especially at many endwall locations. The investigators [6] attribute this behavior to excessive vortices which are generated by the broken ribs. The development of these vortices seems to occur as horseshoe vortices, which are also a major contributor to heat transfer enhancement, additionally continue to be present.

The combined use of rib turbulators, dimples, and protrusions is illustrated by the results which are given in Figures 16 and 17 [4]. The experimental Nusselt number ratio data in the latter figure are presented for the overall arrangement shown in Figure 16 for a Reynolds number of 2,000, with two different protrusion sizes. Here, Reynolds number is based upon bulk spatially averaged air inlet velocity, and a length scale equal to two times channel height. Even though the Reynolds number for this investigation is relatively low, the results presented are representative of fully turbulent flow conditions. Note that the data presented in Figure 17 are given for ratios of protrusion diameter to channel height of 2.54 and 1.70. According to Murata et al. [4], Figure 17 shows that the highest heat transfer augmentations are present near

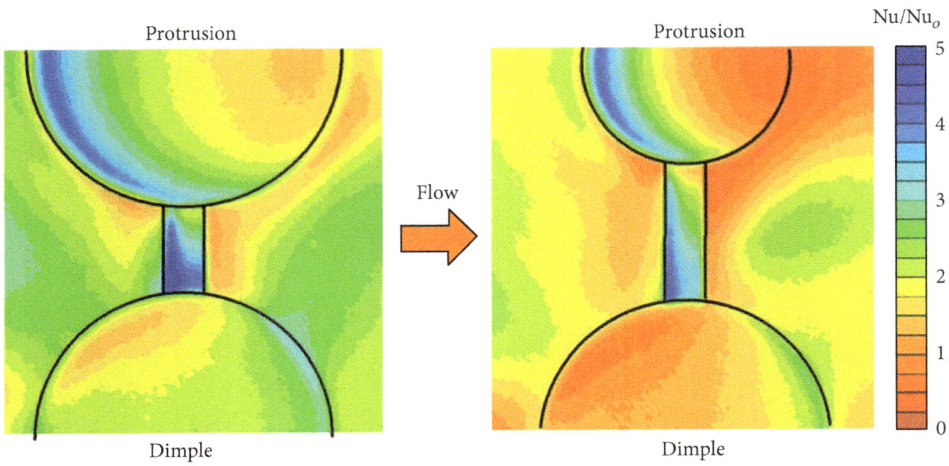

FIGURE 17: Local Nusselt number ratios for the rib, dimple, and protrusion arrangement employed in the investigation by Murata et al. [4] for Re = 2,000 and two different protrusion sizes.

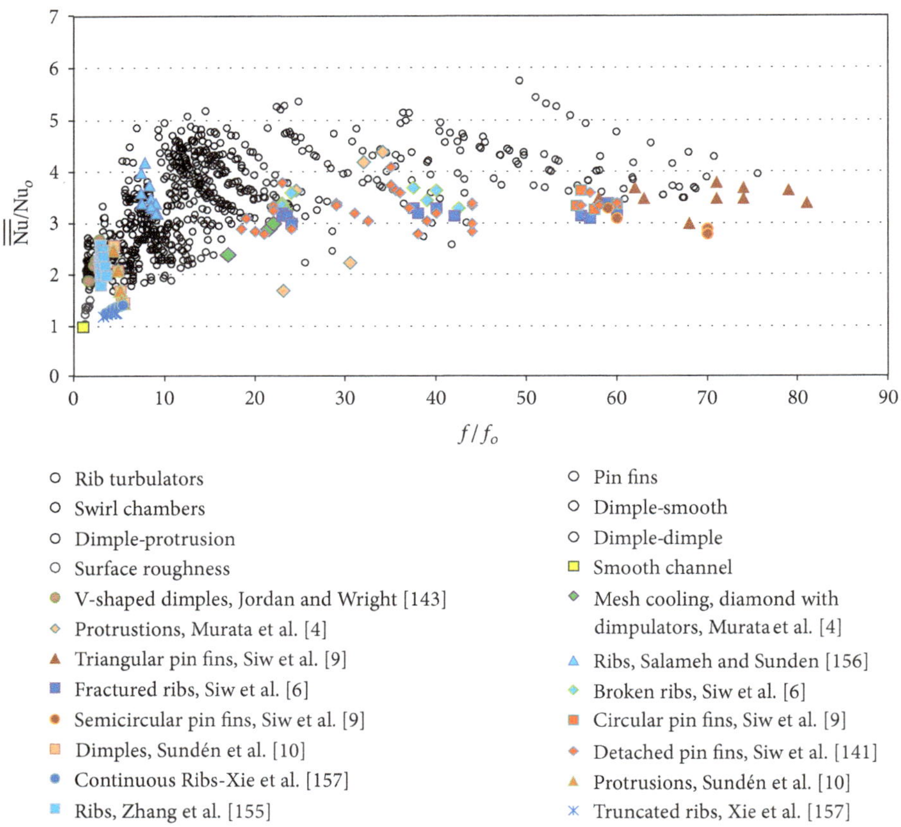

FIGURE 18: Comparison of relative performance of rib turbulators, pin fins, swirl chambers, dimple-smooth arrangements, dimple-protrusion arrangements, dimple-dimple arrangements, surface roughness, and smooth-walled channels, including results from investigations after 2003. Globally averaged Nusselt number ratios $\overline{\overline{\mathrm{Nu}}}/\mathrm{Nu}_o$ are presented as dependent upon friction factor ratios f/f_o.

the leading edge of the protrusion and on the top of the rib surface. Important augmentations are also occurring near the downstream edges of the dimples, as well as on an angled trajectory between adjacent dimples, which is a consequence of the flow blockage produced by the protrusions. When the effects of protrusion diameter are considered, Figure 17 also shows that Nusselt number ratios are higher on the protrusion and ribs when the protrusion diameter is larger. With the smaller protrusion diameter, an isolated high Nusselt number ratio region is observed downstream of the

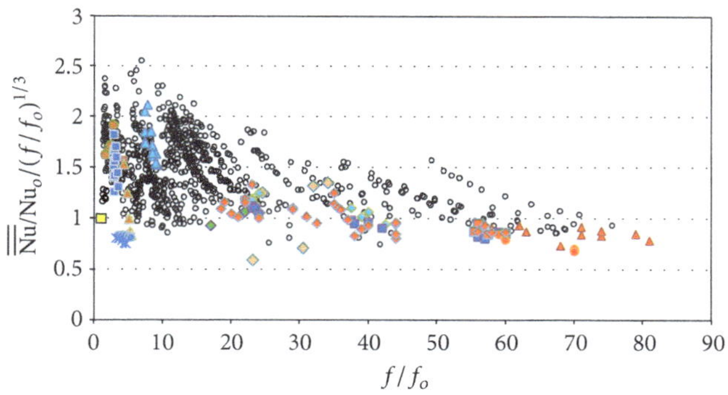

FIGURE 19: Comparison of relative performance of rib turbulators, pin fins, swirl chambers, dimple-smooth arrangements, dimple-protrusion arrangements, dimple-dimple arrangements, surface roughness, and smooth-walled channels, including results from investigations after 2003. Globally averaged thermal performance parameters $\overline{\overline{Nu}}/Nu_o/(f/f_o)^{1/3}$ are presented as dependent upon friction factor ratios f/f_o. Symbol definitions are given in Figure 18.

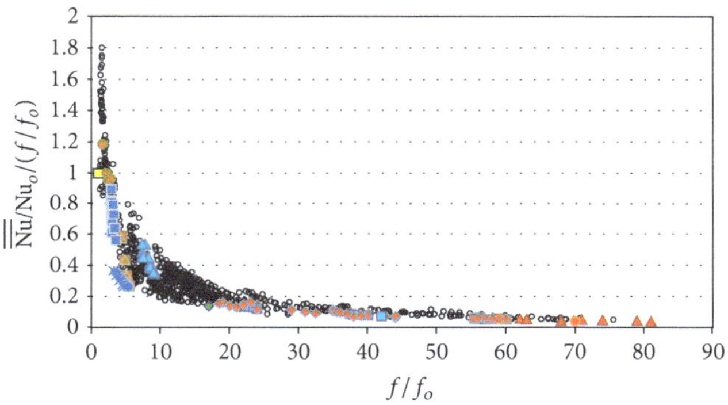

FIGURE 20: Comparison of relative performance of rib turbulators, pin fins, swirl chambers, dimple-smooth arrangements, dimple-protrusion arrangements, dimple-dimple arrangements, surface roughness, and smooth-walled channels, including results from investigations after 2003. Globally averaged thermal performance parameters $\overline{\overline{Nu}}/Nu_o/(f/f_o)$ are presented as dependent upon friction factor ratios f/f_o. Symbol definitions are given in Figure 18.

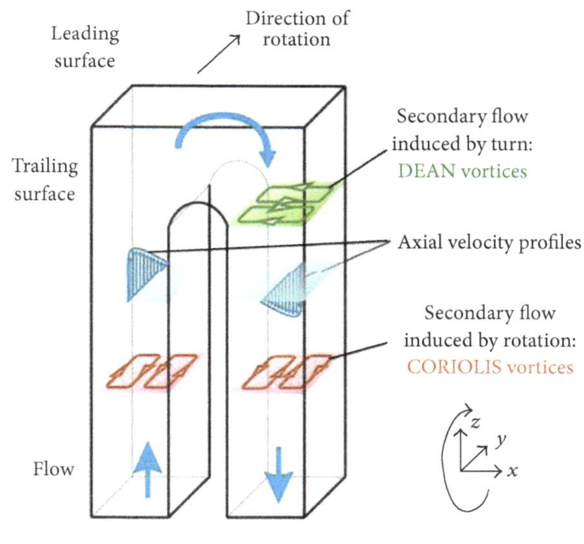

FIGURE 21: Schematic diagram illustrating typical channel arrangement and orientation relative to rotational motion, including resulting secondary flows [50].

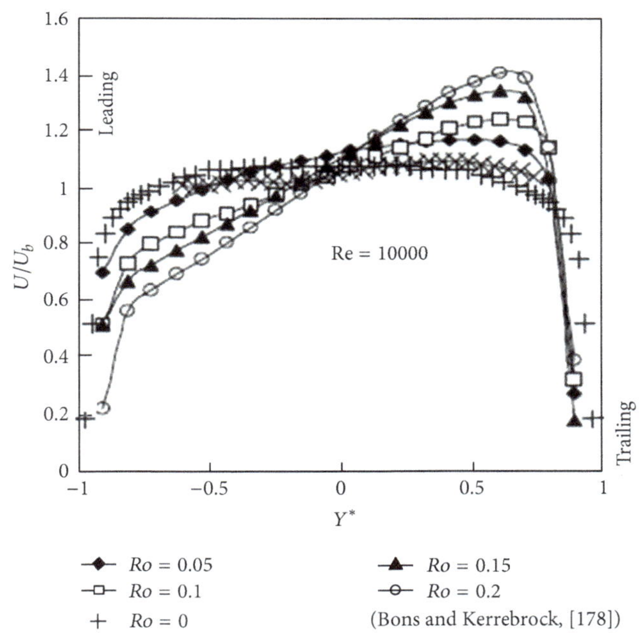

FIGURE 22: Midspan velocity profile for outward radial flow in a rotating smooth-walled channel from Chen and Liou [101].

transverse rib segment, which is due to reattachment of the shear layer, which forms above the flow separation region present just downstream of the rib.

The combined use of inclined pin fins and wavy endwalls is described by Takeishi et al. [174] and Motoda et al. [175]. Typically, pin fins with angles of plus and minus 45 degrees, relative to the flat endwall plane, are employed. Overall, spatially averaged heat transfer rates are higher on endwalls with normal pin fins, compared to endwalls with angled pin fins. However, overall heat transfer increases (for the entire passage surface area) are present for the angled pin fin arrangement because of an increase of overall surface area by a factor of 1.4. Associated pressure drop penalties are then generally lower when inclined pin fins are employed. With wavy endwalls also utilized, the average heat transfer coefficient of the −45 degree inclined pin fins is about the same or a little higher than the heat transfer coefficients of pin fins mounted on a flat endwall. However, the overall pressure loss penalty of the −45 degree inclined pin fins with a wavy endwall is generally less than the penalty when a flat endwall is used.

3.5. Overall Comparisons. Figure 18 presents comparisons of globally averaged Nusselt number ratios $\overline{\overline{Nu}}/Nu_o$ and friction

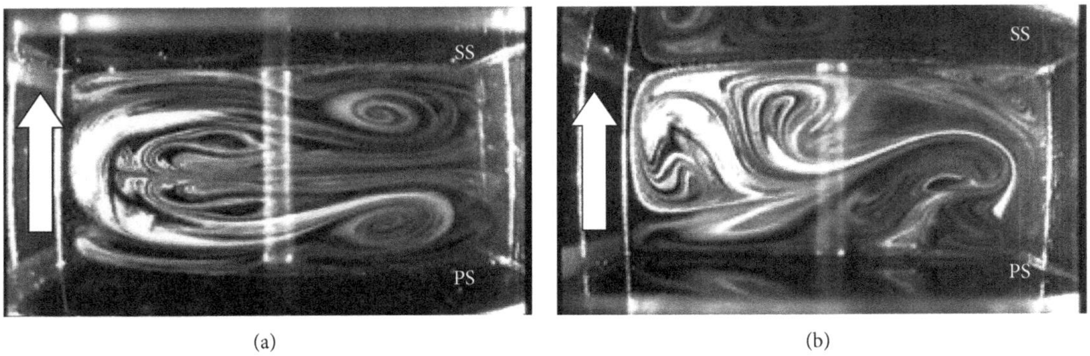

FIGURE 23: Flow visualization images from Elfert et al. [50] for the 180° bend region of the two-pass channel, viewing the flow in the negative z-direction with the bend entrance on the left and the bend exit on the right. (a) Image without rotation. (b) Image with rotation.

FIGURE 24: Flow visualization image from Elfert et al. [50] for the 180° bend region of the two-pass channel, viewing the flow in the positive y-direction with the bend entrance on the left and the bend exit on the right.

factor ratios f/f_o for rib turbulators, pin fins, swirl chambers, different dimple and protrusion arrangements, surface roughness, and smooth-walled channels. Included within Figure 18 are results from investigations since 2003. In regard to these more recent studies, data from pin fin investigations [9, 140] are designated with red symbols, data from rib investigations [6, 154–156] are designated with blue symbols, data from one mesh cooling study [176] are designated with green symbols, and data from investigations which employ dimples and protrusions [4, 10, 142] are designated with brown symbols. Note that some of the combination device results from the previous section [5, 11, 136, 164, 168–172] are not included since friction factor ratio data are not available in these studies. The mesh cooling arrangement, employed by Bunker et al. [176], employs a "small diamond with dimpulators" arrangement. Here, the ratio of dimple print diameter to channel height is 3.0, ratio of protrusion diameter to channel height is 3.5, and ratio of rib height to channel height is 0.15. Description of this mesh cooling arrangement is also provided by Murata et al. [4]. Also included in Figure 18 are a Nusselt number ratio and a friction factor ratio for a smooth channel (designated with a yellow symbol), to provide a baseline comparison. Inspection of Figure 18 reveals that all of the new data, obtained since 2003, collect within scatter of the data obtained from investigations conducted prior to 2003 [1]. The arrangement with the most promising performance characteristics may be the rib investigation of Salameh and Sunden [155], since Nusselt number ratios are as high as 4.2 with a corresponding friction factor ratio of 7.8. The ribs in this investigation are placed transversely to the direction of the main flow on the outer wall of the bend of a U-shaped duct. The ratio of rib height to hydraulic diameter is 0.1, and the rib pitch ratio is 10. Both continuous and truncated ribs are considered.

The data in Figure 18 are also presented again in Figures 19 and 20. In the first of these figures, globally averaged thermal performance parameters $\overline{\overline{\mathrm{Nu}}}/\mathrm{Nu}_o/(f/f_o)^{1/3}$ are given as dependent upon the friction factor ratio f/f_o. The $\overline{\overline{\mathrm{Nu}}}/\mathrm{Nu}_o/(f/f_o)$ parameter, presented in Figure 20, is referred to as the Reynolds analogy performance parameter. Like the results in the previous figure, Figures 19 and 20 show that all of the new data, obtained since 2003, are collected within distributions of the data obtained from investigations conducted prior to 2003 [1].

4. Effects of Rotation

Fluid flows within rotating channels are different from those in stationary channels because of the development of secondary flows, such as Coriolis vortices, from rotational influences. As a result, local surface heat transfer and friction factor augmentation levels are often altered because of the presence of rotation.

Figure 21 shows a typical turbine blade cooling passage arrangement from Elfert et al. [50]. Here, a two-pass system is employed, which is rotating in an orthogonal mode, such that the flat surface which comprises the 180° bend between the two passes is parallel to the axis of rotation. Note that the rotation is directed from the trailing side to the leading side of the passage. In modeling a turbine blade, the leading and trailing surfaces are analogous to the suction and pressure sides, respectively.

According to Servouze and Sturgis [177], the Coriolis force is an apparent force that exists in the noninertial, rotating reference frame. Application of the Coriolis effect on radially outward flow results in a force acting towards the pressure side, whereas application of the Coriolis effect on radially inward flow results in a force acting towards the suction side. As a result, Coriolis forces often induce secondary flows in fluid flows contained within rotating systems. Counter-rotating vortex pairs then develop in both radially outward flows and radially inward flows, with different directions of vortex rotation for different flow directions (relative to pressure and suction sides of the passage). Such vortex pairs are illustrated in Figure 21. Experimental evidence of Coriolis vortex pairs in rotating channels is provided by Elfert et al. [50] and Bons and Kerrebrock [178] from Particle Image Velocimetry (PIV) experiments, as well as by Chen and Liou [101] from Laser-Doppler Velocimetry (LDV) experiments.

The axial velocity profiles, also shown in Figure 21, are altered because of the presence of Coriolis vortices. Midspan velocity profiles for radially outward flow in a rotating smooth-walled channel from Chen and Liou [101] are presented in Figure 22. From these results, it is evident that, as rotation number increases, velocity increases near the pressure (or trailing) side and decreases near the suction (or leading) side relative to the mean stream velocity. Other experimental results from Chen and Liou [39] show that the opposite velocity profile trend is present for radially inward flow. Here, overall deviations and skewness (relative to the mean stream velocity) are quantitatively less as rotation number increases (relative to radially outward flow).

Figure 21 also shows Dean vortices near the exit of the 180° turn and near the beginning of the segment of radially inward channel flow. Dean vortices develop as a result of curvature of flow streamlines around the passage bend. A visualization of such a Dean vortex pair is shown in Figure 23(a) from Elfert et al. [50] for a stationary, nonrotating channel. With this arrangement, the vortex pair is approximately symmetric. With rotation present, the image in Figure 23(b) is obtained with Coriolis vortices on the left side of the image, before the flow enters the turn, and distorted Dean vortices on the right side, just before the flow exits the turn. Here, the influences of rotation cause the vortices near the suction side to become more coherent, whereas the vortices near the pressure side are more distorted and diffuse. This trend continues until local mixing within and around the pressure side vortices dominates the channel, as illustrated by the image in Figure 23(b). Overall, diffuse and convoluted secondary flows and the associated thermal transport mechanisms are most important near the pressure surface.

Another important event associated with turning ducts is formation of a separation flow region on the inner wall of the channel immediately following the bend, as shown in Figure 24. According to Elfert et al. [50], the characteristics and extent of such a separation region depend upon flow characteristics, such as flow speed, channel geometry, and channel surface condition.

4.1. Effects of Rotation on Local Nusselt Numbers, Spatially Averaged Nusselt Numbers, and Overall Thermal Performance. In general, rotation appears to alter heat transfer distributions only over very limited portions of channel surfaces. Existing investigations provide little evidence of overall increases or decreases in overall thermal performance characteristics with rotation, at any value of rotation number, buoyancy parameter, density ratio, or Reynolds number. In many investigations of internal passage flows with heat transfer and rotation, the Reynolds number and rotation number are given by

$$\mathrm{Re} = \frac{UD_h}{\nu}, \qquad \mathrm{Ro} = \frac{\Omega D_h}{U}, \tag{1}$$

respectively. The inlet density ratio is then expressed using an equation of the form

$$DR = T_c\left(\frac{1}{T_w} - \frac{1}{T_c}\right) = \frac{(\rho_w - \rho_c)}{\rho_c}. \tag{2}$$

The buoyancy parameter is then given by the following,

$$Bo = \mathrm{Ro}^2\left(\frac{R}{D_h}\right)DR. \tag{3}$$

The relationships and interdependence of these different parameters are apparent from these definitions.

In regard to overall rotation effects, the investigations of Fu et al. [128], Wright et al. [179], and Acharya et al. [180] generally show that overall thermal performance in smooth channels is affected more by Reynolds number than rotation number. According to Fu et al. [128], pressure losses and Reynolds number are generally insensitive to rotation number. A similar conclusion is reached if ribbed turbulators are present in rotating channels [102, 173, 179, 181, 182].

This is largely because of the competing effects of rotation on the pressure (or trailing) sides and the suction (or leading) sides of internal channel flows. According to Huh et al. [104], differences in local Nusselt number ratios for pressure sides and suction sides are generally a result of the rotation induced secondary flows caused by Coriolis vortices. These investigators indicate that (for radially outward flow channels) the effects of these secondary flows are generally more pronounced on the pressure sides of channels, where larger local heat transfer augmentations are often present (compared to suction sides).

This qualitative trend is often reversed for channels with multiple passes and turns [58, 64, 65, 74, 75, 77, 81, 100,

104, 128, 177, 182–206]. This is because, after the first turn, flow changes from radially outward to radially inward, which alters the application direction of the Coriolis force. The overall result for radially inward flow is generally surface heat transfer augmentations on passage suction sides. However, note that for channels with more than two passes, the quantitative consequences of the change in flow direction often become less pronounced in later passages [74, 185, 188, 189, 194, 202–204, 207].

A number of recent investigations show that rotation effects on Nusselt number ratios are often suppressed at high Reynolds numbers. This is illustrated by Nusselt number ratios which tend toward stationary channel values as Reynolds number increases [58, 64, 65, 74, 82, 104, 128, 182–184, 186, 187, 191, 200, 201, 206, 208–214]. One rationale for this behavior is the inverse dependence of Reynolds number and rotation number, which means that rotation number generally decreases as the associated Reynolds number increases (provided certain experimental parameters are maintained constant). Investigations which maintain constant Reynolds number show that Nusselt number ratios deviate further from stationary values with increasing rotation number [103, 189, 193, 202, 203, 215–218].

According to Wright et al. [219], rotation effects sometimes become more pronounced at lower Reynolds numbers when rib turbulators are employed. In general, according to this study, for Reynolds numbers from 10,000 to 40,000, the higher thermal performance characteristics, with no rotation, do not change significantly when rotation is added. The only exception is observed for Re = 10,000, wherein some improvement in overall thermal performance with rotation number is present [219].

For radially outward flow channels, Nusselt number ratio tends to decrease asymptotically with streamwise development from the channel entrance, with pressure side Nusselt number ratios higher than suction side values [64, 74, 75, 103, 104, 128, 178, 179, 183, 185–193, 196–198, 200, 201, 204–211, 215, 220–230]. Such trends are due to developing thermal boundary layers. When rib turbulators are present, the asymptotic decrease is less pronounced; however, Nusselt number ratios continue to be larger on the pressure side than on the suction side [58, 64–66, 75, 77, 81, 82, 100, 128, 177, 179, 182–184, 186, 189, 194, 199, 208, 209, 212, 213, 220, 229, 231–235].

Many investigations show significant increases in Nusselt number ratio as internal passages turn from radially outward to radially inward flow, and from radially inward flow to radially outward flow [64, 65, 74, 75, 77, 104, 128, 173, 182, 183, 185, 187–189, 191–194, 196–201, 204, 206, 207, 214, 215, 232, 236]. This local augmentation is generally attributed to the flow impinging on the outer walls of the turn, as well as the secondary flows resulting from the presence of Dean vortices. Other investigations show slight decreases in Nusselt number ratio in turn regions, especially when rib turbulators are employed [58, 74, 183, 184, 186, 189, 202, 203]. Such variations may be due to shear, separation, and reattachment phenomena which become more important than Dean vortex effects at high Reynolds numbers [64, 65, 74, 104, 128, 182–184, 186].

A number of investigations consider rotating channels with a variable wall temperature surface boundary condition [128, 186, 194, 201, 207, 215, 217, 218, 227, 228, 234, 236, 237]. In many of these investigations, wall temperature increases along the length of the channel, with slight decreases through turn regions.

Investigations which employ naphthalene sublimation to determine mass transfer generally obtain results which are consistent with heat transfer experiments [216, 238–247]. However, one drawback of such mass transfer methods is the absence of buoyancy within experiments.

Several other experimental investigations of heat transfer and flow behavior in rotating channels consider effects of channel shape and orientation [169, 248–251], pin fins [169, 251, 252], impingement cooling [253], dimpled surfaces [254], arrays of spherical protrusions [255], trailing edge cooling [256–258], and rib turbulators [148, 149, 166, 169, 259, 260]. For example, Tse and Steubner [98] measure different velocity components in a serpentine channel with 45° ribs. Prabhu and Vedula [83] measure surface static pressure variations with different channel aspect ratios, and different rotational speeds in a rectangular channel with tranverse ribs on one wall. Compared with results from nonrotating ribbed channels [59, 87], important effects of rotation are evident in the results from both investigations.

Computational studies of flows and heat transfer in ducts with rib turbulators consider straight single-pass ducts [56, 72, 73, 81, 86, 89, 99], two-pass ducts [57], two-pass ducts with U-shaped channels in between [60, 76, 84, 88, 90], 90° orthogonal ribs [72, 73, 81, 84, 86, 99], 45° angled ribs [56, 57, 60, 76, 88–90], 45° V-shaped ribs [73], and rotation [56, 57, 60, 72, 76, 81, 84, 88, 90]. In one case [89], the angled ribs placed on two opposite walls of the channel are rounded. The results from the most notable of these investigations with rotation show that the secondary flows induced by angled ribs, rotating buoyancy, and Coriolis forces produce alterations to surface heat transfer coefficients, and flow field structure. This includes the development of strong nonisotropic turbulence stresses [88, 90]. Jia et al. [73] and Watanabe and Takahashi [99] include experimentally measured friction factors, Nusselt numbers, and streamwise velocity fluctuations, which are compared to numerically predicted results. Experimental investigation of a complex trailing edge passage of a high pressure turbine blade is described by Bunker et al. [61]. Rigby and Bunker [85] employ a full three-dimensional Navier-Stokes equation solver to investigate the same arrangement.

4.2. Effects of Rotation Number. Figure 25 from Zhou and Acharya [102] shows how spatially averaged pressure and suction side Nusselt number ratios vary with rotation number. Of course, different trends with rotation number are present for the two sides of the rotating channel with radially outward flow. These experimental results from Zhou and Acharya [102] are obtained for both portions of a two-pass channel. As a result, opposite trends with rotation number are also evident in Figure 25 for the suction and pressure

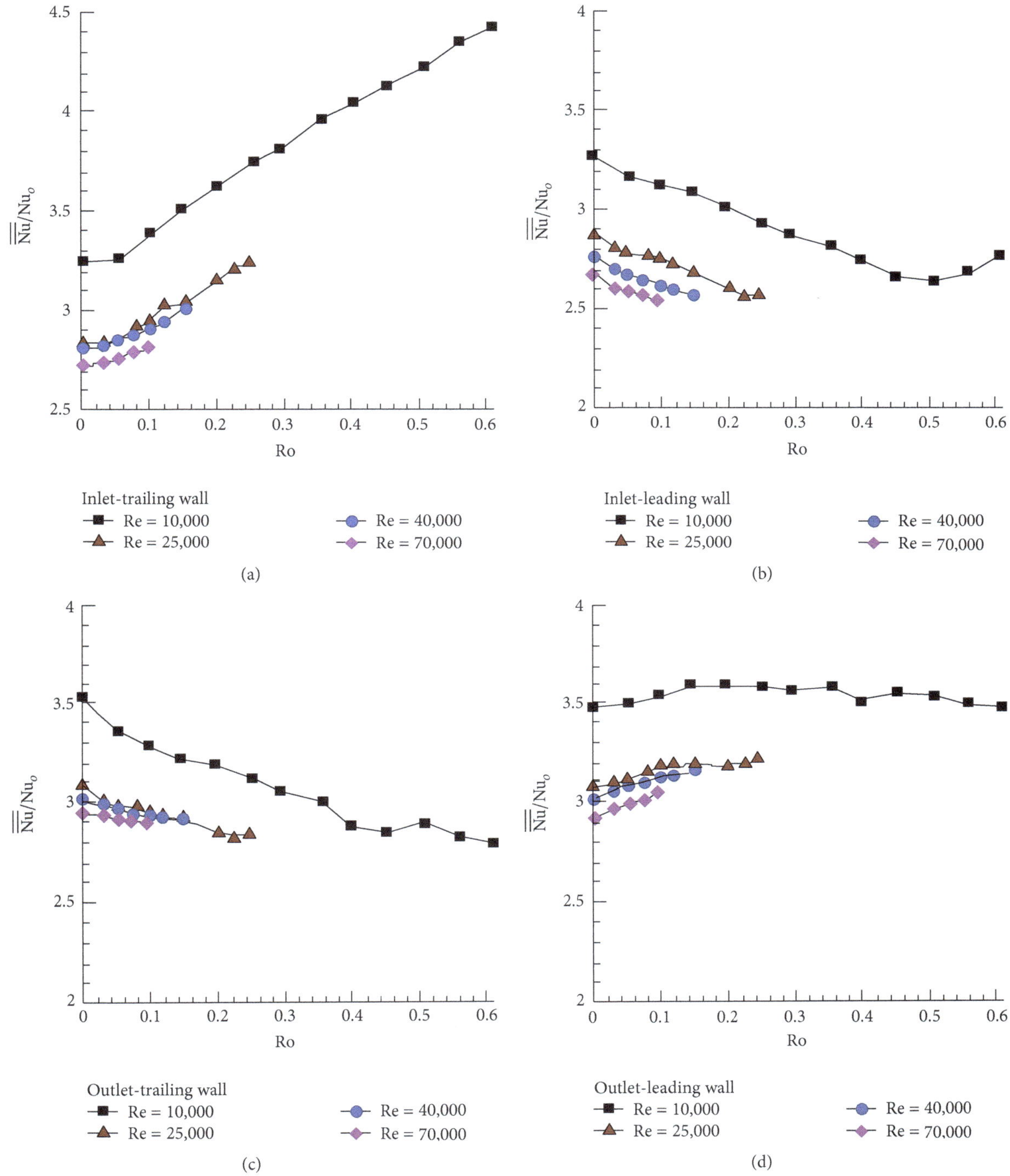

FIGURE 25: Spatially averaged Nusselt number ratio variations as dependent upon rotation number for the rotating channel investigation of Zhou and Acharya [102]. (a) Radial outflow pressure side results. (b) Radial outflow suction side results. (c) Radial inflow pressure side results. (d) Radial inflow suction side results.

sides for radially inward flow [58, 64, 65, 74, 75, 82, 102–104, 179, 180, 184, 185, 188–191, 193, 194, 196, 200, 202–212, 216–218, 221–223, 230, 237, 239, 241, 261–264].

Overall, the results in Figure 25 show that Nusselt number ratios increase with rotation number for the pressure side with radial outward flow, and decrease with rotation number for the suction side with radial outward flow. Decreases with rotation number are then present for the pressure side with radial inward flow, and increases are present with rotation number for the suction side with radial inward flow. Because of such competing effects on each passage portion, overall, spatially averaged Nusselt numbers are generally not significantly different from values, measured in similar configurations, with no rotation. For many turbine thermal

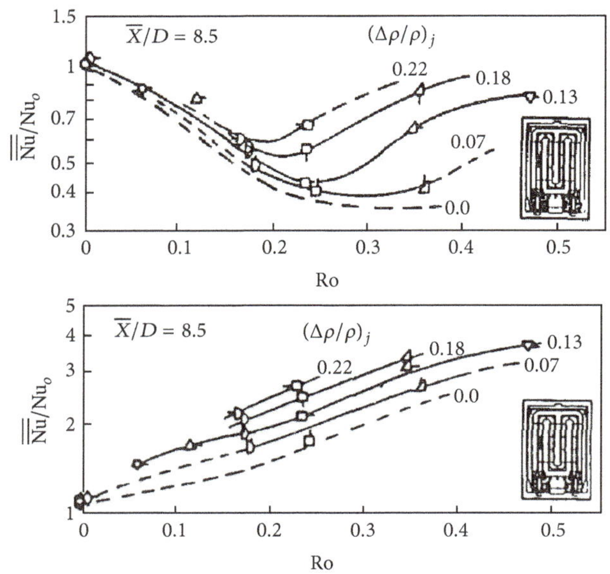

FIGURE 26: Spatially averaged Nusselt number ratio variations as dependent upon inlet density ratio and rotation number for the rotating channel investigation of Wagner et al. [103]. Results for a radial outflow arrangement are presented for suction (top figure) and pressure (bottom figure) sides of the rotating passage.

designers, local Nusselt number reductions are as, or more important, than local augmentations. This is because such degradations of local heat transfer coefficients also must be understood and correctly accommodated within designs in order to insure appropriate engine longevity.

4.3. Effects of Inlet Density Ratio. Figure 26 presents spatially averaged Nusselt number ratio variations as dependent upon inlet density ratio and rotation number for the rotating channel investigation of Wagner et al. [103]. Results for a radial outflow arrangement are presented for suction (top figure) and pressure (bottom figure) sides of the rotating passage. Overall, these data show that increasing the inlet density ratio enhances the Nusselt number ratio for both pressure and suction side surfaces [74, 102, 103, 173, 188, 189, 202–204, 206, 235, 265, 266]. From Figure 26, it is apparent that the density ratio has a significant influence on heat transfer magnitudes over a range of rotation numbers. According to Hajek et al. [189], inlet density ratio has a slightly larger effect on the pressure side than the suction side and diminishing influences in later passages of multipass and serpentine channels. Because the buoyancy parameter is proportional to the product of the inlet density ratio and rotation number squared, alteration of these different parameters results in different data trends.

4.4. Effects of Buoyancy Parameter. Figure 27 from Huh et al. [104] shows that local area-averaged Nusselt number ratios are affected by variations of the buoyancy parameter in the same approximate qualitative manner as when the rotation number changes. Note that the buoyancy parameter

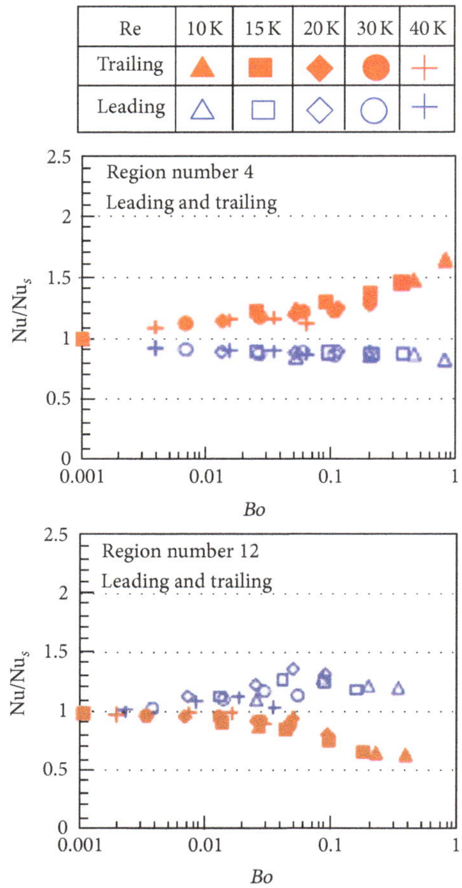

FIGURE 27: Local area-averaged Nusselt number ratio variations as dependent upon buoyancy parameter for the smooth surface rotating channel investigation of Huh et al. [104]. Region 4 is measured with radially outward flow, and region 12 is measured with radially inward flow.

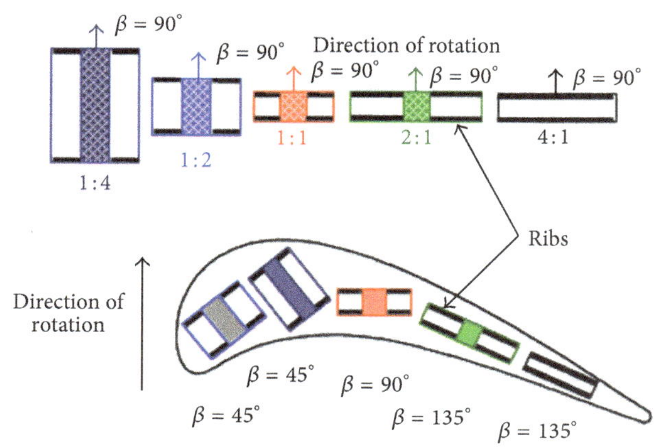

FIGURE 28: Examples of various channel orientations and aspect ratios employed for internal cooling of turbine blades from Fu et al. [128].

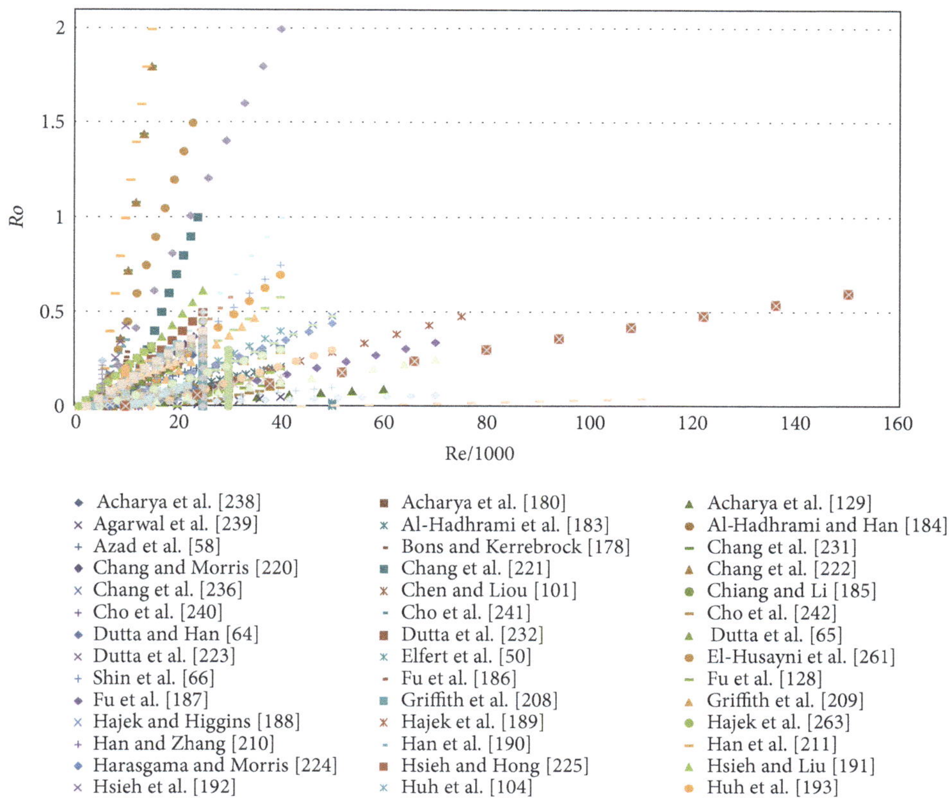

(a)

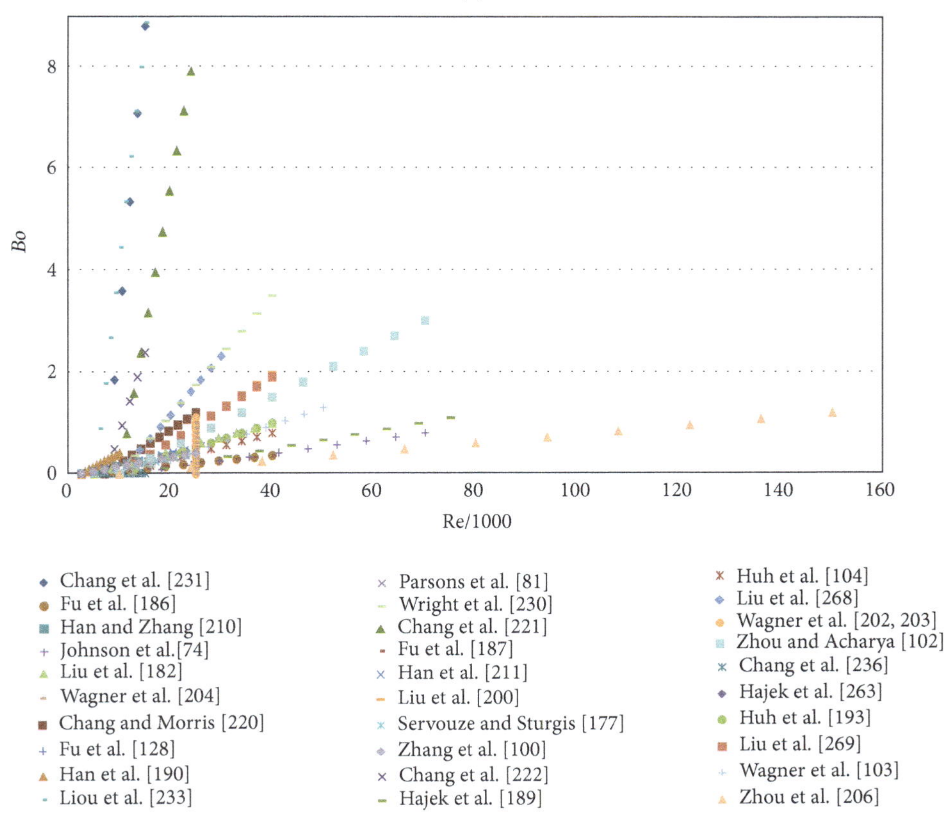

(b)

FIGURE 29: Continued.

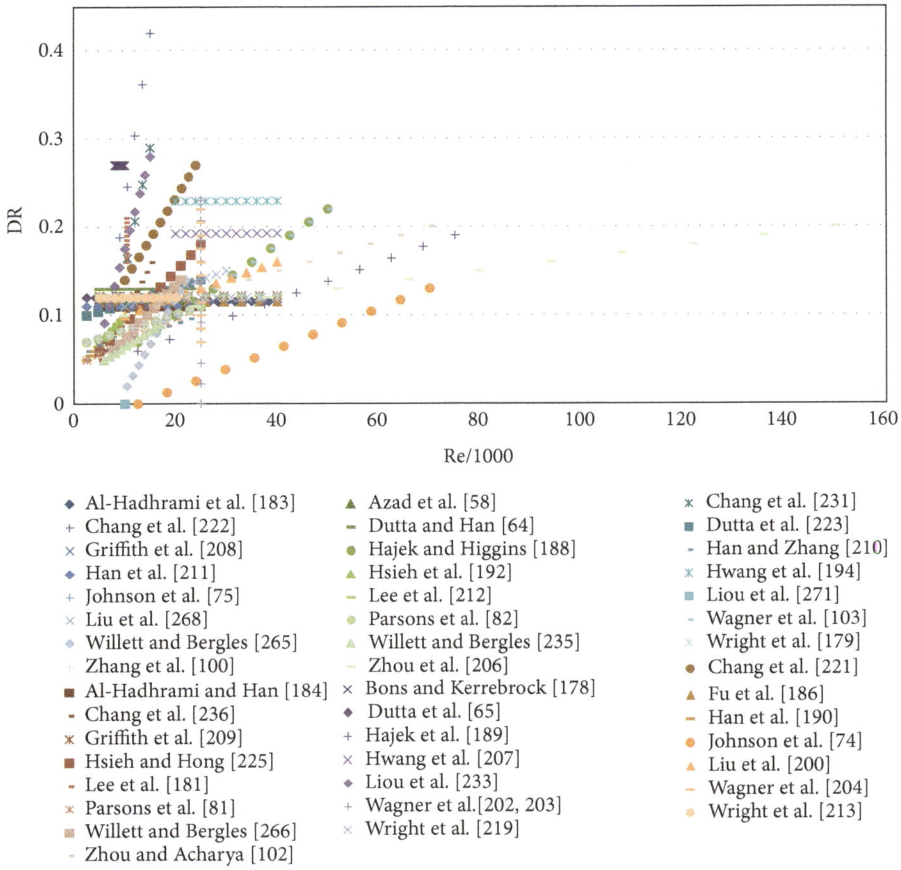

(c)

Figure 29: Ranges of experimental parameters for existing investigations of channel flows with rotation. (a) Rotation number versus Reynolds number. (b) Buoyancy parameter versus Reynolds number. (c) Inlet density ratio versus Reynolds number.

combines the effects of rotation number, density ratio, and a geometric configuration parameter. With radially outward flow, Nusselt number ratios increase with buoyancy parameter on the passage pressure side and decrease with buoyancy parameter on the passage suction side. With radially inward flow, the converse is generally true, with suction side Nusselt numbers increasing with buoyancy parameter, such that values are generally larger than Nusselt numbers measured on the passage pressure side [74, 100, 102–104, 128, 177, 182, 186, 187, 189, 190, 193, 200, 204, 206, 210, 220–222, 263, 265, 267]. The channel cross-section for the Huh et al. [104] investigation has a 12.70 mm height, and 23.81 mm width, resulting in a hydraulic diameter of 16.93 mm.

4.5. Effects of Channel Geometry and Orientation. In regard to aspect ratio, Fu et al. [128, 187] consider a variety of arrangements, and often show similar Nusselt number trends regardless of the aspect ratio, particularly as the buoyancy parameter is changed. In some situations, Nusselt number ratios are slightly higher in channels with low aspect ratios. In addition, pressure losses developing in low aspect ratios are generally significantly less than channels with higher aspect ratios [128, 187].

The results in Figure 28 illustrate the variety of different channel orientations which are sometimes employed within a turbine rotor blade. From these arrangements, the importance of channel orientation with respect to the rotating axis is evident because not all internal cooling channels are oriented perpendicular to the rotating axis. With the arrangement shown in Figure 21, pressure side is synonymous with trailing side, and suction side is synonymous with leading side. Within a 45° oriented channel, for example, designating the suction side and pressure side is not longer so straight-forward. For example, the suction side can still be considered the leading side, or it can be considered to be a combination of a portion of the leading side, a sidewall, and even a small portion of the trailing side. According to several investigations [64, 180, 183, 184, 187, 209], rotation appears to have less influence on heat transfer on the leading and trailing sides when the channel is oriented at 45°, compared to a 90° arrangement.

In regard to channel shape, triangular [65, 181, 232, 268, 269] and circular [226, 270] ducts are also investigated. In regard to triangular arrangements, care is required when considering the pressure and suction sides of the rotating passages. Overall, rotation and buoyancy have less influence

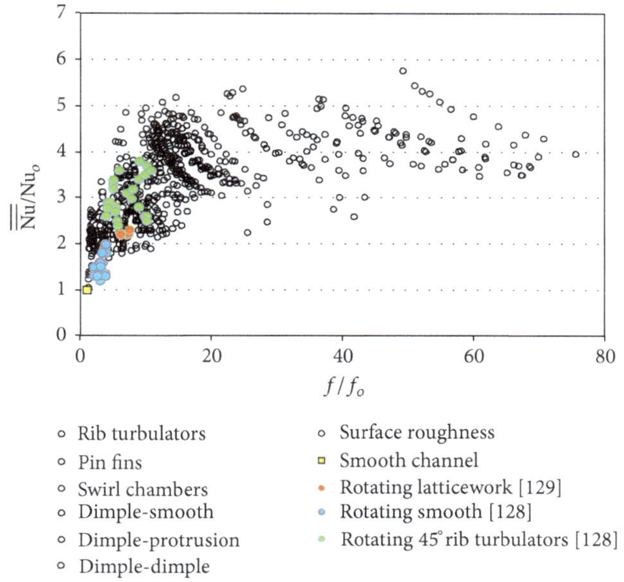

FIGURE 30: Comparisons of relative performance of rib turbulators, pin fins, swirl chambers, dimple-smooth arrangements, dimple-protrusion arrangements, dimple-dimple arrangements, surface roughness, and smooth-walled channels [1], with results from rotation experiments [128, 173]. Globally averaged Nusselt number ratios $\overline{\overline{\mathrm{Nu}}}/\mathrm{Nu}_o$ are presented as dependent upon friction factor ratios f/f_o.

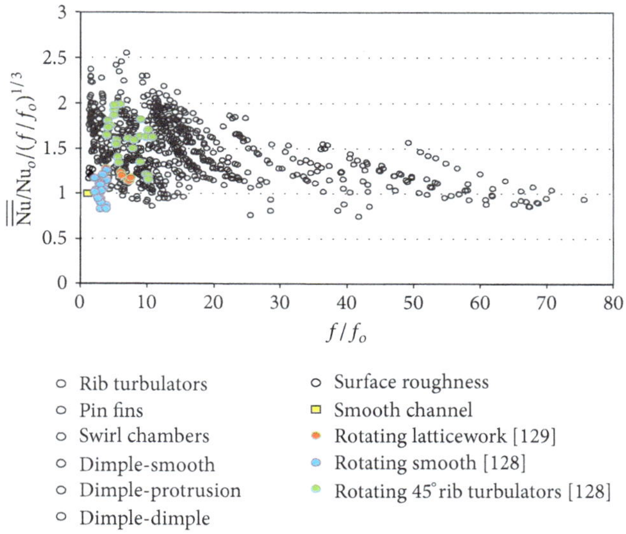

FIGURE 31: Comparisons of relative performance of rib turbulators, pin fins, swirl chambers, dimple-smooth arrangements, dimple-protrusion arrangements, dimple-dimple arrangements, surface roughness, and smooth-walled channels [1], with results from rotation experiments [128, 173]. Globally averaged thermal performance parameters $\overline{\overline{\mathrm{Nu}}}/\mathrm{Nu}_o/(f/f_o)^{1/3}$ are presented as dependent upon friction factor ratios f/f_o.

on heat transfer in ribbed channels than smooth ducts [269]. In addition, different rib orientations and arrangements produce different Nusselt number ratio distributions, especially as either the rotation number or the buoyancy parameter increases [268].

4.6. Effects of Different Wall Heating Arrangements. Several studies examine the effects different wall heating arrangements. For example, El-Husayni et al. [261] apply uniform heat flux to several different rotating channel arrangements: the primary wall only, two opposite walls only, and all four walls together. All of the other non-heated surfaces are maintained adiabatic. Comparisons generally show that the different configurations result in minimal changes to magnitudes and distributions of local Nusselt number ratios. The only exception is an increase in Nusselt number ratio on the pressure side of a ribbed channel with all four walls at uniform heat flux as the rotation number increases. With other conditions and arrangements, Nusselt number ratios either remained constant or decreased. Han and Zhang [210] consider rotating channels with all four walls at the same temperature, four walls with uniform heat flux, and with the pressure side hotter than the suction side. The investigators conclude that heat transfer coefficients on the pressure side are higher with uniform heat flux than with uniform wall temperature. Heat transfer coefficients on the suction side are then much higher with uniform heat flux and hotter pressure side, than when a uniform wall temperature is imposed.

4.7. Ranges of Reynolds Number, Rotation Number, Buoyancy Parameter, and Inlet Density Ratio of Existing Investigations. Investigated ranges of Reynolds number, rotation number, buoyancy parameter, and inlet density ratio for existing investigations of rotating channel flows are shown in Figure 29. The overall objective of the related studies is application of experimental data to the operating conditions of turbine blades within industrial and aeroengine gas turbine engines. However, unfortunately, magnitudes of Reynolds number, rotation number, buoyancy parameter, and inlet density ratio are generally (in some manner) smaller than the actual values which are encountered in operating engines. For example, see investigation of Liou et al. [271].

One reason for these deficiencies is because high rotation numbers can be achieved by maintaining rotational speed and reducing the Reynolds number. As a result, rotation numbers within the range of operating turbines within gas turbine engines can be produced, but at the penalty of a low Reynolds number.

As indicated, *local, spatially resolved* heat transfer augmentations and reductions within rotating channels are dependent upon both Reynolds number and rotation number. As such, additional, high-quality experimental data are needed, with high Reynolds numbers simultaneously with high rotation numbers (to correctly match engine operating conditions), especially when these two parameters are varied independently of each other.

4.8. Rotation and Its Relationship to Thermal Performance Parameters. Here, comparisons are made between results presented by Ligrani et al. [1] for stationary (nonrotating)

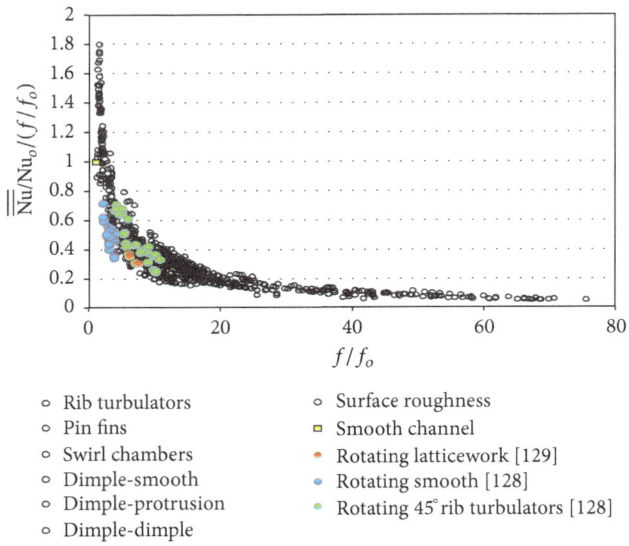

FIGURE 32: Comparisons of relative performance of rib turbulators, pin fins, swirl chambers, dimple-smooth arrangements, dimple-protrusion arrangements, dimple-dimple arrangements, surface roughness, and smooth-walled channels [1], with results from rotation experiments [128, 173]. Globally averaged thermal performance parameters $\overline{\overline{Nu}}/Nu_o/(f/f_o)$ are presented as dependent upon friction factor ratios f/f_o.

channels, and experimental thermal performance data from rotating channels [128, 173]. As discussed previously, Ligrani et al. [1] present information on heat transfer augmentation techniques with respect to internal cooling channels, including rib turbulators, pin fins, dimpled surfaces, surfaces with protrusions, swirl chambers, and surface roughness. Included are spatially averaged thermal performance parameters as they relate to friction factor ratios. The comparisons between the rotating channel experimental data [128, 173] and the Ligrani et al. [1] results are given in Figures 30, 31, and 32. The rotating lattice work arrangement [173] is comprised of a crossing-rib configuration on leading and trailing passage sides to model a coolant passage with inclined serpentine subchannels. Each rib in this arrangement is inclined at 45 degrees relative to the radial direction, such that ribs on opposite walls are arranged with an aligned and crossing arrangement, such that the subchannel on one surface turns smoothly to the subchannel on the opposite surface. This arrangement forces the fluid within the latticework configuration to turn by 90 degrees in a repeated and periodic fashion. As shown by Figures 30, 31, and 32, the rotating thermal performance characteristics are distributed such that they are well within the scatter and distribution of the stationary channel performance data. This indicates that rotation has little effect on overall thermal performance as a function of friction factor. Additional confirmation is also provided that overall, spatially averaged performance characteristics from rotating channel studies are generally insensitive to variations of rotation number.

5. Summary and Conclusions

Many important constraint requirements are imposed upon the development of internal cooling configurations for gas turbine engines. Satisfying these constraints requires simultaneous consideration of multidisciplinary issues, such as manufacturing, airfoil surface cooling requirements, aerodynamic losses, coolant availability, second law losses, available volume space, airfoil shape, and pressure drop penalties, in addition to heat transfer and thermal considerations. As a result, current and future trends for advanced turbine component cooling design require simultaneous development of external thermal protection schemes, like film cooling, with internal cooling technologies, including consideration of the adjacent conjugate conduction variations in adjacent solid components, as well as associated aerodynamic pressure loss penalty issues. To meet these requirements, designers, manufacturers, and researchers must develop internal cooling systems with combinations of different devices together within one coolant passage. Resulting objectives are optimal thermal protection and minimal pressure loss penalties.

To provide an overview of the current state of the art of heat transfer augmentation schemes employed for internal cooling of turbine blades and components, results from an extensive literature review are presented with data from internal cooling channels, both with and without rotation. According to this survey, a very small number of existing investigations consider the use of combination devices for internal passage heat transfer augmentation. Examples are rib turbulators, pin fins and dimples together [4], a combination of pin fins and dimples [5], and rib turbulators and pin fins in combination [6]. The results of such studies are compared with data obtained prior to 2003 without rotation influences, as summarized by Ligrani et al. [1]. Those data [1] are comprised of heat transfer augmentation results for internal cooling channels, with rib turbulators, pin fins, dimpled surfaces, surfaces with protrusions, swirl chambers, or surface roughness.

This comparison reveals that all of the new data, obtained since 2003, are collected within the distribution of data obtained from investigations conducted prior to 2003, without rotation influences [1]. A newer arrangement with the most promising performance characteristics may be the rib investigation of Salameh and Sunden [155], since Nusselt number ratios are as high as 4.2 with a corresponding friction factor ratio of 7.8. The same conclusion in regard to data distributions is also reached in regard to globally averaged thermal performance parameters, $\overline{\overline{Nu}}/Nu_o/(f/f_o)^{1/3}$ and $\overline{\overline{Nu}}/Nu_o/(f/f_o)$, as they vary with friction factor ratio f/f_o. These comparisons, made on the basis of such judgement criteria, lead to the conclusion that improvements in our ability to provide better *overall, spatially averaged* thermal protection have been minimal since 2003. However, in spite of this conclusion, many heat transfer augmentation technologies, developed since 2003, may provide improved *local* thermal protection, compared to many devices developed prior to that time. Such local thermal protection is mostly related to local gradients of temperature and heat transfer, which are also vitally important for the long-term durability and reliability of gas turbine engines. Local temperature gradients are also important regarding local thermal stress magnitudes. The best configurations, with the lowest thermal stresses, are thus, the ones with the lowest local temperature gradients.

When rotation is present, existing investigations provide little evidence of overall increases or decreases in overall thermal performance characteristics with rotation, at any value of rotation number, buoyancy parameter, density ratio, or Reynolds number. In addition, overall thermal performance in smooth channels with rotation is affected more by Reynolds number than rotation number. Comparisons between the rotating channel experimental data [128, 173] and the Ligrani et al. [1] results show that rotating thermal performance characteristics are distributed well within the distribution of the stationary performance data, obtained prior to 2003. This also indicates that rotation has little effect on overall thermal performance as a function of friction factor. Additional confirmation is also provided that overall, spatially averaged performance characteristics from rotating channel studies are generally insensitive to variations of rotation number.

This is largely because of the competing effects of rotation on the pressure (or trailing) sides and the suction (or leading) sides of internal channel flows. Differences in local Nusselt number ratios for pressure sides and suction sides are generally a result of increasing rotation numbers, and the rotation induced secondary flows caused by Coriolis vortices. For radially outward flow channels, the effects of these secondary flows produce local heat transfer augmentations on the pressure (or trailing) sides of channels, and local heat transfer decreases on suction (or leading) sides (compared to similar channels with no rotation). Different trends with increasing rotation number are present for the two sides of a rotating channel with radially inward flow.

Local area-averaged Nusselt number ratios are affected by variations of the buoyancy parameter in the same approximate qualitative manner as when rotation number changes. With radially outward flow, Nusselt number ratios increase with buoyancy parameter on the passage pressure side, and decrease with buoyancy parameter on the passage suction side (compared to similar channels with no rotation). When the inlet density ratio is altered, results for a radial outflow arrangement show that increasing this parameter enhances the Nusselt number ratio for both pressure and suction side surfaces over a range of rotation numbers.

To further advance heat transfer augmentation technologies for internal cooling of turbine components of gas turbine engines, additional high quality experimental data are needed. Each experimental requires complete documentation of relevant characteristics, including *local* and *spatially averaged* distributions of surface Nusselt numbers, as well as overall flow passage pressure drop penalties and friction factor ratios. Relevant engine operating conditions, both with and without rotation, must be matched whenever possible. Additional thermal performance improvements can be obtained with singular devices utilizing detailed and extensive knowledge of the effects and interactions between geometry and configuration details, and the resulting flow and heat transfer characteristics. Use of novel combination devices also offers excellent opportunities for improved internal cooling thermal performance. New high quality rotation experimental data are additionally needed (especially, local, spatially resolved data), particularly with different inflow and outflow arrangements, and a variety of heat transfer augmentation devices.

Nomenclature

Bo: Buoyancy parameter
D: Dimple print diameter
D: Channel length scale
D: Pin fin length scale
D_h: Channel or duct hydraulic diameter
DR: Inlet density ratio, $(\Delta\rho/\rho)_i$
f: Friction factor
f_o: Baseline friction factor in a smooth channel with no augmentation devices
H: Channel length scale
H: Dimple depth
Nu: Local Nusselt number based on channel length scale
Nu_o: Baseline, constant property Nusselt number in a smooth channel with no augmentation devices
Nu_S: Baseline, constant property Nusselt number in a smooth channel with no augmentation devices
R: Rotation radial position
Re: Reynolds number based on channel length scale and channel spatially-averaged inlet velocity
Ro: Rotation number
T: Temperature
U: Channel or duct spatially-averaged inlet velocity
x: Longitudinal coordinate
X: Longitudinal coordinate
$\overline{X}$: Longitudinal coordinate
X^*: Normalized longitudinal coordinate
y: Normal coordinate
Y^*: Normalized normal coordinate
z: Spanwise coordinate.

Greek Symbols

β: Rotation angle
ρ: Density
ν: Kinematic viscosity
Ω: Rotational angular velocity.

Superscripts

=: Globally averaged value.

Subscripts

c: Coolant stagnation value
w: Wall value.

References

[1] P. M. Ligrani, M. M. Oliveira, and T. Blaskovich, "Comparison of heat transfer augmentation techniques," *AIAA Journal*, vol. 41, no. 3, pp. 337–362, 2003.

[2] J. C. Han, S. Dutta, and S. Ekkad, *Gas Turbine Heat Transfer and Cooling Technology*, Taylor and Francis, New York, NY, USA, 2000.

[3] J. C. Han, J. S. Park, and C. K. Lie, "Heat Transfer and pressure drop in blade cooling channels with turbulence promoters," NASA CR-3837, 1984.

[4] A. Murata, S. Nishida, H. Saito, K. Iwamoto, Y. Okita, and C. Nakamata, "Heat transfer enhancement due to combination of dimples, protrusions, and ribs in narrow internal passage of gas turbine blade," ASME Paper GT2011-45356, 2011.

[5] Y. Rao, C. Wan, Y. Xu, and S. Zang, "Local heat transfer characteristics in channels with pin fin and pin fin-dimple arrays," ASME Paper GT2011-45449, 2011.

[6] S. C. Siw, M. K. Chyu, and M. A. Alvin, "Effects of pin detached space on heat transfer in a rib roughened channel," ASME Paper GT2011-46078, 2011.

[7] B. A. Brigham and G. J. VanFossen, "Length to diameter ratio and row number effects in short pin fin heat transfer," *Journal of Engineering for Gas Turbines and Power*, vol. 106, no. 1, pp. 241–245, 1984.

[8] P. M. Ligrani and G. I. Mahmood, "Variable property Nusselt numbers in a channel with pin fins," *Journal of Thermophysics and Heat Transfer*, vol. 17, no. 1, pp. 103–111, 2003.

[9] S. C. Siw, M. K. Chyu, and M. A. Alvin, "Heat transfer enhancement of internal cooling passage with triangular and semi-circular shaped pin-fin arrays," ASME Paper GT2012-69266, 2012.

[10] B. Sundén, G. Xie, and Q. Wang, "Predictions of enhanced heat transfer of an internal blade tip-wall with hemispherical dimples or protrusions," ASME Paper GT2010-22265, 2010.

[11] W. Chen, J. Ren, and H. Jiang, "Effect of turning vane configurations on heat transfer and pressure drop in a ribbed internal cooling system," ASME Paper GT2010-22273, 2010.

[12] D. E. Metzger, R. A. Barry, and J. P. Bronson, "Developing heat transfer in rectangular ducts with staggered arrays of short pin-fins," *Journal of Heat Transfer*, vol. V 104, no. 4, pp. 700–706, 1982.

[13] D. E. Metzger and S. W. Haley, "Heat transfer experiments and flow visualization for arrays of short pin-fins," ASME Paper 82-GT-138, 1982.

[14] E. M. Sparrow, J. W. Ramsey, and C. A. C. Altemani, "Experiments on in-line pin fin arrays and performance comparisons with staggered array," *Journal of Heat Transfer*, vol. 102, no. 1, pp. 44–50, 1980.

[15] A. Zukauskas, "Heat transfer from tubes in cross flow," *Advances in Heat Transfer*, vol. 8, pp. 96–160, 1972.

[16] G. J. VanFossen, "Heat-transfer coefficients for staggered arrays of short pin-fins," *Journal of Engineering for Power*, vol. 104, no. 2, pp. 268–274, 1982.

[17] R. J. Simoneau and G. J. VanFossen, "Effects of location in an array on heat transfer to a short cylinder in cross flow," *Journal of Heat Transfer*, vol. 106, no. 1, pp. 42–48, 1984.

[18] S. C. Lau, J. C. Han, and Y. S. Kim, "Turbulent heat transfer and friction in pin fin channels with lateral flow ejection," *Journal of Heat Transfer*, vol. 111, no. 1, pp. 51–58, 1989.

[19] R. D. McMillin and S. C. Lau, "Effect of trailing-edge ejection on local heat (mass) transfer in pin fin cooling channels in turbine blades," *Journal of Turbomachinery*, vol. 116, no. 1, pp. 159–168, 1994.

[20] M. K. Chyu, "Heat transfer and pressure drop for short pin-fin arrays with pin-endwall fillet," *Journal of Heat Transfer*, vol. 112, no. 4, pp. 926–932, 1990.

[21] M. K. Chyu and R. J. Goldstein, "Influence of an array of wall-mounted cylinders on the mass transfer from a flat surface," *International Journal of Heat and Mass Transfer*, vol. 34, no. 9, pp. 2175–2186, 1991.

[22] M. K. Chyu, Y. C. Hsing, T. I. P. Shih, and V. Natarajan, "Heat transfer contributions of pins and endwall in pin-fin arrays: effects of thermal boundary condition modeling," *Journal of Turbomachinery*, vol. 121, no. 2, pp. 257–263, 1999.

[23] V. B. Grannis and E. M. Sparrow, "Numerical simulation of fluid flow through an array of diamond-shaped pin fins," *Numerical Heat Transfer A*, vol. 19, no. 4, pp. 381–403, 1991.

[24] D. A. Olson, "Heat transfer in thin, compact heat exchangers with circular, rectangular, or pin-fin flow passages," *Journal of Heat Transfer*, vol. 114, no. 2, pp. 373–382, 1992.

[25] M. K. Chyu, Y. C. Hsing, and V. Natarajan, "Convective heat transfer of cubic fin arrays in a narrow channel," *Journal of Turbomachinery*, vol. 120, no. 2, pp. 362–367, 1998.

[26] M. K. Chyu and V. Natarajan, "Heat transfer on the base surface of three-dimensional protruding elements," *International Journal of Heat and Mass Transfer*, vol. 39, no. 14, pp. 2925–2935, 1996.

[27] J. J. Hwang and C. C. Lu, "Lateral-flow effect on endwall heat transfer and pressure drop in a pin-fin trapezoidal duct of various pin shapes," ASME Paper 2000-GT-232, 2000.

[28] O. Uzol and C. Camci, "Elliptical pin fins as an alternative to circular pin fins for gas turbine blade cooling applications. Part 1: endwall heat transfer and total pressure loss characteristics," ASME Paper 2001-GT-180, 2001.

[29] O. Uzol and C. Camci, "Elliptical pin fins as an alternative to circular pin fins for gas turbine blade cooling applications. Part 2: wake flow field measurements and visualization using particle image velocimetry," ASME Paper 2001-GT-181, 2001.

[30] S. Y. Won, G. I. Mahmood, and P. M. Ligrani, "Spatially-resolved heat transfer and flow structure in a rectangular channel with pin fins," *International Journal of Heat and Mass Transfer*, vol. 47, no. 8-9, pp. 1731–1743, 2004.

[31] V. N. Murzin, S. A. Stoklitskii, and A. P. Chebotarev, "Creation of solitary vortices in a flow around shallow spherical depressions," *Soviet Technical Physical Letters*, vol. 12, no. 11, pp. 547–549, 1986.

[32] P. R. Gromov, A. B. Zobnin, M. I. Rabinovich, and M. M. Sushchik, "Creation of solitary vortices in a flow around shallow spherical depressions," *Soviet Technical Physical Letters*, vol. 12, no. 11, pp. 1323–1328, 1986.

[33] V. S. Kesarev and A. P. Kozlov, "Convective heat transfer in turbulized flow past a hemispherical cavity," *Heat Transfer Research*, vol. 25, no. 2, pp. 156–160, 1993.

[34] V. N. Afanasyev, Y. P. Chudnovsky, A. I. Leontiev, and P. S. Roganov, "Turbulent flow friction and heat transfer characteristics for spherical cavities on a flat plate," *Experimental Thermal and Fluid Science*, vol. 7, no. 1, pp. 1–8, 1993.

[35] M. Y. Belen'kiy, M. A. Gotovskiy, B. M. Lekakh, B. S. Fokin, and K. S. Dolgushin, "Heat transfer augmentation using surfaces formed by a system of spherical cavities," *Heat Transfer Research*, vol. 25, no. 2, pp. 196–203, 1993.

[36] V. I. Terekhov, S. V. Kalinina, and Y. M. Mshvidobadze, "Flow structure and heat transfer on a surface with a unit hole depression," *Russian Journal of Engineering Thermophysics*, vol. 5, pp. 11–34, 1995.

[37] V. D. Zhak, "The taylor-goertler vortices and three-dimensional flow evolution in a cavity," *Russian Journal of Engineering Thermophysics*, vol. 5, pp. 165–176, 1995.

[38] A. V. Schukin, A. P. Kozlov, and R. S. Agachev, "Study and application of hemispheric cavities for surface heat transfer augmentation," ASME Paper 95-GT-59, 1995.

[39] P. W. Bearman and J. K. Harvey, "Control of circular cylinder flow by the use of dimples," *AIAA Journal*, vol. 31, no. 10, pp. 1753–1756, 1993.

[40] T. Kimura and M. Tsutahara, "Fluid dynamic effects of grooves on circular cylinder surface," *AIAA Journal*, vol. 29, no. 12, pp. 2062–2068, 1991.

[41] M. E. Kithcart and D. E. Klett, "Heat transfer and skin friction comparison of dimpled versus protrusion roughness," *Journal of Enhanced Heat Transfer*, vol. 3, no. 4, pp. 273–280, 1996.

[42] M. K. Chyu, Y. Yu, H. Ding, J. P. Downs, and F. O. Soechting, "Concavity enhanced heat transfer in an internal cooling passage," ASME Paper 97-GT-437, 1997.

[43] Y. L. Lin, T. I.-P. Shih, and M. K. Chyu, "Computations of flow and heat transfer in a channel with rows of hemispherical cavities," ASME Paper 99-GT-263, 1999.

[44] Y. F. Gortyshov, I. A. Popov, R. D. Amirkhanov, and K. E. Gulitsky, "Studies of hydrodynamics and heat exchange in channels with various types of intensifiers," in *Proceedings of 11th International Heat Transfer Congress (IHTC '98)*, vol. 6, pp. 83–88, 1998.

[45] H. K. Moon, T. O'Connell, and B. Glezer, "Channel height effect on heat transfer and friction in a dimpled passage," ASME Paper 99-GT-163, 1999.

[46] J. Chen, H. Müller-Steinhagen, and G. G. Duffy, "Heat transfer enhancement in dimpled tubes," *Applied Thermal Engineering*, vol. 21, no. 5, pp. 535–547, 2001.

[47] S. W. Moon and S. C. Lau, "Turbulent heat transfer measurements on a wall with concave and cylindrical dimples in a square channel," ASME Paper GT-2002-30208, 2002.

[48] G. I. Mahmood, M. L. Hill, D. L. Nelson, P. M. Ligrani, H. K. Moon, and B. Glezer, "Local heat transfer and flow structure on and above a dimpled surface in a channel," *Journal of Turbomachinery*, vol. 123, no. 1, pp. 115–123, 2001.

[49] G. I. Mahmood and P. M. Ligrani, "Heat transfer in a dimpled channel: combined influences of aspect ratio, temperature ratio, Reynolds number, and flow structure," *International Journal of Heat and Mass Transfer*, vol. 45, no. 10, pp. 2011–2020, 2002.

[50] M. Elfert, M. Voges, and J. Klinner, "Detailed flow investigation using PIV in a rotating square-sectioned two-pass cooling system with ribbed walls," ASME Paper GT2008-51183, 2008.

[51] N. K. Burgess and P. M. Ligrani, "Effects of dimple depth on channel nusselt numbers and friction factors," *Journal of Heat Transfer*, vol. 127, no. 8, pp. 839–847, 2005.

[52] P. M. Ligrani, J. L. Harrison, G. I. Mahmmod, and M. L. Hill, "Flow structure due to dimple depressions on a channel surface," *Physics of Fluids*, vol. 13, no. 11, pp. 3442–3451, 2001.

[53] G. I. Mahmood, M. Z. Sabbagh, and P. M. Ligrani, "Heat transfer in a channel with dimples and protrusions on opposite walls," *Journal of Thermophysics and Heat Transfer*, vol. 15, no. 3, pp. 275–283, 2001.

[54] P. M. Ligrani, G. I. Mahmood, J. L. Harrison, C. M. Clayton, and D. L. Nelson, "Flow structure and local Nusselt number variations in a channel with dimples and protrusions on opposite walls," *International Journal of Heat and Mass Transfer*, vol. 44, no. 23, pp. 4413–4425, 2001.

[55] S. A. Isaev, A. I. Leont'ev, and P. A. Baranov, "Identification of self-organized vortexlike structures in numerically simulated turbulent flow of a viscous incompressible liquid streaming around a well on a plane," *Technical Physics Letters*, vol. 26, no. 1, pp. 15–18, 2000.

[56] M. Al-Qahtani, H.-C. Chen, and J. C. Han, "A numerical study of flow and heat transfer in rotating rectangular channels (AR=4) with 45° Rib turbulators by reynolds stress turbulence model," ASME Paper GT-2002-30216, 2002.

[57] M. Al-Qahtani, Y. Jang, H. C. Chen, and J. C. Han, "Prediction of flow and heat transfer in rotating two-pass rectangular channels with 45° Rib turbulators," ASME Paper 2001-GT-187, 2001.

[58] G. S. Azad, M. J. Uddin, J. C. Han, H. K. Moon, and B. Glezer, "Heat transfer in a two-pass rectangular rotating channel with 45-deg angleb rib turbulators," *Journal of Turbomachinery*, vol. 124, no. 2, pp. 251–259, 2002, ASME Paper 2001-GT-186.

[59] B. Bonhoff, S. Parneix, J. Leusch, B. V. Johnson, J. Schabacker, and A. Bölcs, "Experimental and numerical study of developed flow and heat transfer in coolant channels with 45 degree ribs," *International Journal of Heat and Fluid Flow*, vol. 20, no. 3, pp. 311–319, 1999.

[60] B. Bonhoff, U. Tomm, B. V. Johnson, and I. Jennions, "Heat transfer predictions for rotating U-shaped coolant channels with skewed ribs and with smooth walls," ASME Paper 97-GT-162, 1997.

[61] R. S. Bunker, T. G. Wetzel, and D. L. Rigby, "Heat transfer in a complex trailing edge passage for a high pressure turbine blade—part 1: experimental measurements," ASME Paper GT-2002-30212, 2002.

[62] L. Casarsa, M. Cakan, and T. Arts, "Characterization of the velocity and heat transfer fields in an internal cooling channel with high blockage ratio," ASME Paper GT-2002-30207, 2002.

[63] H. H. Cho, S. Y. Lee, and S. J. Wu, "The combined effects of rib arrangements and discrete ribs on local heat/mass transfer in a square duct," ASME Paper 2001-GT-175, 2001.

[64] S. Dutta and J. C. Han, "Local heat transfer in rotating smooth and ribbed two-pass square channels with three channel orientations," *Journal of Heat Transfer*, vol. 118, no. 3, pp. 578–584, 1996.

[65] S. Dutta, J. C. Han, and C. P. Lee, "Local heat transfer in a rotating two-pass ribbed triangular duct with two model orientations," *International Journal of Heat and Mass Transfer*, vol. 39, no. 4, pp. 707–715, 1996.

[66] F. Shin, Y. Wen-Jei, and Z. Nengli, "Local heat transfer in a rotating serpentine passage with rib-roughened surfaces," *International Journal of Heat and Mass Transfer*, vol. 37, no. 2, pp. 217–228, 1994.

[67] T. S. Griffith, L. Al-Hadhrami, and J. C. Han, "Heat transfer in rotating rectangular cooling channels (AR=4) with angled ribs," *Journal of Heat Transfer*, vol. 124, no. 4, pp. 617–625, 2002.

[68] J. C. Han, L. R. Glicksman, and W. M. Rohsenow, "An investigation of heat transfer and friction for rib-roughened surfaces," *International Journal of Heat and Mass Transfer*, vol. 21, no. 8, pp. 1143–1156, 1978.

[69] J. C. Han, J. J. Huang, and C. P. Lee, "Augmented heat transfer in square channels with wedge-shaped and delta-shaped turbulence promoters," *Journal of Enhanced Heat Transfer*, vol. 1, no. 1, pp. 37–52, 1993.

[70] J. C. Han and J. S. Park, "Developing heat transfer in rectangular channels with rib turbulators," *International Journal of Heat and Mass Transfer*, vol. 31, no. 1, pp. 183–195, 1988.

[71] J. C. Han, Y. M. Zhang, and C. P. Lee, "Augmented heat transfer in square channels with parallel, crossed, and V-shaped angled ribs," *Journal of Heat Transfer*, vol. 113, no. 3, pp. 590–596, 1991.

[72] H. Iacovides, "Computation of flow and heat transfer through rotating ribbed passages," *International Journal of Heat and Fluid Flow*, vol. 19, no. 5, pp. 393–400, 1998.

[73] R. Jia, A. Saidi, and B. Sunden, "Heat transfer enhancement in square ducts with V-shaped Ribs of various angles," ASME Paper GT-2002-30209, 2002.

[74] B. V. Johnson, J. H. Wagner, G. D. Steuber, and F. C. Yeh, "Heat transfer in rotating serpentine passages with trips skewed to the flow," *Journal of Turbomachinery*, vol. 116, no. 1, pp. 113–123, 1994, ASME Paper 92-GT-191.

[75] B. V. Johnson, J. H. Wagner, G. D. Steuber, and F. C. Yeh, "Heat transfer in rotating serpentine passages with selected model orientations for smooth or skewed trip walls," *Journal of Turbomachinery*, vol. 116, no. 4, pp. 738–744, 1994, ASME Paper 93-GT-305.

[76] Y. L. Lin, T. I. P. Shih, M. A. Stephens, and M. K. Chyu, "A numerical study of flow and heat transfer in a smooth and ribbed u-duct with and without rotation," *Journal of Heat Transfer*, vol. 123, no. 2, pp. 219–232, 2001.

[77] T. M. Liou, M. Y. Chen, and Y. M. Wang, "Heat transfer, fluid flow, and pressure measurements inside a rotating two-pass duct with detached 90-deg ribs," *Journal of Turbomachinery*, vol. 125, no. 3, pp. 565–574, 2003, ASME Paper GT-2002-30201.

[78] G. I. Mahmood, P. M. Ligrani, and S. Y. Won, "Spatially-resolved heat transfer and flow structure in a rectangular channel with 45° angled rib turbulators," ASME Paper GT-2002-30215, 2002.

[79] C. W. Park, S. C. Lau, and R. T. Kukreja, "Effect of rib size on heat (Mass) transfer distribution in a rotating channel," ASME Paper 97-AA-126, 1997.

[80] C. W. Park, S. C. Lau, and R. T. Kukreja, "Heat (mass) transfer in a rotating channel with ribs of various sizes on two walls," *Journal of Thermophysics and Heat Transfer*, vol. 12, no. 3, pp. 452–454, 1998.

[81] J. A. Parsons, J. C. Han, and Y. M. Zhang, "Wall heating effect on local heat transfer in a rotating two-pass square channel with 90° rib turbulators," *International Journal of Heat and Mass Transfer*, vol. 37, no. 9, pp. 1411–1420, 1994.

[82] J. A. Parsons, J. C. Han, and Y. Zhang, "Effect of model orientation and wall heating condition on local heat transfer in a rotating two-pass square channel with rib turbulators," *International Journal of Heat and Mass Transfer*, vol. 38, no. 7, pp. 1151–1159, 1995.

[83] S. V. Prabhu and R. P. Vedula, "Pressure drop distribution in a rotating rectangular channel with one ribbed surface," ASME Paper 97-AA-118, 1997.

[84] D. L. Rigby, "Prediction of heat and mass transfer in a rotating ribbed coolant passage with a 180 degree turn," ASME Paper 98-GT-329, 1998.

[85] D. L. Rigby and R. S. Bunker, "Heat transfer in a complex trailing edge passage for a high pressure turbine blade—part 2: simulation results," ASME Paper GT-2002-30213, 2002.

[86] D. L. Rigby, E. Steinthorsson, and A. A. Ameri, "Numerical prediction of heat transfer in a channel with ribs and bleed," ASME Paper 97-GT-431, 1997.

[87] J. Schabacker, A. Bolcs, and B. V. Johnson, "PIV investigation of the flow characteristics in an internal coolant passage with 45 rib arrangement," ASME Paper 99-GT-120, 1999.

[88] T. I.-P. Shih, Y.-L. Lin, M. A. Stephens, and M. K. Chyu, "Flow and heat transfer in a ribbed U-duct under typical engine conditions," ASME Paper 98-GT-213, 1998.

[89] M. A. Stephens, M. K. Chyu, and T. I.-P. Shih, "Computations of convective heat transfer in a square duct with inclined ribs of rounded cross section," ASME Paper 96-WA/HT-12, 1996.

[90] M. A. Stephens and T. I.-P. Shih, "Computation of compressible flow and heat transfer in a rotating duct with inclined ribs and 180 degree bend," ASME Paper 97-GT-192, 1997.

[91] M. E. Taslim, L. A. Bondi, and D. M. Kercher, "Experimental investigation of heat transfer in an orthogonally rotating channel roughened with 45 deg criss-cross ribs on two opposite walls," *Journal of Turbomachinery*, vol. 113, no. 3, pp. 346–353, 1991, ASME Paper 90-GT-332.

[92] M. E. Taslim, T. Li, and D. M. Kercher, "Experimental heat transfer and friction in channels roughened with angled, V-shaped, and discrete ribs on two opposite walls," *Journal of Turbomachinery*, vol. 118, no. 1, pp. 20–28, 1996.

[93] M. E. Taslim, T. Li, and S. D. Spring, "Measurements of heat transfer coefficients and friction factors in passages rib-roughened on all walls," *Journal of Turbomachinery*, vol. 120, no. 3, pp. 564–570, 1998.

[94] M. E. Taslim and A. Lengkong, "45° staggered rib heat transfer coefficient measurements in a square channel," *Journal of Turbomachinery*, vol. 120, no. 3, pp. 571–580, 1998.

[95] M. E. Taslim, A. Rahman, and S. D. Spring, "Experimental investigation of heat transfer coefficients in a spanwise rotating channel with two opposite rib-roughened walls," *Journal of Turbomachinery*, vol. 113, no. 1, pp. 75–82, 1991, ASME Paper 89-GT-150.

[96] M. E. Taslim and C. M. Wadsworth, "An experimental investigation of the rib surface-averaged heat transfer coefficient in a rib-roughened square passage," *Journal of Turbomachinery*, vol. 119, no. 2, pp. 381–389, 1997.

[97] D. Thurman and P. Poinsatte, "Experimental heat transfer and bulk air temperature measurements for a multipass internal cooling model with ribs and bleed," ASME Paper 2000-GT-233, 2000.

[98] D. G. N. Tse and G. D. Steubner, "Flow in a rotating square serpentine coolant passage with skewed trips," ASME Paper 97-GT-529, 1997.

[99] K. Watanabe and T. Takahashi, "LES simulation and experimental measurement of fully developed ribbed channel flow and heat transfer," ASME Paper GT-2002-30203, 2002.

[100] Y. M. Zhang, J. C. Han, J. A. Parsons, and C. P. Lee, "Surface heating effect on local heat transfer in a rotating two-pass square channel with 60 deg angled rib turbulators," *Journal of Turbomachinery*, vol. 117, no. 2, pp. 272–280, 1995, ASME Paper 93-GT-336.

[101] C. C. Chen and T. M. Liou, "Rotating effect on fluid flow in a smooth duct with a 180-deg sharp turn," ASME Paper 2000-GT-0228, 2000.

[102] F. Zhou and S. Acharya, "Heat transfer at high rotation numbers in a two-pass 4:1 aspect ratio rectangular channel with 45 deg skewed ribs," *Journal of Turbomachinery*, vol. 130, no. 2, Article ID 021019, 2008.

[103] J. H. Wagner, B. V. Johnson, and T. J. Hajek, "Heat transfer in rotating passages with smooth walls and radial outward flow," *Journal of Turbomachinery*, vol. 113, no. 1, pp. 42–51, 1991, ASME Paper 89-GT-272.

[104] M. Huh, J. Lei, Y. H. Liu, and J. C. Han, "High rotation number effects on heat transfer in a rectangular (AR=2:1) two pass channel," ASME Paper GT2009-59421, 2009.

[105] A. Ogawa, *Vortex Flow*, CRC Press, New York, NY, USA, 1993.

[106] B. Glezer, H.-K. Moon, and T. O'Connell, "A novel technoique for the internal blade cooling," ASME Paper 96-GT-181, 1996.

[107] B. Glezer, H.-K. Moon, J. Kerrebrock, J. Bons, and G. Guenette, "Heat transfer in a rotating radial channel with swirling internal flow," ASME Paper 98-GT-214, 1998.

[108] A. E. Bergles, "Survey and evaluation of techniques to augment convective heat and mass transfer," *International Journal of Heat and Mass Transfer*, vol. 1, pp. 331–413, 1969.

[109] H. H. Bruun, A. Fitouri, and M. K. Khan, "Use of a multiposition single yawed hot-wire probe for measurements in swirling flow," in *Proceedings of the Fluids Engineering Conference*, vol. 167, pp. 57–65, June 1993.

[110] F. Chang and V. K. Dhir, "Turbulent flow field in tangentially injected swirl flows in tubes," *International Journal of Heat and Fluid Flow*, vol. 15, no. 5, pp. 346–356, 1994.

[111] A. W. Date, "Prediction of fully-developed flow in a tube containing a twisted-tape," *International Journal of Heat and Mass Transfer*, vol. 17, no. 8, pp. 845–859, 1974.

[112] M. Dong and D. G. Lilley, "Parameter effects on flow patterns in confined turbulent swirling flows," in *Proceedings of the International Joint Power Generation Conference*, vol. 17, October 1993.

[113] A. Fitouri, M. K. Khan, and H. H. Bruun, "A multiposition hot-wire technique for the study of swirling flows in vortex chambers," *Experimental Thermal and Fluid Science*, vol. 10, no. 1, pp. 142–151, 1995.

[114] W. R. Gambill and R. D. Bundy, "An evaluation of the present status of swirl-flow heat transfer," ASME Paper 62-HT-42, 1962.

[115] C. R. Hedlund and P. M. Ligrani, "Local swirl chamber heat transfer and flow structure at different Reynolds numbers," *Journal of Turbomachinery*, vol. 122, no. 2, pp. 375–385, 2000.

[116] C. R. Hedlund, P. M. Ligrani, H. K. Moon, and B. Glezer, "Heat transfer and flow phenomena in a swirl chamber simulating turbine blade internal cooling," *Journal of Turbomachinery*, vol. 121, no. 4, pp. 804–813, 1999.

[117] S. W. Hong and A. E. Bergles, "Augmentation of laminar flow heat transfer in tubes by means of twisted-tape inserts," *Journal of Heat Transfer*, vol. 98, no. 2, pp. 251–256, 1976.

[118] J. J. Hwang and C. S. Cheng, "Augmented heat transfer in a triangular duct by using multiple swirling jets," *Journal of Heat Transfer*, vol. 121, no. 3, pp. 683–690, 1999.

[119] J. B. W. Kok, F. J. J. Rosendal, and J. J. H. Brouwers, "LDA-measurements on swirling flows in tubes," in *Proceedings of the 5th International Conference on Laser Anemometry*, vol. 2052 of *Proceedings of SPIE*, pp. 721–728, August 1993.

[120] F. Kreith and D. Margolis, "Heat transfer and friction in turbulent vortex flow," *Applied Scientific Research*, vol. 8, no. 1, pp. 457–473, 1959.

[121] R. Kumar and T. Conover, "Flow visualization studies of a swirling flow in a cylinder," *Experimental Thermal and Fluid Science*, vol. 7, no. 3, pp. 254–262, 1993.

[122] H. Li and Y. Tomita, "Characteristics of swirling flow in a circular pipe," *Journal of Fluids Engineering*, vol. 116, no. 2, pp. 370–373, 1994.

[123] P. M. Ligrani, C. R. Hedlund, B. T. Babinchak, R. Thambu, H. K. Moon, and B. Glezer, "Flow phenomena in swirl chambers," *Experiments in Fluids*, vol. 24, no. 3, pp. 254–264, 1998.

[124] H.-K. Moon, T. O'Connell, and B. Glezer, "Heat transfer enhancement in a circular channel using lengthwise continuous tangential injection," in *Proceedings of the International Heat Transfer Congress (IHTC '98)*, Seoul, South Korea, 1998.

[125] P. Papadopoulos, D. M. France, and W. J. Minkowycz, "Heat transfer to dispersed swirl flow of high-pressure water with low wall superheat," *Experimental Heat Transfer*, vol. 4, no. 2, pp. 153–169, 1991.

[126] R. Razgaitis and J. P. Holman, "A survey of heat transfer in confined swirling flows," in *Future Energy Production Systems, Heat and Mass Transfer Processes*, vol. 2, pp. 831–866, Academic Press, New York, NY, USA, 1976.

[127] W. F. J. Sampers, A. P. G. G. Lamers, and A. A. van Steenhoven, "Experimental and numerical analysis of a turbulent swirling flow in a tube," in *Proceedings of the 3rd UK National Conference incorporating 1st European Conference on Thermal Sciences*, vol. 2 of *ICHEME Symposium Series no. 129*, pp. 765–771, September 1992.

[128] W. L. Fu, L. M. Wright, and J. C. Han, "Rotational buoyancy effects on heat transfer in five different aspect-ratio rectangular channels with smooth walls and 45 degree ribbed walls," *Journal of Heat Transfer*, vol. 128, no. 11, pp. 1130–1141, 2006, ASME Paper GT 2005-68493.

[129] B. J. Belnap, J. A. Van Rij, and P. M. Ligrani, "A Reynolds analogy for real component surface roughness," *International Journal of Heat and Mass Transfer*, vol. 45, no. 15, pp. 3089–3099, 2002.

[130] J. A. Van Rij, B. J. Belnap, and P. M. Ligrani, "Analysis and experiments on three-dimensional, irregular surface roughness," *Journal of Fluids Engineering, Transactions of the ASME*, vol. 124, no. 3, pp. 671–677, 2002.

[131] D. L. Gee and R. L. Webb, "Forced convection heat transfer in helically Rib-roughened tubes," *International Journal of Heat and Mass Transfer*, vol. 23, no. 8, pp. 1127–1136, 1980.

[132] X. Chi, T. I.-P. Shihb, K. M. Bryden et al., "Effects of pin-fin height on flow and heat transfer in a rectangular duct," ASME Paper GT2011-46014, 2011.

[133] J. K. Ostanek and K. A. Thole, "Flow field measurements in a single row of low aspect ratio pin-fins," ASME Paper GT2011-45757, 2011.

[134] J. K. Ostanek and K. A. Thole, "Effects of varying steamwise and spanwise spacing in pin-fin arrays," ASME Paper GT2012-68127, 2012.

[135] X. Wang, X. Bai, J. Wu, R. Liu, D. Zhu, and G. Liao, "Numerical study on heat transfer characteristics in rectangular ducts with pin-fins," ASME Paper GT2011-45286, 2011.

[136] Y. Rao, C. Wan, and S. Zang, "Comparisons of flow friction and heat transfer performance in rectangular channels with pin fin-dimple arrays," ASME Paper GT2010-22442, 2010.

[137] C. Bianchini, B. Facchini, S. Zacchi, F. Simonetti, and L. Tachi, "Numerical and experimental investigation of turning flow effects of innovative pin fin arrangements for trailing edge cooling configurations," ASME Paper GT2010-23536, 2010.

[138] M. L. Busche, L. P. Moualeu, N. Chowdhury, C. Tang, and F. E. Ames, "Heat transfer and pressure drop measurements in high solidity pin fin cooling arrays with incremental replenishment," ASME Paper GT2012-69289, 2012.

[139] J. B. Cowan, D. K. Tafti, and A. Kohli, "Investigation of sand particle deposition and erosion characteristics within a short pin fin array," ASME Paper GT2010-22362, 2010.

[140] S. C. Siw, M. K. Chyu, T. I.-P. Shih, and M. A. Alvin, "Effects of pin detached space on heat transfer and from pin fin arrays," ASME Paper GT2010-23227, 2010.

[141] L. V. Tran, M. I. Valentino, A. Saha et al., "PIV study on the dimple mid-plane of a narrow rectangular channel with dimples applied to one wall," ASME Paper GT2011-46132, 2011.

[142] C. N. Jordan and L. M. Wright, "Heat transfer enhancement in a rectangular (AR=3:1) channel with V-shaped dimples," ASME Paper GT2011-46128, 2011.

[143] R. Amano, K. Guntur, J. M. Lucci, and Y. Ashitaka, "Study of flow through a stationary ribbed channel for blade cooling," ASME Paper GT2010-23031, 2010.

[144] J. E. Dees, G. A. Ledezma, A. K. Tolpadi, G. M. Lashowski, and D. G. Bogard, "Experimental measurements and computational predictions for an internally cooled simulated turbine vane with 90 degree Rib turbulators," ASME Paper GT2010-23004, 2010.

[145] T. Hagari, K. Ishida, K.-I. Takeishi, Y. Oda, and H. Kitada, "Investigation on heat transfer characteristics of a cooling channel With dense array of angled Rib turbulator," ASME Paper GT2012-68949, 2012.

[146] C. LeBlanc, S. V. Ekkad, T. Lambert, and V. Rajendran, "Detailed heat transfer distributions in engine similar cooling channels for a turbine rotor blade with different Rib orientations," ASME Paper GT2011-45254, 2011.

[147] G. Ledezma and R. Bunker, "The optimal distribution of chordwise Rib Fin arrays for blade Tip Cap underside cooling," ASME Paper GT2012-68432, 2012.

[148] J. Lei, M. Huh, and J.-C. Han, "Effect of rib spacing on heat transfer in a two pass rectangular channel (AR=2:1) at high rotation numbers," ASME Paper GT2011-45926, 2011.

[149] J. Lei, S.-J. Li, J.-C. Han, L. Zhang, and H.-K. Moon, "Heat transfer in rotating multi-pass rectangular ribbed channel with and without a turning vane," ASME Paper GT2012-69139, 2012.

[150] S. Patil and D. Tafti, "Large-eddy simulation with zonal near wall treatment of flow and heat transfer in a ribbed duct for the internal cooling of turbine blades," ASME Paper GT2011-45749, 2011.

[151] B. V. N. Ramakumar, V. R. Kollati, S. Goswami, V. Kulkarni, J. S. Liu, and D. C. Crites, "Numerical validation of heat transfer augmentation factor in serpentine passages with ribbed walls," ASME Paper GT2010-22636, 2010.

[152] K. Saha and S. Acharya, "Effect of entrance geometry on heat transfer in a narrow (AR=1:4) rectangular two pass channel with smooth and ribbed walls," ASME Paper GT2011-46076, 2011.

[153] T. Salameh and B. Sunden, "Effects of ribs on internal blade-tip cooling," ASME Paper GT2011-45118, 2011.

[154] H. Zhang, J. Wang, X. Wu, and H. Lu, "A simplified approach to design transverse ribs which array alternately in rectangular channel," *ASME Paper*, no. GT2010-22562, 2010.

[155] T. Salameh and B. Sunden, "Comparison of continuous and truncated ribs on internal blade tip cooling," ASME Paper GT2012-68028, 2012.

[156] G. Xie, S. Li, W. Zhang, and B. Sundén, "Computational analysis of side-wall heat transfer of a turbine blade internal cooling passage with truncated ribs on opposite walls," ASME Paper GT2012-68073, 2012.

[157] J. Wang, H. Liu, M. Mao, X. Li, and Z. Zhang, "Numerical investigations of fluid flow and heat transfer performances of semiattached rib channel design," ASME Paper GT2010-22563, 2010.

[158] D. M. Lee, J. S. Park, D. H. Lee, B. S. Kim, and H. H. Cho, "Heat transfer characteristics of a non-rotating two-pass rectangular duct with various guide vanes in the tip turn region," ASME Paper GT2011-45962, 2011.

[159] B. Cukurel, C. Selcan, and T. Arts, "Film cooling extraction effects on the aero-thermal characteristics of rib roughened cooling channel flow," ASME Paper GT2012-68680, 2012.

[160] T. Guo, H. Zhu, and D. Xu, "Heat transfer investigation of the channels with Rib turbulators and double-row bleed holes," ASME Paper GT2011-46319, 2011.

[161] Y. Xing and B. Weigand, "Experimental investigation on staggered impingement heat transfer on a rib roughened plate with different crossflow schemes," ASME Paper GT2010-22043, 2010.

[162] F. Coletti, M. Scialanga, and T. Arts, "Experiemental investigation of conjugate heat transfer in a rib-roughened trailing edge channel with crossing-jets," ASME Paper GT2010-22432, 2010.

[163] F. Hoefler, S. Schueren, J. von Wolfersdorf, and S. Naik, "Heat transfer characteristics of an oblique jet impingement configuration in a passage with ribbed surfaces," ASME Paper GT2010-22288, 2010.

[164] Q. Yang, Y. Lin, C. Zhang, C.-J. Sung, and Q. Xu, "Cooling effectiveness of impingement/effusion cooling with and turbulence promoter ribs," ASME Paper GT2012-69209, 2012.

[165] J. Liu, J. Gao, and T. Gao, "An experimental investigation of heat transfer characteristics in a steam-cooled square channel with rib turbulators," ASME Paper GT2011-46134, 2011.

[166] L. Shui, J. Gao, L. Xu, and X. Wang, "Numerical investigation of heat transfer and flow characteristics in a steam-cooled square ribbed duct," ASME Paper GT2010-22407, 2010.

[167] T. S. Dhanasekaran and T. Wang, "Mist/air cooling in a two-pass rectangular rotating channel with 45-deg angled Rib turbulators," ASME Paper GT2011-45954, 2011.

[168] R. S. Amano, K. S. Guntur, J. Martinez-Lucci, and S. Kumar, "Numerical comparison of heat transfer and pressure drop in gas turbine blade cooling channels with dimples and rib-turbulators," ASME Paper GT2011-45277, 2011.

[169] L. Qui, H. Deng, and Z. Tao, "Effect of channel orientation in a rotating wedge-shaped cooling channel with pin fins and ribs," ASME Paper GT2012-68439, 2012.

[170] J. Lan, Y. Xie, and D. Zhang, "Heat transfer enhancement in a rectangular channel with the combination of ribs, dimples and protrusions," ASME Paper GT2011-46031, 2011.

[171] Z. Feng, Z. Dou, J. Wang, S. Ma, and Z. Zhang, "Numerical investigations of cooling enhancement with internal ribs and external coolant film," ASME Paper GT2012-68682, 2012.

[172] L. Wang, B. Sunden, A. Borg, and H. Abrahamsson, "Heat transfer in a channel under the effects of a shallow-angle jet impingement and a rib," ASME Paper GT2011-46333, 2011.

[173] S. Acharya, F. Zhou, J. Lagrone, G. Mahmood, and R. S. Bunker, "Latticework (Vortex) cooling effectiveness: rotating channel experiments," *Journal of Turbomachinery*, vol. 127, no. 3, pp. 471–478, 2005, ASME Paper 2004-GT-53983.

[174] K. Takeishi, Y. Oda, Y. Miyake, and Y. Motoda, "Experimental and numerical study on the convective heat transfer and pressure loss in rectangular ducts with inclined pin-fins on a wavy endwall," ASME Paper 2012-69625, 2012.

[175] Y. Motoda, K. Takeishi, Y. Oda, and Y. Miyake, "A study of the effect of angled pin fins on endwall heat transfer," in *Proceedings of the International Conference on Power Engineering (ICOPE '09)*, Kobe, Japan, 2009.

[176] R. S. Bunker, J. C. Bailey, C.-P. Lee, and C. W. Stevens, "In-wall netweork (Mesh) cooling augmentation of gas turbine airfoils," ASME Paper 2004-54260, 2004.

[177] Y. Servouze and J. C. Sturgis, "Heat transfer and flow field measurements in a rib-roughened branch of a rotating two-pass duct," ASME Paper GT2003-38048, 2003.

[178] J. P. Bons and J. L. Kerrebrock, "Complementary velocity and heat transfer measurements in a rotating cooling passage with smooth walls," *Journal of Turbomachinery*, vol. 121, no. 4, pp. 651–662, 1999, ASME Paper 98-GT-464.

[179] L. M. Wright, W. L. Fu, and J. C. Han, "Influence of entrance geometry on heat transfer in rotating rectangular cooling channels (AR=4:1) with angled ribs," *Journal of Heat Transfer*, vol. 127, no. 4, pp. 378–387, 2005.

[180] S. Acharya, D. E. Nikitopoulos, and P. Agarwal, "Heat/mass transfer in a 4:1 AR smooth and ribbed coolant passage with rotation in 90-degree and 45-degree orientations," ASME Paper GT2004-53928, 2004.

[181] D. H. Lee, D. H. Rhee, and H. H. Cho, "Heat transfer measurements in a rotating equilateral triangular channel with various rib arrangements," ASME Paper GT2006-90973, 2006.

[182] Y. H. Liu, L. M. Wright, W. L. Fu, and J. C. Han, "Rib spacing effect on heat transfer and pressure loss in a rotating two-pass rectangular channel (AR=1:2) with 45-degree angled ribs," *AIAA Journal of Thermophysics and Heat Transfer*, vol. 21, no. 3, pp. 582–595, 2006, ASME Paper GT2006-90368.

[183] L. Al-Hadhrami, T. Griffith, and J. C. Han, "Heat transfer in two-pass rotating rectangular channels (AR=2) with five different orientations of 45 deg v-shaped rib turbulators," *Journal of Heat Transfer*, vol. 125, no. 2, pp. 232–242, 2003, AIAA Paper 2002-0789.

[184] L. Al-Hadhrami and J. C. Han, "Effect of rotation on heat transfer in two-pass square channels with five different orientations of 45° angled rib turbulators," *International Journal of Heat and Mass Transfer*, vol. 46, no. 4, pp. 653–669, 2003.

[185] H. D. Chiang and H. L. Li, "Jet impingement and forced convection cooling experimental study in rotating turbine blades," ASME Paper GT2009-59795, 2009.

[186] W. L. Fu, L. M. Wright, and J. C. Han, "Heat transfer in two-pass rotating rectangular channels (AR = 1:2 and AR = 1:4) with 45° angled Rib turbulators," *Journal of Turbomachinery*, vol. 127, no. 1, pp. 164–174, 2004, ASME Paper GT2004-53261.

[187] W. L. Fu, L. M. Wright, and J. C. Han, "Heat transfer in two-pass rotating rectangular channels (AR=1:2 and AR=1:4) with smooth walls," *Journal of Heat Transfer*, vol. 127, no. 3, pp. 265–277, 2005.

[188] T. J. Hajek and A. W. Higgins, "Coolant passage heat transfer with rotation," NASA N88-11160, Lewis Research Center, Turbine Engine Hot Section Technology, 1985.

[189] T. J. Hajek, J. H. Wagner, and B. V. Johnson, "Coolant passage heat transfer with rotation," NASA N89-17314, Lewis Research Center, Turbine Engine Hot Section Technology, 1986.

[190] J. C. Han, Y. M. Zhang, and K. Kalkuehler, "Uneven wall temperature effect on local heat transfer in a rotating two-pass square channel with smooth walls," *Journal of Heat Transfer*, vol. 115, no. 4, pp. 912–920, 1993.

[191] S. S. Hsieh and W. J. Liu, "Uneven wall heat flux effect on local heat transfer in rotating two-pass channels with two opposite ribbed walls," *Journal of Heat Transfer*, vol. 118, no. 4, pp. 864–876, 1996.

[192] S. S. Hsieh, Y. S. Wang, and M. H. Chiang, "Local heat transfer and velocity measurements in a rotating ribbed two-pass square channel with uneven wall temperatures," *Journal of Heat Transfer*, vol. 119, no. 4, pp. 843–848, 1997.

[193] M. Huh, Y. H. Liu, J. C. Han, and S. Chopra, "Effect of rib spacing on heat transfer in a two-pass rectangular channel (AR=1:4) with a sharp entrance at high rotation numbers," ASME Paper GT2008-50311, 2008.

[194] G. J. Hwang, S. C. Tzeng, C. P. Mao, and C. Y. Soong, "Heat transfer in a radially rotating four-pass serpentine channel with staggered half-V rib turbulators," *Journal of Heat Transfer*, vol. 123, no. 1, pp. 39–50, 2001.

[195] H. Ito and K. Nanbu, "Flow in rotating straight pipes of circular cross section," *Journal of Basic Engineering*, vol. 93, no. 3, pp. 383–394, 1971.

[196] T. M. Liou and C. C. Chen, "Heat transfer in a rotating two-pass smooth passage with a 180° rectangular turn," *International Journal of Heat and Mass Transfer*, vol. 42, no. 2, pp. 231–247, 1998.

[197] T. M. Liou, C. C. Chen, and M. Y. Chen, "TLCT and LDV measurements of heat transfer and fluid flow in a rotating sharp turning duct," *International Journal of Heat and Mass Transfer*, vol. 44, no. 9, pp. 1777–1787, 2001.

[198] T. M. Liou, M. Y. Chen, and M. H. Tsai, "Fluid flow and heat transfer in a rotating two-pass square duct with in-line 90-deg ribs," *Journal of Turbomachinery*, vol. 124, no. 2, pp. 260–268, 2002, ASME Paper 2001-GT-185.

[199] T. M. Liou, Y. S. Hwang, and M. Y. Chen, "Heat transfer improvement by arranging detached ribs on suction surfaces of rotating internal coolant passages," *International Journal of Heat and Mass Transfer*, vol. 50, no. 11-12, pp. 2414–2424, 2007.

[200] Y. H. Liu, M. Huh, J. C. Han, and S. Chopra, "Heat transfer in a two-pass rectangular channel (AR=1:4) under high rotation numbers," *Journal of Heat Transfer*, vol. 130, no. 8, Article ID 081701, 9 pages, 2008, ASME Paper GT2007-27067.

[201] S. Mochizuki, J. Takamura, S. Yamawaki, and W. J. Yang, "Heat transfer in serpentine flow passages with rotation," *Journal of Turbomachinery*, vol. 116, no. 1, pp. 133–140, 1994, ASME Paper 92-GT-190.

[202] J. H. Wagner, B. V. Johnson, R. A. Graziani, and F. C. Yeh, "Heat transfer in rotating serpentine passages with trips normal to the flow," NASA TM 103758, 1991.

[203] J. H. Wagner, B. V. Johnson, R. A. Graziani, and F. C. Yeh, "Heat transfer in rotating serpentine passages with trips normal to the flow," *Journal of Turbomachinery*, vol. 114, no. 4, pp. 847–857, 1991.

[204] J. H. Wagner, B. V. Johnson, and F. C. Kopper, "Heat transfer in rotating serpentine passages with smooth walls," *Journal of Turbomachinery*, vol. 113, no. 3, pp. 321–330, 1991, ASME Paper 90-GT-331.

[205] W. J. Yang, N. Zhang, and J. Chiou, "Local heat transfer in a rotating serpentine flow passage," *Journal of Heat Transfer*, vol. 114, no. 2, pp. 354–361, 1992.

[206] F. Zhou, J. Lagrone, and S. Acharya, "Internal cooling in 4:1 AR passages at high rotation numbers," *Journal of Heat Transfer*, vol. 129, no. 12, pp. 1666–1675, 2007.

[207] G. J. Hwang, S. C. Tzeng, and C. P. Mao, "Heat transfer of compressed air flow in a spanwise rotating four-pass serpentine channel," *Journal of Heat Transfer*, vol. 121, no. 3, pp. 583–591, 1999.

[208] T. S. Griffith, L. Al-Hadhrami, and J. C. Han, "Heat transfer in rotating rectangular cooling channels (AR=4) with angled ribs," *Journal of Heat Transfer*, vol. 124, no. 4, pp. 617–625, 2002.

[209] T. S. Griffith, L. Al-Hadhrami, and J. C. Han, "Heat transfer in rotating rectangular cooling channels (AR = 4) with dimples," *Journal of Turbomachinery*, vol. 125, no. 3, pp. 555–563, 2003, ASME Paper 2002-GT-30220.

[210] J. C. Han and Y. M. Zhang, "Effect of uneven wall temperature on local heat transfer in a rotating square channel with smooth walls and radial outward flow," *Journal of Heat Transfer*, vol. 114, no. 4, pp. 850–858, 1992.

[211] J. C. Han, Y. M. Zhang, and C. P. Lee, "Influence of surface heating condition on local heat transfer in a rotating square channel with smooth walls and radial outward flow," *Journal of Turbomachinery*, vol. 116, no. 1, pp. 149–158, 1994, ASME Paper 92-GT-188.

[212] E. Lee, L. M. Wright, and J. C. Han, "Heat transfer in rotating rectangular channels (AR=4:1) with V-shaped and angled rib turbulators with and without gaps," ASME Paper GT2003-38900, 2003.

[213] L. M. Wright, E. Lee, and J. C. Han, "Effect of rotation on heat transfer in rectangular channels with pin fins," *Journal of Thermophysics and Heat Transfer*, vol. 18, no. 2, pp. 263–272, 2004.

[214] N. Zhang, J. Chiou, S. Fann, and W. J. Yang, "Local heat transfer distribution in a rotating serpentine rib-roughened flow passage," *Journal of Heat Transfer*, vol. 115, no. 3, pp. 560–567, 1993.

[215] G. J. Hwang and C. R. Kuo, "Experimental studies and correlations of convective heat transfer in a radially rotating serpentine passage," *Journal of Heat Transfer*, vol. 119, no. 3, pp. 460–466, 1997.

[216] K. M. Kim, Y. Y. Kim, D. H. Lee, D. H. Rhee, and H. H. Cho, "Influence of duct aspect ratio on heat/mass transfer in coolant passages with rotation," *International Journal of Heat and Fluid Flow*, vol. 28, no. 3, pp. 357–373, 2007.

[217] C. R. Kuo and G. J. Hwang, "Experimental studies and correlations of radially outward and inward air-flow heat transfer in a rotating square duct," *Journal of Heat Transfer*, vol. 118, no. 1, pp. 23–30, 1996.

[218] C. R. Kuo and G. J. Hwang, "Aspect ratio effect on convective heat transfer of radially outward flow in rotating rectangular ducts," *International Journal of Rotating Machinery*, vol. 1, no. 1, pp. 1–18, 1994.

[219] L. M. Wright, W. L. Fu, and J. C. Han, "Thermal performance of angled, V-shaped, and W-shaped rib turbulators in rotating rectangular cooling channels (AR=4:1)," *Journal of Turbomachinery*, vol. 126, no. 4, pp. 604–614, 2004, ASME Paper 2004-GT-54073.

[220] S. W. Chang and W. D. Morris, "Heat transfer in a radially rotating square duct fitted with in-line transverse ribs," *International Journal of Thermal Sciences*, vol. 42, no. 3, pp. 267–282, 2003.

[221] S. W. Chang, T. L. Yang, T. M. Liou, and G. F. Hong, "Heat transfer of rotating rectangular duct with compound scaled roughness and V-ribs at high rotation numbers," *International Journal of Thermal Sciences*, vol. 48, no. 1, pp. 174–187, 2009.

[222] S. W. Chang, T. L. Yang, T. M. Liou, and H. G. Fang, "Heat transfer in rotating scale-roughened trapezoidal duct at high rotation numbers," *Applied Thermal Engineering*, vol. 29, no. 8-9, pp. 1682–1693, 2009.

[223] S. Dutta, J. C. Han, Y. Zhang, and C. P. Lee, "Local heat transfer in a rotating two-pass triangular duct with smooth walls," *Journal of Turbomachinery*, vol. 118, no. 3, pp. 435–443, 1996, ASME Paper 94-GT-337.

[224] S. P. Harasgama and W. D. Morris, "The influence of rotation on the heat transfer characteristics of circular, triangular, and square-sectioned coolant passages of gas turbine rotor blades," *Journal of Turbomachinery*, vol. 110, no. 1, pp. 44–50, 1988.

[225] S. S. Hsieh and Y. J. Hong, "Heat transfer coefficients in an orthogonally rotating duct with turbulators," *Journal of Heat Transfer*, vol. 117, no. 1, pp. 69–78, 1995.

[226] W. D. Morris and T. Ayhan, "Observations on the influence of rotation on heat transfer in the coolant channels of gas turbine rotor blades," *Proceedings of the Institution of Mechanical Engineers*, vol. 193, pp. 303–311, 1979.

[227] W. D. Morris and S. W. Chang, "An experimental study of heat transfer in a simulated turbine blade cooling passage," *International Journal of Heat and Mass Transfer*, vol. 40, no. 15, pp. 3703–3716, 1997.

[228] W. D. Morris and R. Salemi, "An attempt to uncouple the effect of Coriolis and buoyancy forces experimentally on heat transfer in smooth circular tubes that rotate in the orthogonal mode," *Journal of Turbomachinery*, vol. 114, no. 4, pp. 858–864, 1992, ASME Paper 91-GT-17.

[229] A. Murata, S. Mochizuki, and T. Takahashi, "Local heat transfer measurements of an orthogonally rotating square duct with angled rib turbulators," *International Journal of Heat and Mass Transfer*, vol. 42, no. 16, pp. 3047–3056, 1999.

[230] L. M. Wright, Y. H. Liu, J. C. Han, and S. Chopra, "Heat transfer in trailing edge, wedge-shaped cooling channels under high rotation numbers," *Journal of Heat Transfer*, vol. 130, no. 7, Article ID 071701, 11 pages, 2007, ASME Paper GT2007-27093.

[231] S. W. Chang, T. M. Liou, J. H. Hung, and W. H. Yeh, "Heat transfer in a radially rotating square-sectioned duct with two opposite walls roughened by 45 deg staggered ribs at high rotation numbers," *Journal of Heat Transfer*, vol. 129, no. 2, pp. 188–199, 2007.

[232] S. Dutta, J. C. Han, and C. P. Lee, "Experimental heat transfer in a rotating triangular duct: effect of model orientation," *Journal of Heat Transfer*, vol. 117, no. 4, pp. 1058–1061, 1995.

[233] T. M. Liou, S. W. Chang, J. H. Hung, and S. F. Chiou, "High rotation number heat transfer of a 45° rib-roughened rectangular duct with two channel orientations," *International Journal of Heat and Mass Transfer*, vol. 50, no. 19-20, pp. 4063–4078, 2007.

[234] W. D. Morris and K. F. Rahmat-Abadi, "Convective heat transfer in rotating ribbed tubes," *International Journal of Heat and Mass Transfer*, vol. 39, no. 11, pp. 2253–2266, 1996.

[235] F. T. Willett and A. E. Bergles, "Heat transfer in rotating narrow rectangular pin-fin ducts," *Experimental Thermal and Fluid Science*, vol. 25, no. 7, pp. 573–582, 2002.

[236] S. W. Chang, T. L. Yang, and W. J. Wang, "Heat transfer in a rotating twin-pass trapezoidal-sectioned passage roughened by skewed ribs on two opposite walls," *Heat Transfer Engineering*, vol. 27, no. 10, pp. 63–79, 2006.

[237] W. D. Morris and G. Ghavami-Nasr, "Heat transfer measurements in rectangular channels with orthogonal mode rotation," *Journal of Turbomachinery*, vol. 113, no. 3, pp. 339–345, 1991, ASME Paper 90-GT-138.

[238] S. Acharya, V. Eliades, and D. E. Nikitopoulos, "Heat transfer enhancements in rotating two-pass coolant channels with profiled ribs: part 1-average results," *Journal of Turbomachinery*, vol. 123, no. 1, pp. 97–106, 2001, ASME Paper 2000-GT-227.

[239] P. Agarwal, S. Acharya, and D. E. Nikitopoulos, "Heat transfer in 1:4 rectangular passages with rotation," *Journal of Turbomachinery*, vol. 125, no. 4, pp. 726–733, 2003.

[240] H. H. Cho, Y. Y. Kim, K. Kim, and D. H. Rhee, "Effects of rib arrangements and rotation speed on heat transfer in a two-pass duct," ASME Paper GT-2003-38609, 2003.

[241] H. H. Cho, S. Y. Lee, and D. H. Rhee, "Effects of cross ribs on heat/mass transfer in a two-pass rotating duct," *Heat and Mass Transfer*, vol. 40, no. 10, pp. 743–755, 2004.

[242] H. H. Cho, S. Y. Lee, J. H. Won, and D. H. Rhee, "Heat/mass transfer in a two-pass rotating rectangular duct with and without 70°-angled ribs," *Heat and Mass Transfer*, vol. 40, no. 6-7, pp. 467–475, 2004.

[243] K. M. Kim, D. H. Lee, and H. H. Cho, "Detailed measurement of heat/mass transfer and pressure drop in a rotating two-pass duct with transverse ribs," *Heat and Mass Transfer*, vol. 43, no. 8, pp. 801–815, 2007.

[244] C. W. Park and S. C. Lau, "Effect of channel orientation of local heat (mass) transfer distributions in a rotating two-pass square channel with smooth walls," *Journal of Heat Transfer*, vol. 120, no. 3, pp. 624–632, 1998.

[245] C. W. Park, S. C. Lau, and R. T. Kukreja, "Heat (mass) transfer in a rotating channel with ribs of various sizes on two walls," *Journal of Thermophysics and Heat Transfer*, vol. 12, no. 3, pp. 452–454, 1998.

[246] C. W. Park, S. C. Lau, and R. T. Kukreja, "Heat/mass transfer in a rotating two-pass channel with transverse ribs," *Journal of Thermophysics and Heat Transfer*, vol. 12, no. 1, pp. 80–86, 1998.

[247] C. W. Park, C. Yoon, and S. C. Lau, "Heat (mass) transfer in a diagonally oriented rotating two-pass channel with rib-roughened walls," *Journal of Heat Transfer*, vol. 122, no. 1, pp. 208–211, 2000.

[248] S. W. Chang, T.-M. Liou, and T.-H. Lee, "Detailed heat transfer distributions and pressure drop measurements for a rotating parallelogram channel with radially outward flow," ASME Paper GT2011-45456, 2011.

[249] S.-Y. Hu, T. I.-P. Shih, M. K. Chyu et al., "Flow and heat transfer in the tip-turn region of a U-duct under rotating and non-rotating conditions," ASME Paper GT2011-46013, 2011.

[250] M. Huh, J. Lei, and J.-C. Han, "Influence of channel orientation on heat transfer in a two-pass smooth and ribbed rectangular channel (AR=2:1) under large rotation numbers," ASME Paper GT2010-22190, 2010.

[251] A. P. Rallabandi, Y.-H. Liu, and J.-C. Han, "Heat transfer in trailing edge wedge-shaped pin-fin channels with slot ejection under high rotation numbers," ASME Paper GT2010-22832, 2010.

[252] S. W. Chang, T.-M. Liou, and T.-H. Lee, "Heat transfer of rotating rectangular channel with diamond shaped pin-fin array at high rotation numbers," ASME Paper GT2012-68676, 2012.

[253] J. A. Lamont and S. V. Ekkad, "Effects of rotation on jet impingement channel heat transfer," ASME Paper GT2011-45744, 2011.

[254] M. A. Elyyan and D. K. Tafti, "Investigation of coriolis forces effect of flow structure and heat transfer distribution in a rotating dimpled channel," ASME Paper GT2010-22657, 2010.

[255] S. W. Chang, T.-M. Liou, and W. C. Chen, "Heat transfer in a rotating rectangular channel with two opposite walls roughened with spherical protrusions at high rotation numbers," ASME Paper GT2010-22609, 2010.

[256] L. Bonanni, C. Carcasci, B. Facchini, and L. Tarchi, "Experiemental survey on heat transfer in a trailing edge cooling system: effects of rotation in internal cooling ducts," ASME Paper GT2012-69638, 2012.

[257] D. Borello, G. Delibra, C. Bianchini, and A. Andreini, "Unsteady CFD analysis of turbulent flow and heat transfer in a gas turbine blade trailing edge subjected to rotation," ASME Paper GT2012-69903, 2012.

[258] M. Pascotto, A. Armellini, L. Casarsa, P. Giannattasio, and C. Mucignat, "Effects of rotation and channel orientation on the flow field inside a trailing edge internal cooling channel," ASME Paper GT2012-68050, 2012.

[259] M. Schroll, M. Elfert, and L. Lange, "Investigation of the effect of rotation on the flow in a two-pass cooling system with smooth and ribbed walls using PIV," ASME Paper GT2011-46427, 2011.

[260] M. Schüler, Dreher, H. -M, M. Elfert, S. O. Neumann, and B. Weigand, "Numerical predictions of the effect of rotation on fluid flow and heat transfer in an engine-similar two-pass internal cooling channel with smooth and ribbed walls," ASME Paper GT2010-22870, 2010.

[261] H. A. El-Husayni, M. E. Taslim, and D. M. Kercher, "Experimental heat transfer investigation of stationary and orthogonally rotating asymmetric and symmetric heated smooth and turbulated channels," *Journal of Turbomachinery*, vol. 116, no. 1, pp. 124–132, 1994, ASME Paper 92-GT-189.

[262] J. Guidez, "Study of the convective heat transfer in a rotating coolant channel," *Journal of Turbomachinery*, vol. 111, no. 1, pp. 43–50, 1989.

[263] T. J. Hajek, J. H. Wagner, and B. V. Johnson, "Coolant passage heat transfer with rotation," NASA, Lewis Research Center, Turbine Engine Hot Section Technology, 1987.

[264] E. Sethuraman, S. Acharya, and D. E. Nikitopoulos, "Mass/heat transfer in rotating, smooth, high-aspect ratio (4:1) coolant channels with curved walls," *Journal of Turbomachinery*, vol. 131, no. 2, Article ID 021002, 9 pages, 2009.

[265] F. T. Willett and A. E. Bergles, "Heat transfer in rotating narrow rectangular ducts with heated sides parallel to the r-z plane," *Journal of Heat Transfer*, vol. 124, no. 1, pp. 1–7, 2002.

[266] F. T. Willett and A. E. Bergles, "Heat transfer in rotating narrow rectangular ducts with heated sides oriented at 60° to the r-z plane," *Journal of Turbomachinery*, vol. 123, no. 2, pp. 288–295, 2001, ASME Paper 2000-GT-224.

[267] R. J. Clifford, W. D. Morris, and S. P. Harasgama, "An experimental study of local and mean heat transfer in a triangular-sectioned duct rotating in the orthogonal mode," *Journal of Engineering for Gas Turbines and Power*, vol. 106, no. 3, pp. 661–667, 1984.

[268] Y. H. Liu, M. Huh, J. C. Han, and H. K. Moon, "High rotation number effects on heat transfer in a triangular channel with 45°, inverted 45°, and 90° ribs," ASME Paper GT2009-59216, 2009.

[269] Y. H. Liu, M. Huh, D. H. Rhee, J. C. Han, and H. K. Moon, "Heat transfer in leading edge, triangular shaped cooling channels with angled ribs under high rotation numbers," *Journal of Turbomachinery*, vol. 131, no. 4, Article ID 041017, 12 pages, 2009, ASME Paper GT2008-50344.

[270] Y. Mori, T. Fukada, and W. Nakayama, "Convective heat transfer in a rotating radial circular pipe (2nd report)," *International*

Journal of Heat and Mass Transfer, vol. 14, no. 11, pp. 1807–1824, 1971.

[271] T. M. Liou, C. C. Chen, and M. Y. Chen, "Rotating effect on fluid flow in two smooth ducts Connected by a 180-degree bend," *ASME Journal of Fluids Engineering*, vol. 125, no. 1, pp. 138–148, 2003.

A Rotor Flux and Speed Observer for Sensorless Single-Phase Induction Motor Applications

Massimo Caruso,[1] Vittorio Cecconi,[1] Antonino O. Di Tommaso,[1] and Ronilson Rocha[2]

[1] *Department of Electrical, Electronics and Telecommunications Engineering, University of Palermo, Vialle delle Scienze, Edificio 9, 90128 Palermo, Italy*
[2] *Department of Control and Automation Engineering, School of Mines, Federal University of Ouro Preto, Campus Morro do Cruzeiro, 35400-000 Ouro Preto, MG, Brazil*

Correspondence should be addressed to Ronilson Rocha, rocha@decat.em.ufop.br

Academic Editor: Hyeong Joon Ahn

It is usual to find single-phase induction motor (SPIM) in several house, office, shopping, farm, and industry applications, which are become each time more sophisticated and requiring the development of efficient alternatives to improve the operational performance of this machine. Although the rotor flux and rotational speed are essential variables in order to optimize the operation of a SPIM, the use of conventional sensors to measure them is not a viable option. Thus, the adoption of sensorless strategies is the more reasonable proposal for these cases. This paper presents a rotor flux and rotational speed observer for sensorless applications involving SPIMs. Computer simulations and the experimental results are used to verify the performance of the proposed observer.

1. Introduction

The single-phase induction motor (SPIM) has a very robust design and can be mass produced with relatively low cost. Traditionally used in fractional and subfractional horsepower applications, it is usual to find this motor in several house, office, shopping, farm, and industry appliances such as air conditioning systems, mixers, washers, blowers, dryers, fans, refrigerators, vacuum cleaners, compressors, pumps, and so forth. Due to introduction of low-cost static power converters and the development of more sophisticated appliances, the traditional SPIM applications have required the development of efficient alternatives to improve the performance of this machine.

The SPIM is basically constituted by a squirrel-cage rotor and two stator windings displaced 90° in space, known as main and auxiliary windings. This machine is usually unbalanced, and its windings present different impedances. When the main winding is fed by a single-phase source, the produced magnetic field is pulsating and stationary, and a starting torque is not developed. To start a SPIM, the auxiliary winding must emulate a second phase. Although the auxiliary winding is often disconnected by a centrifugal switch when the SPIM rotor reaches 60% to 80% of rated speed, it can be maintained at running operation in order to provide a higher and less pulsating torque. In general, a series capacitor is connected with the auxiliary winding in order to improve the starting and the SPIM performance.

Many proposals for the better performance of a SPIM are based in the use of an electronically switched series capacitor in the auxiliary winding, which can be controlled to improve the machine performance at different operating conditions [1–4]. The variable speed drives are also proposed in several SPIM applications in order to provide energy savings and torque improvement [5–13]. Although the magnetic flux and/or rotational speed of this machine are essential variables in order to optimize the performance of a SPIM, the use of sensors to measure them cannot be justified in fractional and low-cost SPIM applications since then implies in more electronics, higher cost, lower reliability, assembling difficulties, and the increase in weight, size, and electrical susceptibility. Considering these problems, the adoption of a sensorless strategy is a more reasonable option for high-performance SPIM applications, where the magnetic flux and/or rotational speed are estimated from stator voltage and current measurements [4, 12, 14–17].

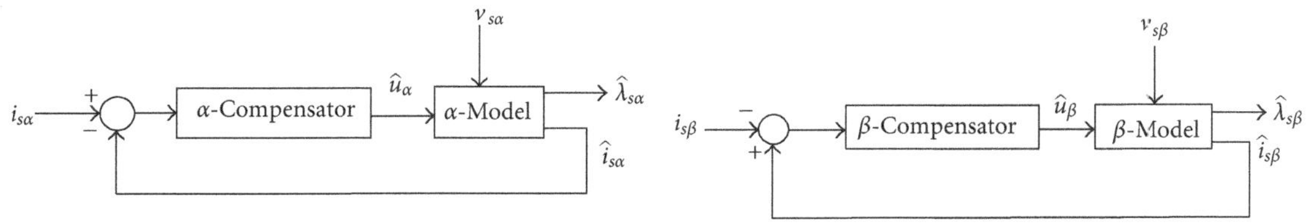

FIGURE 1: α- and β-Flux observer structures.

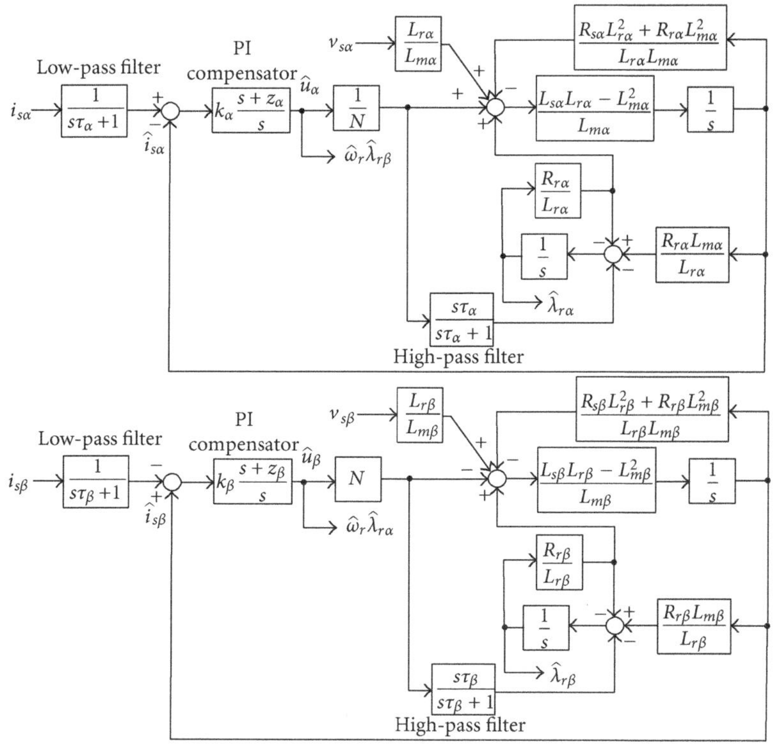

FIGURE 2: Block diagram of α- and β-flux observers.

This paper presents an original proposal of a rotor flux observer for SPIM, from which the rotor speed of this motor can be estimated. This observer can be used in applications involving SPIMs that require the estimation of rotor flux and/or rotor speed, such as optimization strategies, sensorless vector control of the machine based on the philosophy of the direct flux orientation, and so forth. Two half flux observers are designed using the stationary reference frame SPIM model. From these half flux observers, data are obtained to estimate the rotor speed of the SPIM. The performance of the proposed observer is satisfactorily demonstrated from computer simulations and experimental tests considering two situations: a step variation in the stator frequency of the no load SPIM and the application of mechanical load to SPIM shaft.

2. Flux Observer

Traditionally, the dynamic model of a SPIM is described considering a stationary reference frame α-β fixed in the stator, where α and β denote the auxiliary and main windings respectively [18]. If the stator currents i_s and the rotor flux λ_r vectors are admitted as the state variables, this dynamic model can be written as follows [17]:

(i) α-model

$$\dot{i}_{s\alpha} = \frac{L_{m\alpha}}{L_{\sigma\alpha}}\left(\frac{L_{r\alpha}}{L_{m\alpha}}v_{s\alpha} + \frac{R_{r\alpha}}{L_{r\alpha}}\lambda_{r\alpha} - \frac{R_{\phi\alpha}}{L_{r\alpha}L_{m\alpha}}i_{s\alpha} + \frac{u_\alpha}{N}\right),$$
$$\dot{\lambda}_{r\alpha} = -\frac{R_{r\alpha}}{L_{r\alpha}}\left(\lambda_{r\alpha} - L_{m\alpha}i_{s\alpha}\right) - \frac{u_\alpha}{N}, \quad (1)$$

(ii) β-model

$$\dot{i}_{s\beta} = \frac{L_{m\beta}}{L_{\sigma\beta}}\left(\frac{L_{r\beta}}{L_{m\beta}}v_{s\beta} + \frac{R_{r\beta}}{L_{r\beta}}\lambda_{r\beta} - \frac{R_{\phi\beta}}{L_{r\beta}L_{m\beta}}i_{s\beta} - Nu_\beta\right),$$
$$\dot{\lambda}_{r\beta} = -\frac{R_{r\beta}}{L_{r\beta}}\left(\lambda_{r\beta} - L_{m\beta}i_{s\beta}\right) + Nu_\beta, \quad (2)$$

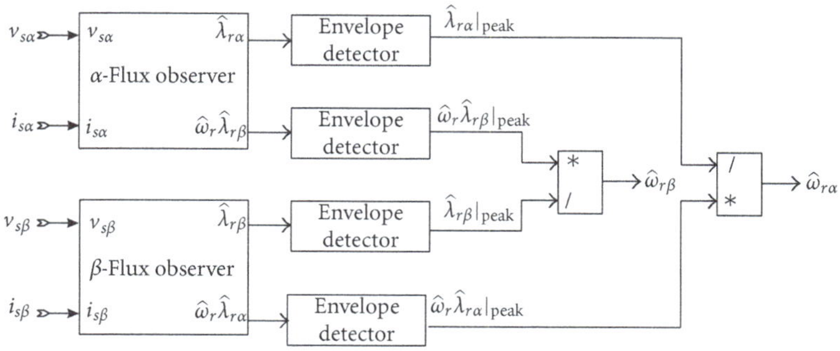

FIGURE 3: Speed estimation using flux observers and envelope detectors.

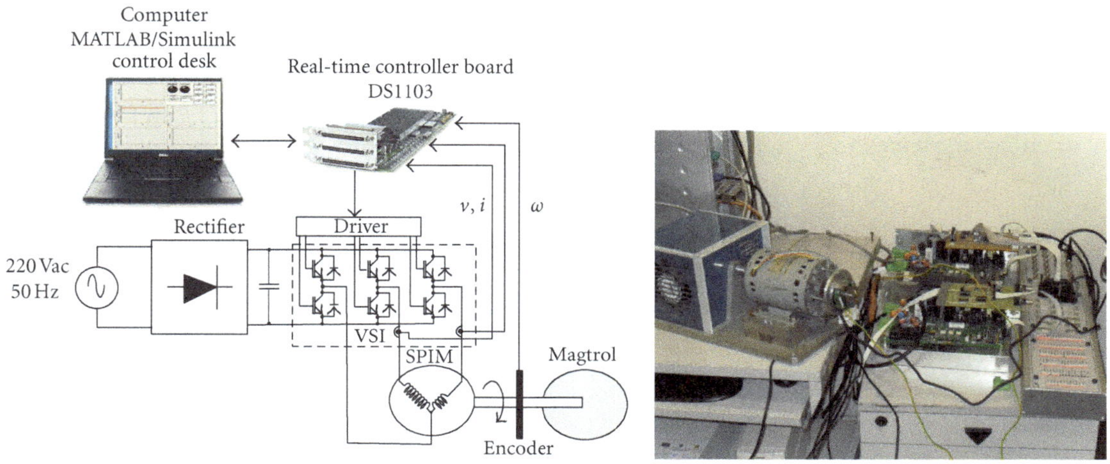

FIGURE 4: SPIM test bench.

TABLE 1: SPIM data.

Rated values			
Voltage		110 V	
Frequency		50/60 Hz	
Main winding current		2.3 A	
Speed		2700 RPM	
No. of poles		2	
Power		0.25 HP (180 W)	
Main winding (β)		Auxiliary winding (α)	
$R_{s\beta}$	5.2 Ω	$R_{s\alpha}$	29 Ω
$R_{r\beta}$	9.4 Ω	$R_{r\alpha}$	35.9 Ω
$L_{m\beta}$	0.3 H	$L_{m\alpha}$	0.45 H
$L_{s\beta}$	0.3068 H	$L_{s\alpha}$	0.55 H
$L_{r\beta}$	0.3068 H	$L_{r\alpha}$	0.55 H
$N = N_\beta/N_\alpha$		0.67	

TABLE 2: Electrical characteristics of the IGBT power converter.

Input voltage (single phase)	230 VCA
Motor supply rail	310 VDC
Output power peak	6.5 kW
Phase current peak (crest)	30 A
Phase current cont.	21 A(RMS)
Switching frequency	5 to 20 kHz
Shunt power peak	3200 W

where v_s = stator voltage, ω_r = rotational speed, $u_\alpha = \omega_r\lambda_{r\beta}$, $u_\beta = \omega_r\lambda_{r\alpha}$, R = resistance, L = winding inductance, L_m = mutual inductance, $R_\phi = R_sL_r^2 + R_rL_m^2$, $L_\sigma = L_sL_r - L_m^2$, and N = transformer ratio between both stator windings. The subscripts s and r are related, respectively, to stator and rotor variables and parameters.

In order to obtain the convergence of the observed variables to its real value in conventional state observers, a correctional term is usually inserted in the system model to reduce the error between the measurable variables and their respective estimated values. If this principle is applied to observe the rotor fluxes of a SPIM, the resulting observer is nonlinear, since the terms u_α and u_β consist in the product of two state variables of the system. However, if these nonlinear terms u_α and u_β of SPIM model are considered to correct the observation error, the flux observer can be decoupled in two linear and independent closed-loop systems as shown in Figure 1, which can be designed and analyzed using linear techniques. Another advantage of this proposed structure for

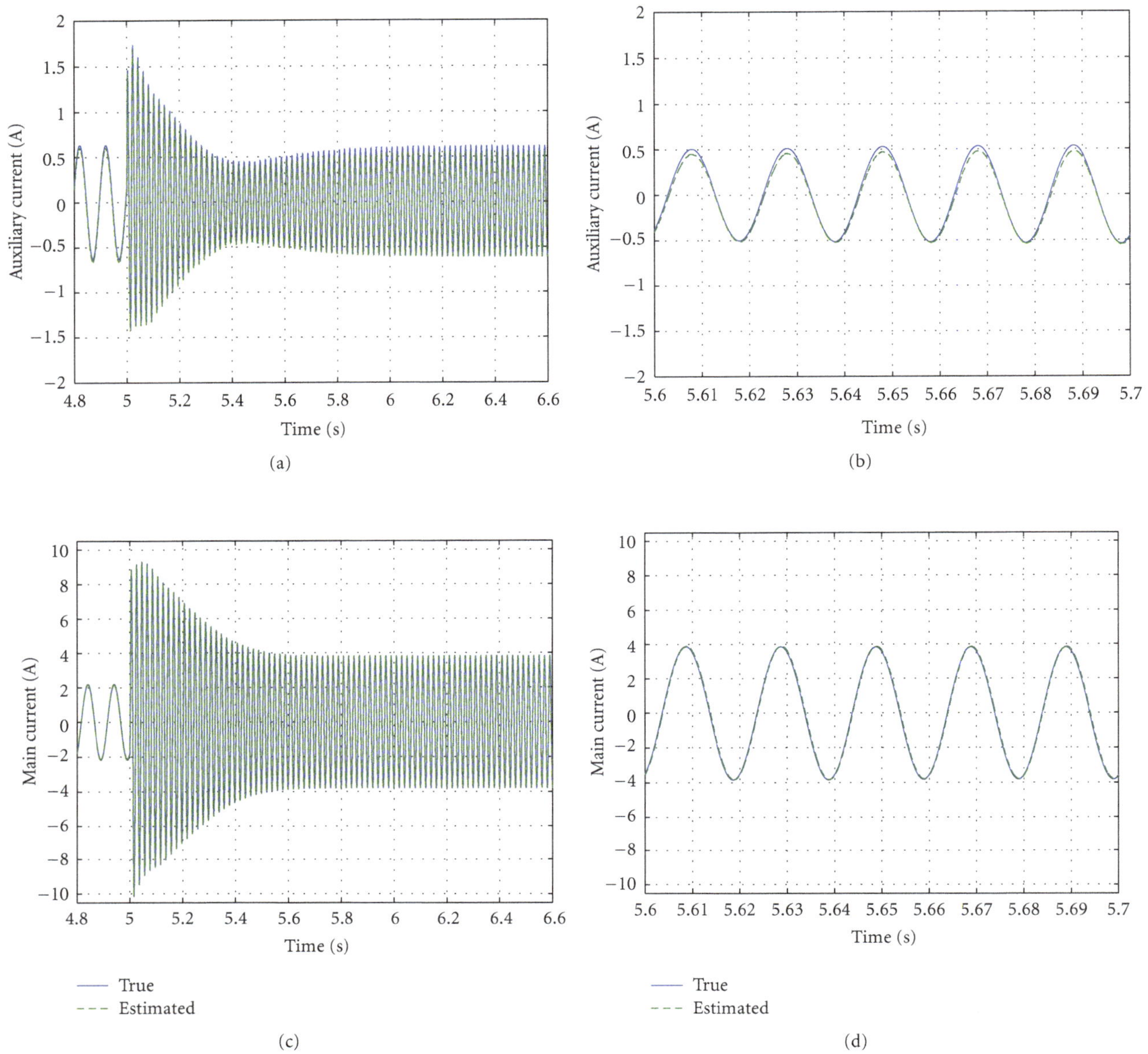

Figure 5: Stator currents of the no-load SPIM and their estimated values during a step variation from 10 Hz to 50 Hz in the stator frequency (computational simulation): (a) auxiliary winding $i_{s\alpha}$, (b) details of $i_{s\alpha}$, (c) main winding $i_{s\beta}$, (d) details of $i_{s\beta}$.

rotor flux observer is the possibility to evaluate the rotor speed of the SPIM, as will be seen in the next section.

Considering only the structure of β-flux observer shown in Figure 1, the open-loop transfer function between $\hat{i}_{s\beta}$ and $\hat{u}_{\beta}$ admitting $v_{s\beta}$ as system disturbance is given by

$$\frac{\hat{I}_{s\beta}(s)}{\hat{U}_{\beta}(s)} = \frac{-sNL_{m\beta}}{s^2 L_{\sigma\beta} + sR_{\sigma\beta} + R_{s\beta}R_{r\beta}}, \tag{3}$$

where $R_{\sigma\beta} = R_{r\beta}L_{s\beta} + R_{s\beta}L_{r\beta}$. This transfer function has a zero in the origin and two real poles. This zero provides a derivative action, which does not contribute to reduce the current error. Furthermore, this derivative action increases the observer sensitive to noises and disturbances. In order to prevent the effects of derivative action, a PI structure is considered for β-compensator:

$$\frac{\hat{U}_{\beta}(s)}{\hat{I}_{s\beta}(s) - I_{s\beta}(s)} = k_{\beta}\frac{s + z_{\beta}}{s}, \tag{4}$$

where the pure integrator of β-compensator eliminates the derivative action of β-model. The zero z_{β} of the β-compensator can be chosen to cancel the slowest pole of the open-loop transfer function of (3). Since the PI compensator and the SPIM model are inherently stable, the Bode criteria can be

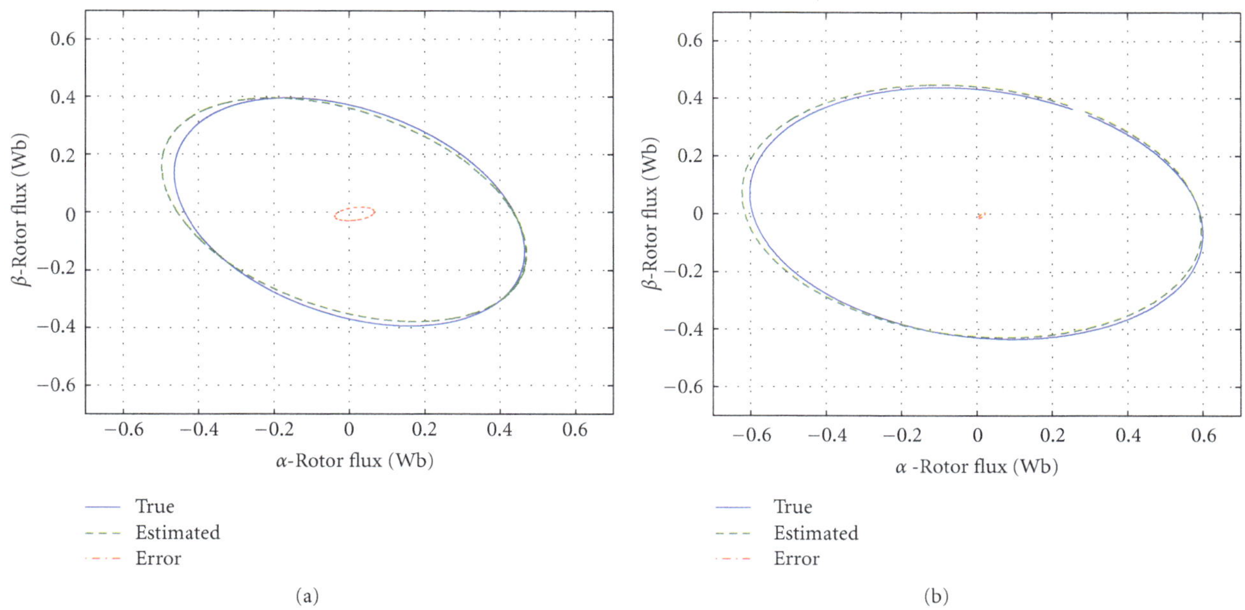

FIGURE 6: Phase portrait $\lambda_{r\alpha} \times \lambda_{r\beta}$ of the vector rotor flux of no-load SPIM (computational simulation): (a) 10 Hz, (b) 50 Hz.

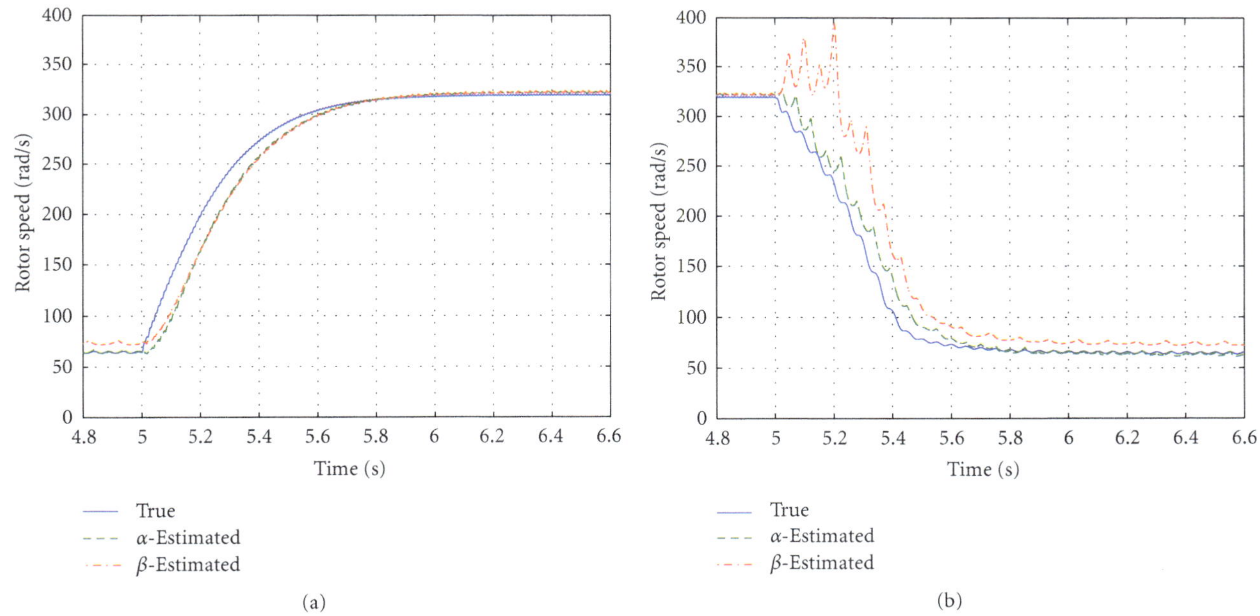

FIGURE 7: Rotor speed ω_r of the SPIM and its estimations $\hat{\omega}_{r\alpha}$ and $\hat{\omega}_{r\beta}$ during steps in the frequency stator (computational simulation): (a) from 10 Hz to 50 Hz, (b) from 50 Hz to 10 Hz.

used in order to verify the stability of the proposed structure for rotor flux observation. The proposed design of the PI compensator imposes a pole-zero cancellation in the open-loop transfer function $\hat{I}_{s\beta}/(\hat{I}_{s\beta} - I_{s\beta})$, assuring that the slope of the amplitude diagram will be always -20 db/dec and the phase angle is always smaller than 135° in the cross-over frequency. Thus, the proposed flux observer will be always stable in these circumstances independently of the gain k_β. This compensator provides an effective proportional action and, consequently, an offset error will always be present in the observation of β-flux. The performance of this structure is determined by the gain k_β, which can be chosen in order to obtain a fastest response and minimize the influence of the parameter variations over the flux estimation.

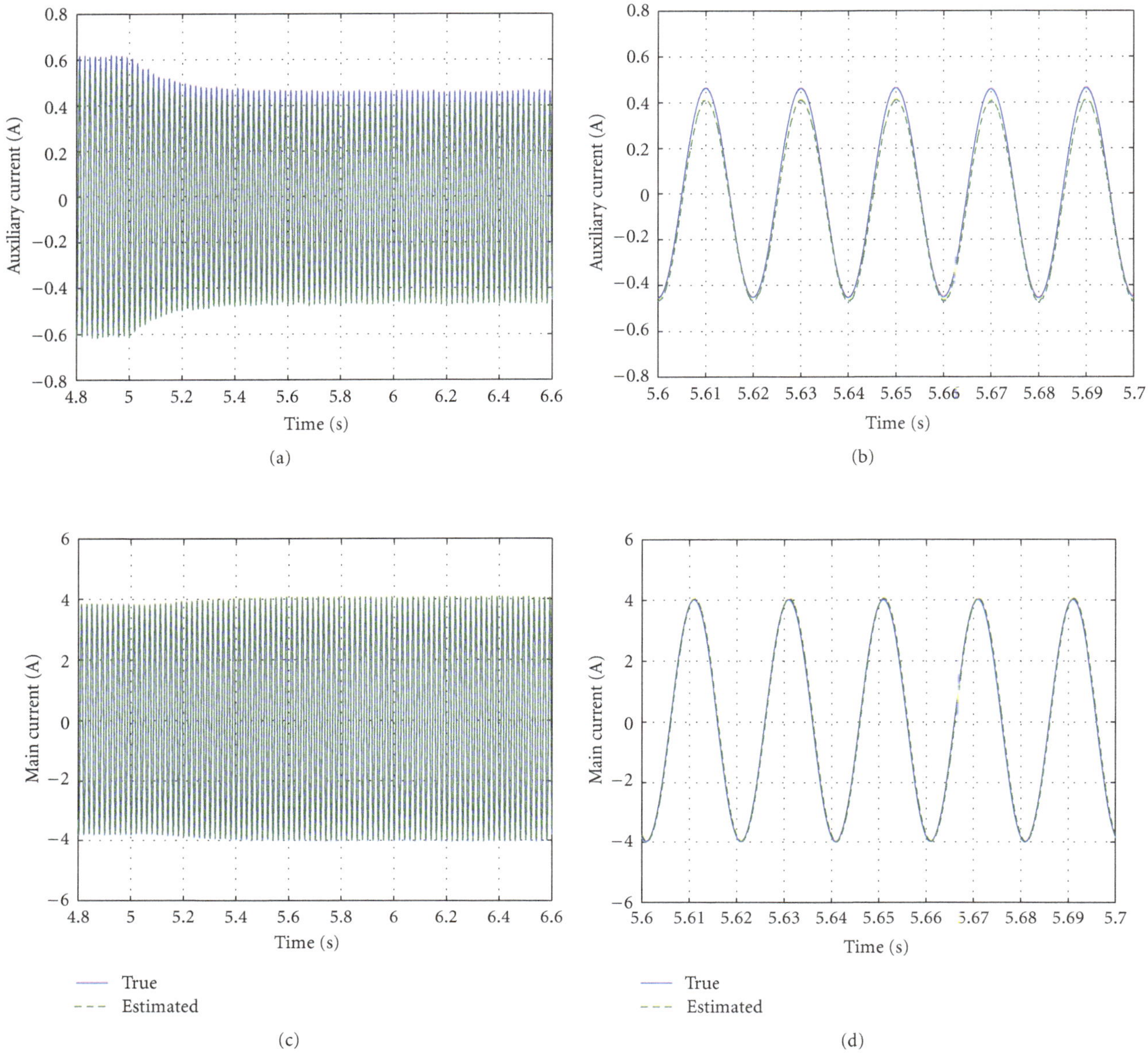

Figure 8: Stator currents of the SPIM and their estimated values for 50 Hz during a load step (computational simulation): (a) auxiliary winding $i_{s\alpha}$, (b) details of $i_{s\alpha}$, (c) main winding $i_{s\beta}$, (d) details of i_{s_β}.

Considering the close-loop β-observer with a PI compensator, the transfer function between the estimated rotor flux $\hat{\lambda}_{r\beta}$ and the measurement $i_{s\beta}$ is given by

$$\frac{\hat{\Lambda}_{r\beta}(s)}{I_{s\beta}(s)} = -\frac{sNk_\beta + Nk_\beta z_\beta}{s} \times \frac{sL_{\sigma\beta} + R_{s\beta}L_{r\beta}}{s^2 L_{\sigma\beta} + s\left(R_{\sigma\beta} + Nk_\beta L_{m\beta}\right) + \left(R_{s\beta}R_{r\beta} + Nk_\beta z_\beta L_{m\beta}\right)}, \quad (5)$$

where it is observed that the PI compensator introduces a pure integrator in the path of the estimated flux. This fact can compromise the performance of the proposed observer, since any DC offset in the current sensors will be continuously integrated until the saturation of the integrators. In order to prevent this problem, a first-order high-pass filter with a very low cut-off frequency (around Hz) must be inserted in the path of the estimated flux to cancel the effects of the pure integrator introduced by PI compensator. Although this solution can deteriorate the dynamical performance of the proposed flux observer at very low rotor speed, it is satisfactory for the most of SPIM applications which usually do not work in this speed range. In order to reduce the effect of measurement noises in the observer performance, a first-order low-pass filter with an adequate cut-off frequency could be inserted in the measured current input to filter high-frequency signals.

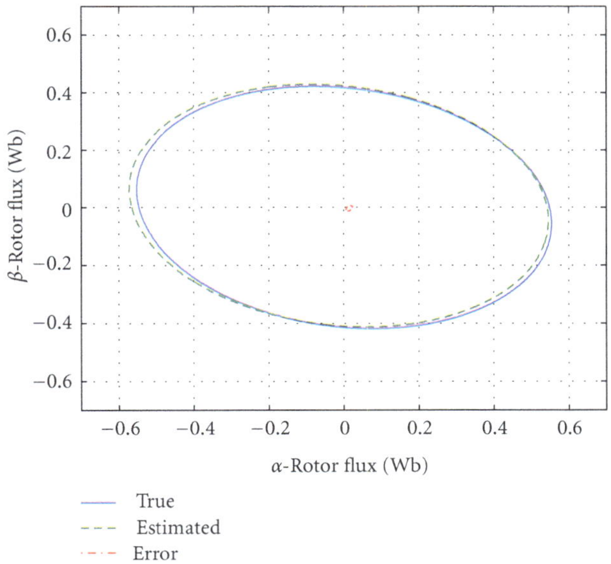

FIGURE 9: Phase portrait $\lambda_{r\alpha} \times \lambda_{r\beta}$ of rotor flux for 50 Hz considering the loaded SPIM (computational simulation).

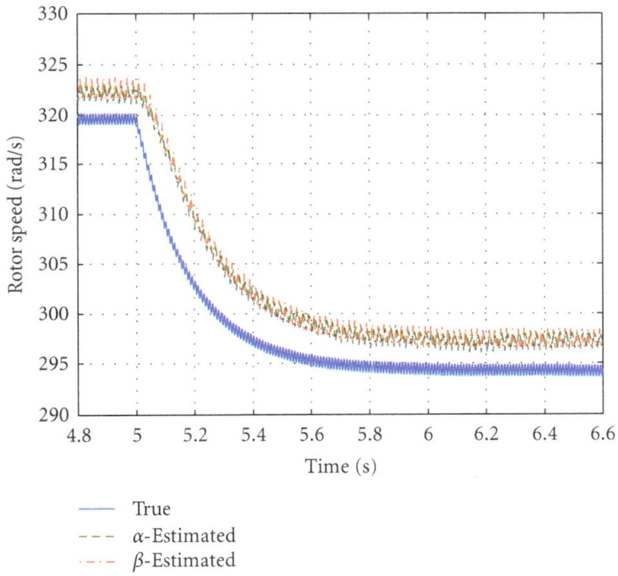

FIGURE 10: Rotor speed ω_r and its estimations $\widehat{\omega}_{r\alpha}$ and $\widehat{\omega}_{r\beta}$ for 50 Hz during the application of a load step (computational simulation).

The design of close-loop α-flux observer is analogous to close-loop β-flux observer. The complete block diagram of the proposed flux observer is presented in the Figure 2. The basic difference between both observers is the form as the error signal must be computed, since the close-loop β-flux observer presents a reverse action.

3. Rotor Speed Estimation

If the observation of the rotor fluxes is satisfactory, an estimative of the rotor speed $\widehat{\omega}_r$ can be obtained from the proposed flux observers for sensorless applications. Since the outputs of the PI compensators correspond to nonlinear terms, $\widehat{u}_\alpha = \widehat{\omega}_r \widehat{\lambda}_{r\beta}$ and $\widehat{u}_\beta = \widehat{\omega}_r \widehat{\lambda}_{r\alpha}$, $\widehat{\omega}_r$ can be estimated dividing $\widehat{u}_\alpha$ by $\widehat{\lambda}_{r\beta}$ or $\widehat{u}_\beta$ by $\widehat{\lambda}_{r\alpha}$. Since $\widehat{\lambda}_{r\alpha}$ and $\widehat{\lambda}_{r\beta}$ are AC values, the computation of $\widehat{\omega}_r$ is subject to a singularity due to zero division. In order to avoid this problem, the solution adopted in this work is only to compute $\widehat{\omega}_r$ using the peak values of the signals u_α (or u_β) and $\lambda_{r\beta}$ (or $\lambda_{r\alpha}$), which occur simultaneously and can be captured using envelope detectors as shown in Figure 3.

4. Experimental Test Bench

The experimental apparatus shown in Figure 4 is used to verify the performance of the proposed rotor flux/speed observer. The data of the SPIM used in test bench are in Table 1,

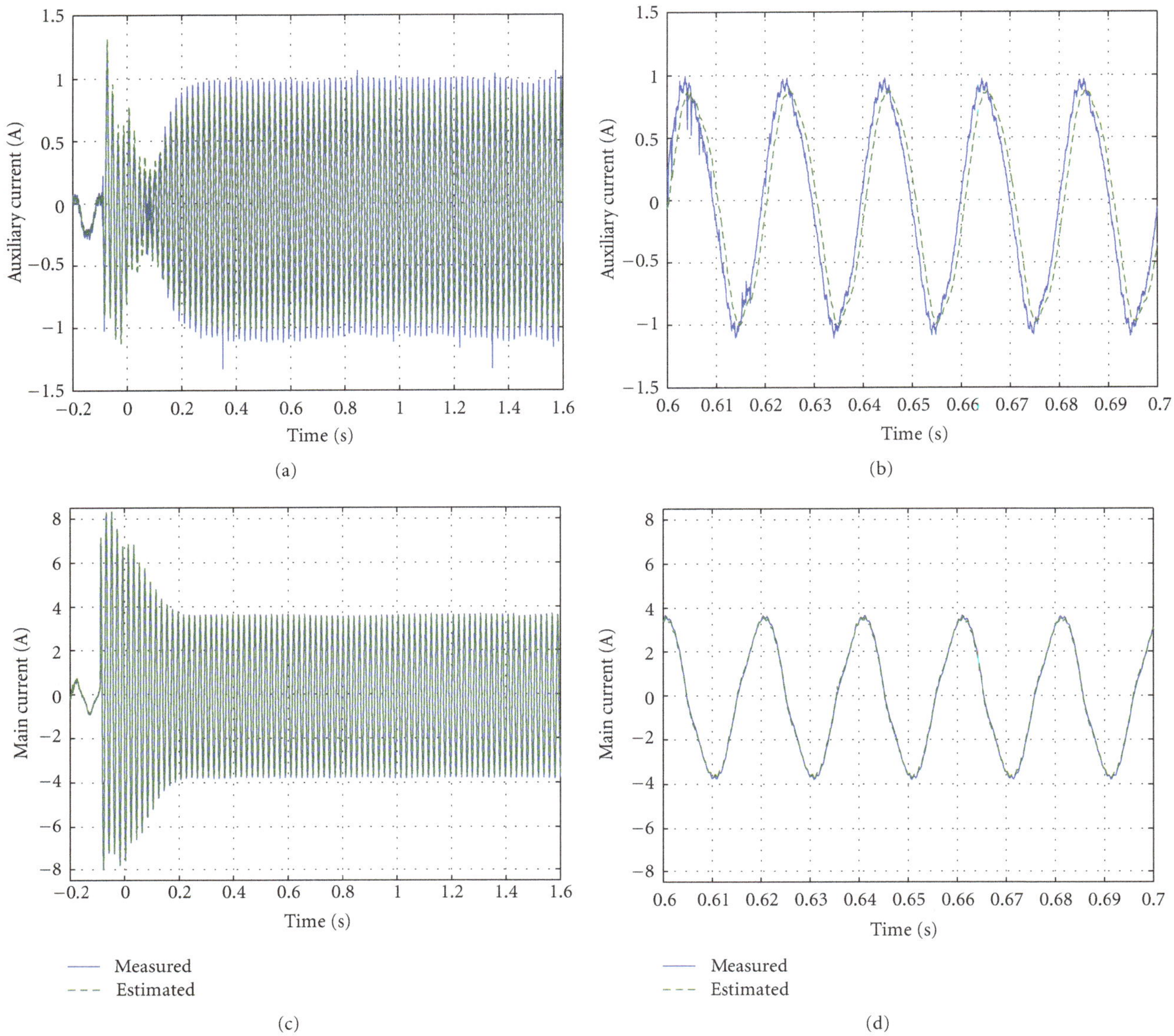

FIGURE 11: Measured and estimated stator currents of the SPIM during the step from 10 Hz to 50 Hz in the frequency stator (experimental results): (a) auxiliary winding $i_{s\alpha}$, (b) details of $i_{s\alpha}$ at 50 Hz, (c) main winding $i_{s\beta}$, (d) details of $i_{s\beta}$ at 50 Hz.

where the parameters of the equivalent circuit are evaluated from the experimental DC, locked rotor, and no-load tests [19]. An optical incremental encoder (SUMTAK Corporation, type LDA-001-1000 CE, 1024 pulses per revolution) is coupled to SPIM shaft in order to measure the rotational speed and compare it to its estimative. A magnetic powder brake Magtrol emulates the mechanical load. The SPIM windings are fed by two quadrature AC voltages with constant V/f ratio generated by an IGBT-based three-leg voltage source inverter (VSI) whose electrical characteristics are summarized in Table 2. The use of a three-leg inverter presents many advantages over a two-leg topology such as generate a zero voltage vector, avoid the circulation of an AC current in the DC link capacitor, and improve the output voltage quality in terms of harmonic distortion [6, 7, 12, 13]. The integrated sensors to VSI are used to measure the currents of the main and auxiliary stator windings. The experimental implementation of the proposed observer is realized using a dSPACE DS1103 controller board, which allows a rapid real-time prototyping [20, 21]. Since this hardware is linked to MATLAB/Simulink/Stateflow development software from MathWorks, a friendly interface can be used in order to set up and tune up the electrical drive and the parameters of the proposed observer. In this philosophy, the model of the proposed observer is converted into a discrete-time block diagram, which is introduced in Simulink, adequately compiled, and inserted in the DS1103, where the algorithm is executed to compute the rotor flux and rotational speed estimations based in the measured values acquired by A/D converters. This controller board also generates the PWM signals for the VSI driver and acquires the signal of an incremental encoder for speed measurement. The control desk provides a virtual

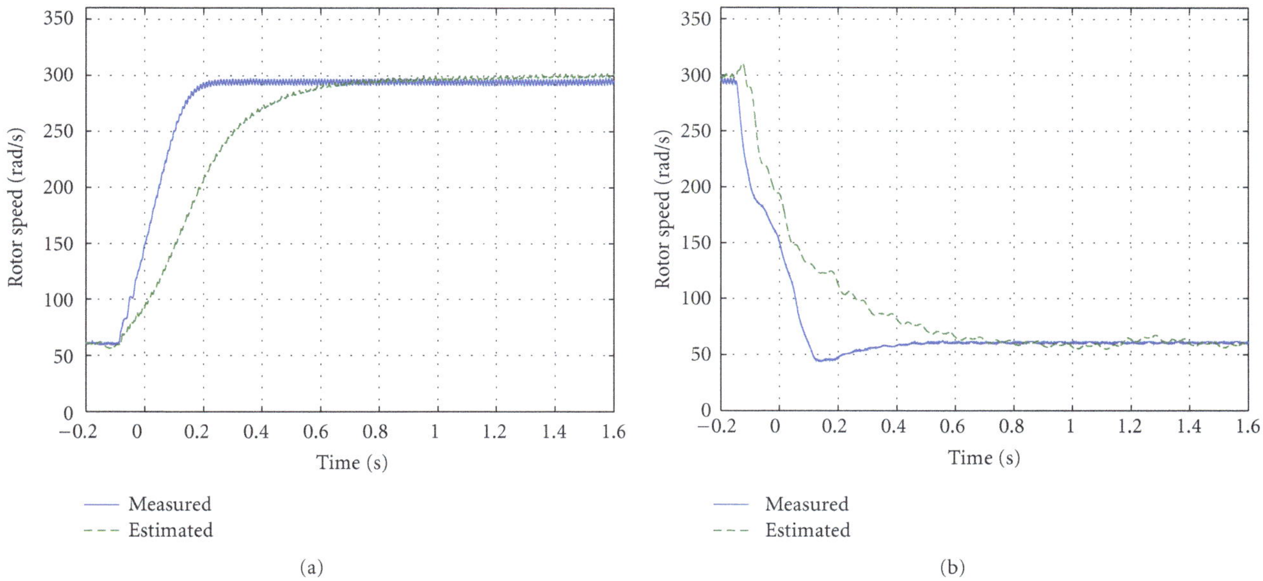

FIGURE 12: Estimated and measured rotor speed ω_r of the SPIM during the step in the frequency stator (experimental results): (a) from 10 Hz to 50 Hz, (b) from 50 Hz to 10 Hz.

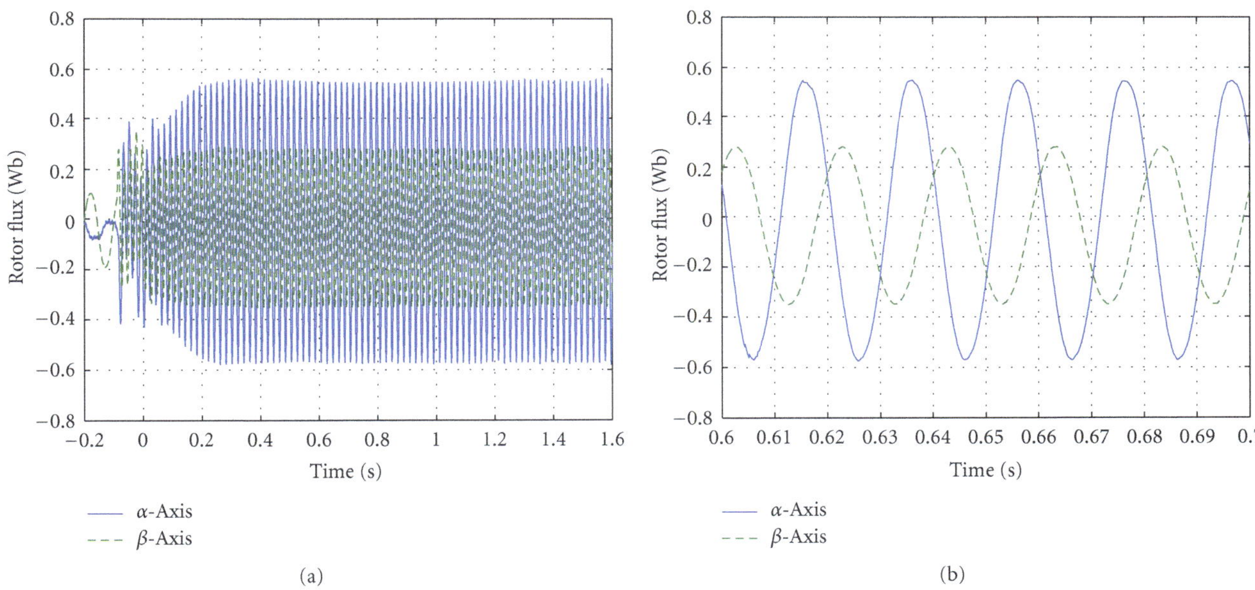

FIGURE 13: Estimated rotor fluxes λ_r of the SPIM during the step from 10 to 50 Hz in the frequency stator (experimental results): (a) estimated rotor fluxes, (b) details of the estimated rotor fluxes at 50 Hz.

instrument panel that enables the monitoring of measured and estimated variables and the adjustment of system values.

5. Observer Performance

5.1. Simulation Results. Considering the experimental apparatus presented in the last section, a computer simulation was done using MATLAB/simulink in order to verify the performance of the proposed observer. In the first test, the stator frequency of the no-load SPIM is varied from 10 to 50 Hz and from 50 Hz to 10 Hz. The stator currents of the SPIM are shown in Figure 5. The auxiliary winding current $i_{s\alpha}$ is $0.40\,A_{\text{rms}}$ at 10 Hz, which is practically maintained at 50 Hz. The main winding current $i_{s\beta}$ is $1.50\,A_{\text{rms}}$, and increases to $2.70\,A_{\text{rms}}$ at 50 Hz. Although a delay is noted in the estimation, the proposed observer presents a good accuracy to estimate the stator currents of the SPIM with an error smaller than 8%. The phase diagrams of the vector flux $\lambda_{r\alpha} \times \lambda_{r\beta}$ for 10 Hz and 50 Hz are presented in Figure 6, where

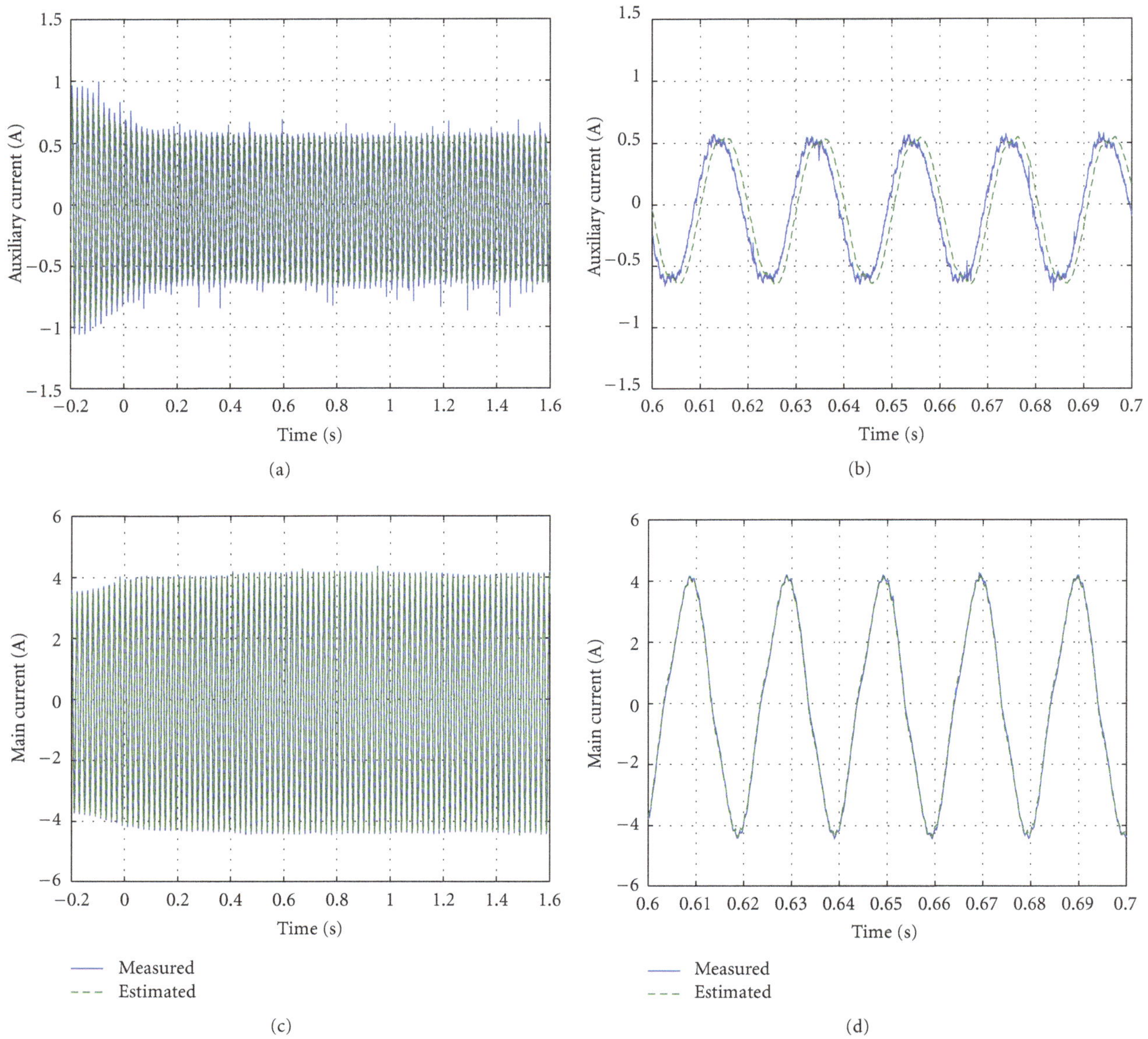

FIGURE 14: Measured and estimated stator currents of the SPIM when the rated mechanical load is applied (experimental results): (a) auxiliary winding $i_{s\alpha}$, (b) details of $i_{s\alpha}$, (c) main winding $i_{s\beta}$, (d) details of $i_{s\beta}$.

the maximum error $\Delta\lambda_r$ in the flux estimation is smaller than 8% and it is observed when the SPIM operates at 10 Hz. This error in the flux estimative at lower frequencies can be justified by the presence of the high-pass filter in the observer structure, whose function is to prevent problems related to DC offset in the current sensors. The acceleration and deceleration of the SPIM are shown in the Figure 7, where a good agreement can be noted in steady state condition between the speed estimative $\widehat{\omega}_{r\alpha}$ obtained from $\widehat{\lambda}_{r\alpha}$ and the rotor speed of the SPIM ω_r. Although $\widehat{\omega}_{r\beta}$ obtained from $\widehat{\lambda}_{r\beta}$ also has a good agreement with ω_r at high frequencies, this speed estimative presents an error at low frequencies which can be attributed to presence of the high-pass filter in the observer structure. The dynamic response of $\widehat{\omega}_{r\beta}$ is more oscillatory than $\widehat{\omega}_{r\alpha}$. It is also observed that the estimated values $\widehat{\omega}_{r\alpha}$ and $\widehat{\omega}_{r\beta}$ present a lag in relation to ω_r, which is caused principally by the low-pass filters of the envelope detectors.

In the second simulated test, the rated mechanical load is applied to SPIM shaft when the a stator frequency is 50 Hz. In this simulated situation, the application of mechanical load reduces the auxiliary winding current $i_{s\alpha}$ to $0.30\,A_{\text{rms}}$ while the main winding current $i_{s\beta}$ is slightly increased to $2.80\,A_{\text{rms}}$ as presented in Figure 8. The error observed in the estimation of both currents is lower than 6%. The phase diagram of the vector flux considering the loaded SPIM is presented in Figure 9, where it can be verified that the error in the flux estimations is smaller than 4%. Considering Figure 10, the proposed observer provides accurate estimations for rotor speed ω_r in this simulated situation, with little delay and an error smallest than 3.75 rad/s.

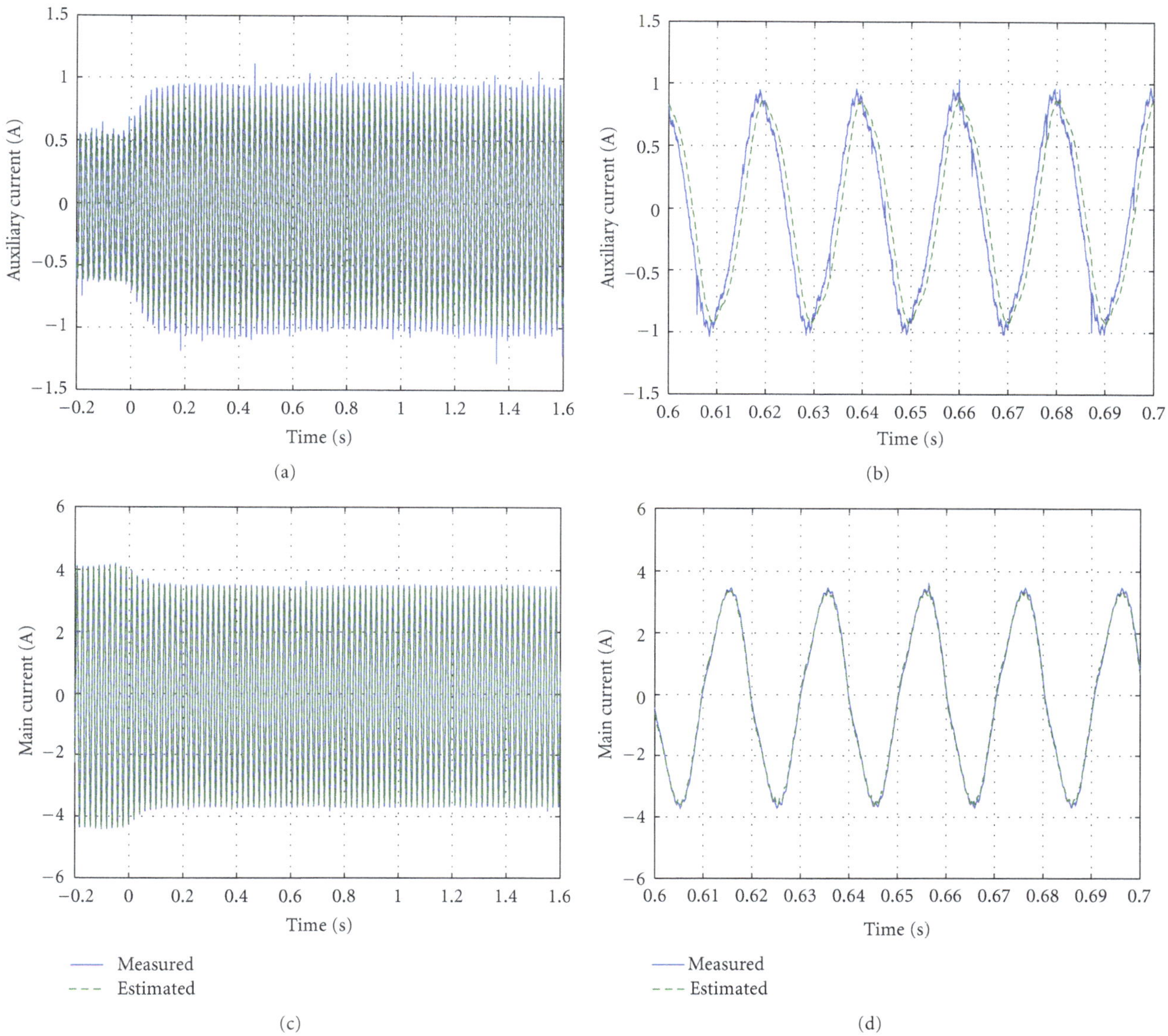

Figure 15: Measured and estimated currents of the SPIM when the rated mechanical load is removed (experimental results): (a) auxiliary winding $i_{s\alpha}$, (b) details of $i_{s\alpha}$, (c) main winding $i_{s\beta}$, (d) details of $i_{s\beta}$.

5.2. Experimental Results. In order to acquire the electrical and mechanical measurements, the "trigger" function of the virtual oscilloscope offered by the dSPACE system is used to register an event occurrence, causing the registrations with negative value for time intervals before the trigger event.

The first experimental test consists of the application of a step variation from 10 to 50 Hz in the stator frequency of the no-load SPIM. Both time response of the auxiliary winding current $i_{s\alpha}$ and main winding current $i_{s\beta}$ obtained in this test are shown in Figure 11, where it can be observed that the estimated currents are very close to the measured values, demonstrating a good accuracy of the proposed observer to estimate the stator currents. It is observed a delay in the estimation of the stator currents, which is negligible for $\hat{i}_{s\beta}$. The SPIM speed response stepwise is reported in Figure 12, where the estimated speed $\hat{\omega}_{r\beta}$ and the measured speed ω_r clearly present a good agreement in steady-state condition in the experimental acceleration and deceleration of the SPIM. The delay in the estimation of the speed can be attributed to the presence of the low-pass filters in the envelope detectors. Figure 13 shows the estimated fluxes $\hat{\lambda}_{r\alpha}$ and $\hat{\lambda}_{r\beta}$, which can be considered accurate since the proposed observer has obtained good estimations for stator currents and rotor speed in the acceleration and deceleration of the SPIM.

The second experimental test consists in to apply and remove the rated mechanical load of the SPIM, which is operating with a stator frequency of 50 Hz. The auxiliary winding current $i_{s\alpha}$ is reduced while the main winding current $i_{s\beta}$ increases with the load application, as shown in Figure 14. The estimation error in the estimation of the stator currents disappears when the mechanical load is removed as presented in Figure 15. These figures also show that the

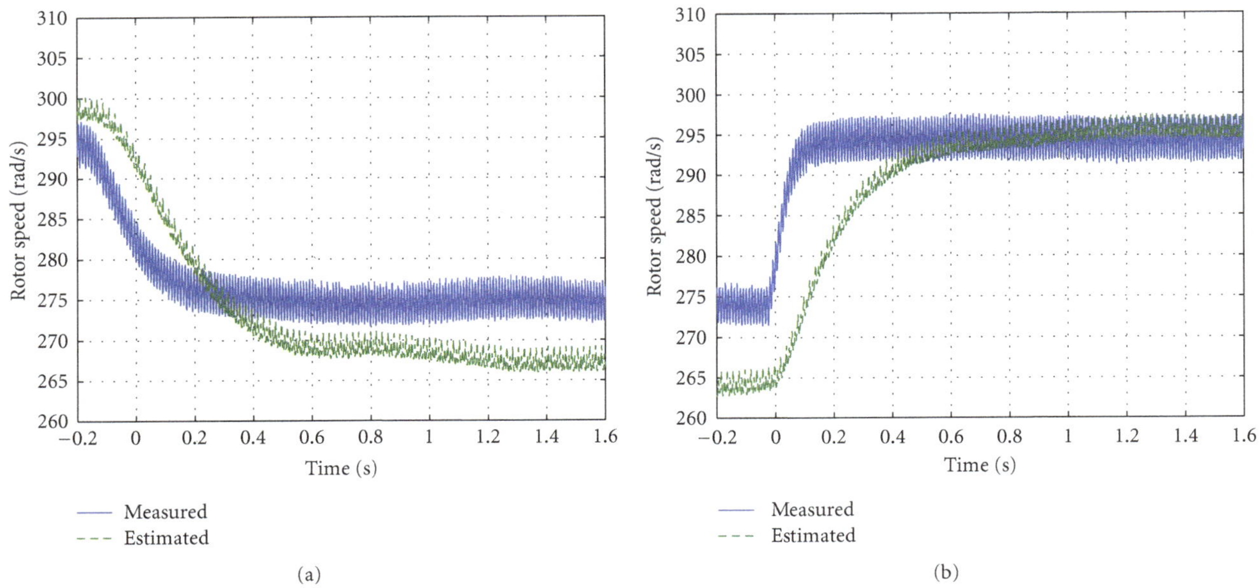

FIGURE 16: Measured and estimated rotor speed ω_r (experimental results): (a) when a mechanical load is applied to SPIM shaft, (b) when a mechanical load is removed of the SPIM shaft.

current estimations of the proposed observer are accurate and present a small delay, which is negligible for $\hat{i}_{s\beta}$. The effect of the mechanical load over the estimated speed $\omega_{r\beta}$ and the measured speed ω_r is shown in Figure 16, where is observed the presence of a little delay during the transient behavior and a small error between the estimated speed $\hat{\omega}_{r\beta}$ and the measured speed ω_r in steady-state, which is acceptable for the majority of the SPIM applications.

It is noted that the experimental results practically reproduce the computational simulations, whose differences can be justified by the uncertainties in the parameter estimation of the SPIM, which are greatest for the auxiliary winding and mechanical parameters. The experiments and the simulations demonstrate that the estimations of the stator currents and the rotor speed are satisfactory, indicating that the rotor flux estimation must be correct. Thus, the experimental performance of the proposed observer corroborates the computational simulations, demonstrating that it provides a satisfactory estimation for the rotor flux and the rotor speed of the SPIM with good accuracy within the usual frequency range considered for most SPIM application.

6. Conclusion

This paper presents the design, simulation, and implementation of a rotor flux and speed observer for the development of sensorless applications involving SPIMs. This observer is based in two half flux observers, which are designed using the stationary reference frame SPIM model. From these half flux observers, data are obtained in order to estimate the rotor speed of the SPIM. The performance of the proposed observer is computationally and experimentally evaluated considering two situations: a step variation in the stator frequency of the no-load SPIM and the application of mechanical load to SPIM shaft. The computational simulations and the experimental results present good agreement, demonstrating that the proposed observer provides satisfactory estimations of the rotor flux and rotor speed of the SPIM, with good accuracy within of the usual frequency range considered for most SPIM applications. In this context, the observer presented in this paper consists in an interesting option to estimate the rotor flux and/or rotational speed for sensorless applications involving the control of variable speed drives and/or development of optimization strategies for SPIMs.

Acknowledgment

The authors gratefully thank SDESLab (Sustainable Development and Energy Savings Laboratory), University of Palermo, for the support given during the experimental tests, and the financial support of MIUR (Ministero dell'Istruzione dell'Universitá e della Ricerca, Italy), CNPq (Conselho Nacional de Desenvolvimento Cientifico e Tecnológico, Brazil), CAPES (Coordenação de Aperfeiçoamento de Pessoal de Nivel Superior, Brazil), FAPEMIG (Fundação de Amparo a Pesquisa do Estado de Minas Gerais-Brazil) and Gorceix foundation (Brazil).

References

[1] T. A. Lettenmaier, D. W. Novotny, and T. A. Lipo, "Single-phase induction motor with an electronically controlled capacitor," *IEEE Transactions on Industry Applications*, vol. 27, no. 1, pp. 38–43, 1991.

[2] E. Muljadi, Y. Zhao, T. H. Liu, and T. A. Lipo, "Adjustable ac capacitor for a single-phase induction motor," *IEEE Transactions on Industry Applications*, vol. 29, no. 3, pp. 479–485, 1993.

[3] H. M. B. Metwally, "New method for speed control of single phase induction motor with improved motor performance," *Energy Conversion and Management*, vol. 42, no. 8, pp. 941–950, 2001.

[4] S. Vaez-Zadeh and A. Payman, "Design and analysis of sensorless torque optimization for single phase induction motors," *Energy Conversion and Management*, vol. 47, no. 11-12, pp. 1464–1477, 2006.

[5] E. R. Collins, "Torque and slip behavior of single-phase induction motors driven from variable-frequency supplies," *IEEE Transactions on Industry Applications*, vol. 28, no. 3, pp. 710–715, 1992.

[6] A. S. Ba-thunya, R. Khopkar, K. Wei, and H. A. Toliyat, "Single phase induction motor drives-a literature survey," in *Proceedings of the IEEE International on Electric Machines and Drives Conference (IEMDC'01)*, pp. 911–916, Cambridge, Mass, USA, June 2001.

[7] M. B. de Rossiter Corrêa, C. B. Jacobina, A. M. N. Lima, and E. R. C. Silva, "A three-leg voltage source inverter for two-phase AC motor drive systems," *IEEE Transactions on Power Electronics*, vol. 17, no. 4, pp. 517–523, 2002.

[8] D.-H. Jang and D.-Y. Yoon, "Space-vector PWM technique for two-phase inverter-fed two-phase induction motors," *IEEE Transactions on Industry Applications*, vol. 39, no. 2, pp. 542–549, 2003.

[9] M. B. de Rossiter Corrêa, C. B. Jacobina, E. R. C. Silva, and A. M. N. Lima, "Vector control strategies for single-phase induction motor drive systems," *IEEE Transactions on Industrial Electronics*, vol. 51, no. 5, pp. 1073–1080, 2004.

[10] R. Rocha, L. S. Martins Filho, and J. C. D. de Melo, "A speed control for variable-speed single-phase induction motor drives," in *Proceedings of the IEEE International Symposium on Industrial Electronics (ISIE'05)*, pp. 43–48, June 2005.

[11] M. Jemli, H. B. Azza, and M. Gossa, "Real-time implementation of IRFOC for Single-Phase Induction Motor drive using dSpace DS 1104 control board," *Simulation Modelling Practice and Theory*, vol. 17, no. 6, pp. 1071–1080, 2009.

[12] M. Jemli, H. B. Azza, M. Boussak, and M. Gossa, "Sensorless indirect stator field orientation speed control for single-phase induction motor drive," *IEEE Transactions on Power Electronics*, vol. 24, no. 6, pp. 1618–1627, 2009.

[13] A. H. Rajaei, M. Mohamadian, S. M. Dehghan, and A. Yazdian, "Single-phase induction motor drive system using z-source inverter," *IET Electric Power Applications*, vol. 4, no. 1, pp. 17–25, 2010.

[14] B. Bose, *Power Electronics and Variable Frequency Drives—Technology and Applications*, IEEE Press, 1997.

[15] J. Holtz, "Sensorless position control of induction motors—an emerging technology," *IEEE Transactions on Industrial Electronics*, vol. 45, no. 6, pp. 840–851, 1998.

[16] E. A. Vendrusculo and J. A. Pomilio, "An adaptive speed estimator for single-phase induction motors," in *Proceedings of the International Symposium on Power Electronics, Electrical Drives, Automation and Motion (SPEEDAM'08)*, pp. 1248–1252, Ischia Naples, Italy, June 2008.

[17] M. Caruso, V. Cecconi, A. O. Di Tommaso, and R. Rocha, "A flux and speed observer for sensorless SPIM applications," in *Proceedings of the 19th International Conference on Electrical Machines (ICEM'10)*, pp. 1–6, Rome, Italy, September 2010.

[18] P. Krause, *Analysis of Electric Machinery*, McGraw-Hill, New York, NY, USA, 1987.

[19] IEEE Standard test procedure for single-phase induction motors, IEEE Std 114—1982 (Revision of ANSI/IEEE Std 114-19691), December, 1982.

[20] A. O. Di Tommaso and R. Miceli, "A new high accuracy software based resolver-to-digital converter," in *Proceedings of the 29th Annual Conference of the IEEE Industrial Electronics Society (IECON'03)*, pp. 2435–2440, Industrial Electronics Society, November 2003.

[21] C. Cavallaro, A. O. Di Tommaso, R. Miceli, A. Raciti, G. Ricco Galluzzo, and M. Trapanese, "Analysis a DSP implementation and experimental validation of a loss minimization algorithm applied to permanent magnet synchronous motor drives," in *Proceedings of the 29th Annual Conference of the IEEE Industrial Electronics Conference (IECON'03)*, vol. 1, pp. 312–317, 2003.

Permissions

The contributors of this book come from diverse backgrounds, making this book a truly international effort. This book will bring forth new frontiers with its revolutionizing research information and detailed analysis of the nascent developments around the world.

We would like to thank all the contributing authors for lending their expertise to make the book truly unique. They have played a crucial role in the development of this book. Without their invaluable contributions this book wouldn't have been possible. They have made vital efforts to compile up to date information on the varied aspects of this subject to make this book a valuable addition to the collection of many professionals and students.

This book was conceptualized with the vision of imparting up-to-date information and advanced data in this field. To ensure the same, a matchless editorial board was set up. Every individual on the board went through rigorous rounds of assessment to prove their worth. After which they invested a large part of their time researching and compiling the most relevant data for our readers. Conferences and sessions were held from time to time between the editorial board and the contributing authors to present the data in the most comprehensible form. The editorial team has worked tirelessly to provide valuable and valid information to help people across the globe.

Every chapter published in this book has been scrutinized by our experts. Their significance has been extensively debated. The topics covered herein carry significant findings which will fuel the growth of the discipline. They may even be implemented as practical applications or may be referred to as a beginning point for another development. Chapters in this book were first published by Hindawi Publishing Corporation; hereby published with permission under the Creative Commons Attribution License or equivalent.

The editorial board has been involved in producing this book since its inception. They have spent rigorous hours researching and exploring the diverse topics which have resulted in the successful publishing of this book. They have passed on their knowledge of decades through this book. To expedite this challenging task, the publisher supported the team at every step. A small team of assistant editors was also appointed to further simplify the editing procedure and attain best results for the readers.

Our editorial team has been hand-picked from every corner of the world. Their multi-ethnicity adds dynamic inputs to the discussions which result in innovative outcomes. These outcomes are then further discussed with the researchers and contributors who give their valuable feedback and opinion regarding the same. The feedback is then collaborated with the researches and they are edited in a comprehensive manner to aid the understanding of the subject.

Apart from the editorial board, the designing team has also invested a significant amount of their time in understanding the subject and creating the most relevant covers. They scrutinized every image to scout for the most suitable representation of the subject and create an appropriate cover for the book.

The publishing team has been involved in this book since its early stages. They were actively engaged in every process, be it collecting the data, connecting with the contributors or procuring relevant information. The team has been an ardent support to the editorial, designing and production team. Their endless efforts to recruit the best for this project, has resulted in the accomplishment of this book. They are a veteran in the field of academics and their pool of knowledge is as vast as their experience in printing. Their expertise and guidance has proved useful at every step. Their uncompromising quality standards have made this book an exceptional effort. Their encouragement from time to time has been an inspiration for everyone.

The publisher and the editorial board hope that this book will prove to be a valuable piece of knowledge for researchers, students, practitioners and scholars across the globe.

List of Contributors

Riccardo Da Soghe, Bruno Facchini, Mirko Micio and Antonio Andreini
Energy Engineering Department "S. Stecco", University of Florence, Street S. Marta 3, 50139 Florence, Italy

Konstantinos G. Barmpalias, Ndaona Chokani and Reza S. Abhari
Laboratory for Energy Conversion (LEC), Swiss Federal Institute of Technology, ETH Zurich, 8092 Zurich, Switzerland

Anestis I. Kalfas
Department of Mechanical Engineering, Aristotle University of Thessaloniki, GR-54124 Thessaloniki, Greece

Yang Sun-Sheng, Kong Fan-Yu and Xue Ling
Research Center of Fluid Machinery Engineering and Technology, Jiangsu University, Zhenjiang, Jiangsu 212013, China

Fu Jian-Hui
Zhejiang Jiali Technology Co., Ltd, Hangzhou, Zhejiang 311241, China

Melissa Wilcox and Klaus Brun
Machinery, Southwest Research Institute, San Antonio, Tx, USA

Rainer Kurz
Systems Analysis, Solar Turbines Incorporated, San Diego, CA, USA

Michele Pinelli, Pier Ruggero Spina and Mauro Venturini
Dipartimento di Ingegneria, Universita degli Studi di Ferrara, Via G. Saragat, 1-44122 Ferrara, Italy

Wenhai Du
Turbomachinery Group, University of Liege, Chemin des Chevreuils 1, 4000 Liege, Belgium
Institute of Engineering Thermophysics, Chinese Academy of Sciences, Beijing 100190, China

Olivier Leonard
Turbomachinery Group, University of Liege, Chemin des Chevreuils 1, 4000 Liege, Belgium

Cheng Xu and Ryoichi S. Amano
Department of Mechanical Engineering, University of Wisconsin-Milwaukee, Milwaukee, WI 53212, USA

Silvia M. Zanoli and Giacomo Astolfi
D.I.I., Universita Politecnica delle Marche, Ancona, Italy

O. Coutier-Delgosha, A. Dazin, G. Caignaert and G. Bois
Arts et Metiers ParisTech/LML Laboratory, 8 boulevard Louis XIV, 59046 Lille Cedex, France

Antoine Ducoin and Yin Lu Young
Department of Naval Architecture and Marine Engineering, University of Michigan, 2600 Draper Drive, Ann Arbor, MI 48109, USA

Biao Huang
Department of Vehicle Engineering, School of Mechanical Engineering, Beijing Institute of Technology, Beijing 100081, China

Osama N. Alshroof and Robert B. Randall
School of Mechanical and Manufacturing Engineering, The University of New South Wales, Kensington, Sydney, NSW 2052, Australia

Gareth L. Forbes
Department of Mechanical Engineering, Curtin University, Bentley, Perth, WA 6102, Australia

Nader Sawalhi
Mechanical Engineering Department, Prince Mohammad Bin Fahd University (PMU), Al Khobar 31952, Saudi Arabia

Guan H. Yeoh
School of Mechanical and Manufacturing Engineering, The University of New South Wales, Kensington, Sydney, NSW 2052, Australia
Australian Nuclear Science Technology Organisation (ANSTO), Locked Bag 2001, Kirrawee DC, NSW 2233, Australia

Spyridon D. Kyparissis and Dionissios P. Margaris
Fluid Mechanics Laboratory, Mechanical Engineering and Aeronautics Department, University of Patras, 26500 Patras, Greece

Spyros A. Kinnas, Ye Tian and Abhinav Sharma
Ocean Engineering Group, Department of Civil, Architectural and Environmental Engineering, University of Texas at Austin, Austin, TX 78712, USA

Toshiyuki Hirano
Production Systems Engineering Course, Tokyo Metropolitan College of Industrial Technology, 1-10-40 Higashiohi, Shinagawa-ku, Tokyo 140-0011, Japan

Takanori Uchida
Graduate School of Engineering, Hosei University, 3-7-2 Kajinocho, Koganei-shi, Tokyo 184-8584, Japan

Hoshio Tsujita
Department of Mechanical Engineering, Faculty of Science and Engineering, Hosei University, Tokyo, Japan

Phil Ligrani
Parks College of Engineering, Aviation, and Technology, Saint Louis University, 3450 Lindell Boulevard, McDonnell Douglas Hall Room 1033A, St. Louis, MO 63103, USA

Massimo Caruso, Vittorio Cecconi and Antonino O. Di Tommaso
Department of Electrical, Electronics and Telecommunications Engineering, University of Palermo, Vialle delle Scienze, Edificio 9, 90128 Palermo, Italy

Ronilson Rocha
Department of Control and Automation Engineering, School of Mines, Federal University of Ouro Preto, Campus Morro do Cruzeiro, 35400-000 Ouro Preto, MG, Brazil